From Music to Mathematics

From Music to Mathematics

Exploring the Connections

GARETH E. ROBERTS

College of the Holy Cross
Worcester, Massachusetts

Johns Hopkins University Press | *Baltimore*

© 2016 Johns Hopkins University Press
All rights reserved. Published 2016
Printed in the United States of America on acid-free paper
9 8 7 6 5 4 3 2

Johns Hopkins University Press
2715 North Charles Street
Baltimore, Maryland 21218-4363
www.press.jhu.edu

ISBN-13: 987-1-4214-1918-3 (hardcover)
ISBN-10: 1-4214-1918-1 (hardcover)
ISBN-13: 978-1-4214-1919-0 (ebook)
ISBN-10: 1-4214-1919-X (ebook)

Library of Congress Control Number: 2015943423

A catalog record for this book is available from the British Library.

*Special discounts are available for bulk purchases of this book. For more
information, please contact Special Sales at 410-516-6936 or specialsales@
press.jhu.edu.*

Johns Hopkins University Press uses environmentally friendly book mate-
rials, including recycled text paper that is composed of at least 30 percent
post-consumer waste, whenever possible.

In Memory of Abigail Eaton Roberts (1939–2013),

dedicated educator, valued friend, and loving mother

Contents

Preface

This book is the product of an introductory course in math and music which I have taught several times at the College of the Holy Cross. It combines two of my passions and most favorite subjects, math and music, a love that was fostered during my undergraduate days studying at Oberlin College & Conservatory.

My primary intention for writing this book was to create a usable textbook for both students and instructors, one that invites the reader to explore the many fascinating connections between math and music, while also fostering the growth of the mathematical (and musical) abilities of the student. I have tried to create a text that provides a sufficiently comprehensive and workable source for instructors who are currently teaching a college-level course on math and music, or have always wanted to give such a course a try. Individual chapters may also be useful for "liberal arts" math courses or advanced high school classes, where the goal is to expose students to some interesting mathematics, but not necessarily prepare them for future study.

The content for this book has been culled from many excellent sources on math and music. References to these sources, as well as to relevant articles, music, and websites, are given at the end of the particular chapter they apply to. I am grateful to these authors for taking the time to contribute their expertise and insights to these combined fields. Some topics are fairly new or less familiar (e.g, the discussion and examples about rhythm, or the Bartók controversy). In choosing which material to use, I have selected the topics that my students seem to have found the most appealing, while still containing some serious mathematics. The text features several musical examples from jazz and popular music, genres typically more familiar to our students.

Throughout the book, ideas and applications are presented in detail, modeling the type of in-depth analysis I ask of my students. Sample problems are included as part of the exposition in nearly every section, with carefully written solutions provided to assist the reader. Each section closes with a variety of exercises, many of which are directly related to the exposition. In total, there are more than 200 exercises in the text covering a wide range of skill levels. Many exercises have been previously vetted by past students. In addition, two chapters contain sample projects that have worked well in my courses.

The level of this text is "introductory" in the sense that no advanced mathematical or musical training is assumed. A few sections in Chapter 3 require some calculus to fully understand, but the reader can still discern the key ideas even without any knowledge of calculus. Any exercise that requires calculus is labeled accordingly. In terms of musical ability, it is not assumed that the reader knows how to read music. Part of the goal in writing this book was to provide enough foundational material (e.g., basic music theory, trigonometry, group theory, etc.) to welcome the untrained musician or terrified mathematics student. By offering a substantial amount of fundamental exercises and explanations, it is hoped that this text will ease the burden on the math/music instructor.

One of the misconceptions about students who are not math or science majors, students who do not feel comfortable or confident in their mathematical abilities, is that these mitigating circumstances imply they are incapable of, or uninterested in, learning any higher-level mathematics. Fortunately, my experiences teaching this material have often been the exact opposite, to the point where students who didn't realize they enjoyed math or know they had any proclivity toward the subject found themselves considering a math major. For example, some students respond very well to learning group theory. In secondary school they struggled through limits, algebra, and precalculus, but completing the group table for the symmetries of the square and seeing its connection to an extent on four bells kindles a newfound interest in abstract algebra.

In support of this pedagogical credo, the book presents some fairly sophisticated mathematics using music as the motivating subject. In most parts of the book, the musical topic is presented first to capture the readers interest; then, the requisite mathematics is developed to explore that particular topic. A good example of this is the notion of symmetry in music (Chapter 5), where composers use techniques such as transposition (mathematical translation), retrograde (vertical reflection), and inversion (horizontal reflection) to help develop their musical themes. The mathematical structure behind the symmetry can be described by a subgroup of the symmetries of the square. Hence, it is natural to study group theory and learn how key examples such as $\mathbb{Z}_{12}$ and the dihedral group play an important role in understanding music. In short, the music motivates the mathematics.

I have also found that any college student can, and should, learn to follow and construct their own proofs. For example, the fact that the $\sqrt{2}$ is irrational is interesting, historically significant, and important in the study of how we tune our instruments. Students need to understand this context and its significance while they try to comprehend the mathematical details. In addition to group theory and analysis (e.g., proofs), number theory makes a significant appearance through such topics as equivalence relations, the least common multiple, the greatest common divisor, and continued fractions. There is also some combinatorics discussed in a few sections. Rather than shy away from these challenging mathematical topics, the text strives to embrace them and then teach them.

Suggestions for using the text

I have been fortunate to have a yearlong course in math and music as part of a special first-year program (called *Montserrat*) at Holy Cross. During the year, I cover Chapters 1–4 (rhythm, music theory, sound, and tuning and temperament) in the first semester and Chapters 5–8 (musical group theory, change ringing, 12-tone music, and modern mathematical music) in the second. This allows sufficient time for students to master the basic concepts from the first few chapters so they can appreciate and comprehend the material in the second half of the book. For instance, a student who struggles with musical intervals in Chapter 2 will have little chance of following the intricacies of 12-tone music in Chapter 7.

In a one-semester course, the chapters on rhythm, music theory, tuning and temperament, and musical group theory are critical to cover. Understanding sound is certainly also important, but it can be covered in less detail than presented in the text. Selections from the final three chapters can be chosen as time allows, but it is important to show how composers have used mathematical ideas in their compositions. As an illustrative example, students enjoy analyzing the patterns in Steve Reich's *Clapping Music*, and even performing it (or at least trying to) in front of their peers.

One of the projects in the text which has been fun for instructor and students alike is the monochord lab in Section 3.5. This project requires the use of a few monochords that have rulers attached in order to study the relationship between pitch (frequency) and the length of a plucked string. The lab is a great way to motivate the Pythagorean scale, and my class always enjoys playing a major scale together, or simple tunes such as *Twinkle Twinkle Little Star* (in Pythagorean tuning), once the correct lengths have been calculated. Another project that I have really enjoyed is a final assignment asking students to compose and perform (or find someone else to perform) their own piece of mathematical music. Although it can be challenging to adequately prepare students to undertake such an assignment, I have been pleasantly surprised by some of the artistic and mathematical works that students have produced (some memorable examples are described in Section 8.4). Assigning a mathematical composition is a great way to blend students' artistic and analytic skills, not to mention a fun way to end the semester (or year), in lieu of a final exam.

The website www.frommusictomath.com has been designed to accompany the text. It contains some useful resources for instructors, including worksheets, sample labs, a discography, and slides to accompany lectures. Both PDF and LaTeX files are provided so that instructors may tailor them for their own courses. Solutions to many of the exercises are available for any instructor who adopts the book for their course. Please contact me at groberts@holycross.edu should you desire these solutions. Hopefully, these additional resources will be helpful for teaching your math and music course.

Ultimately, my goal in writing this text has been to share some of the wonderful connections between these two great fields and to use those links to teach some important and interesting mathematics, math that a nonmajor is not typically exposed to. For students, I believe that this text will help develop your analytical skills, enhance your appreciation of music, and foster a love of, and respect for, mathematics. I hope you enjoy reading it as much as I have enjoyed writing it.

Acknowledgments

First and foremost, this book would never have been completed without the steadfast support of my wife, Julie, and son, Owen. Their love, encouragement, and willingness to hear every last detail about this book have helped me persevere throughout this project. My father, George, and mother-in-law, Kathy, have also been a valuable source of strength and assistance.

Many friends and colleagues have contributed through their enthusiasm, understanding, praise, and good counsel. Others were there to listen well and help me overcome the inevitable obstacles faced when writing a textbook. This special list includes Jeff Bernstein, Gina Cranford, Alisa DeStefano, Kate Ford, Chris Gengarelli, Sean Gonick, Dick Hall, Rich Kramer, John Little, Catherine Roberts, Grace Ross, and Ben Stumpf. My jazz quintet Blue Champagne was a source of inspiration, as well as a welcome musical respite.

I am grateful for the support of both the Mathematics and Computer Science Department and the Music Department (my second home) at Holy Cross (HC). Special thanks to Shirish Korde in music and Alan Karass of the HC music library for the generous amount of time they spent helping me gather the material for this book. Their knowledge and musical expertise were invaluable.

My writing has been inspired in part by my valued time as a graduate student at Boston University working on the *Differential Equations* textbook written by Paul Blanchard, Bob Devaney, and my thesis advisor, Dick Hall. The ongoing support and mentoring of these three BU legends have had a wonderful impact on my career.

Special thanks are also due to Alex Barnett, John Little, Arthur White, and an anonymous reviewer for their many helpful comments and suggestions that have undoubtedly improved this text. I would also like to thank my editor, Vince Burke, for his steadfast support and belief in my ability to write this book, as well as Jeremy Horsefield, for his meticulous copy-editing. To Carrie Peck I owe a debt of gratitude for her hard work and attention to detail on many of the figures and illustrations in the text.

This book would not have been possible without the many students who first tested its material and contributed to its shape. I am grateful to those students who encouraged me to write this book and who participated with such joy and enthusiasm in my classes. Specific contributions to the text came from former students Christina Catalano, Alexandra Gitto, Michael Greco, John Kane, Joseph Kramkowski, Julia Lam, Matthew McIvor, Jacob Miller, Emely Ventura, Megan Whitacre, and Greg Wood.

All of the music in the text was typeset using the open-source software LilyPond, version 2.18.0. The online assistance and learning manuals provided by the LilyPond team were a huge help in producing such high-quality musical notation. The mathematical graphs and images were created using Maple, version 15.

Introduction

Why do some collections of notes sound better together than others? Why does a major chord sound "happy" and a minor chord sound "sad"? Why does the 12-note Western musical scale work so well? What happens if the octave is divided into a number of notes other than 12? What happens in our ears and brains that allows us to perceive and enjoy music? Where do we locate the frets on a guitar? Do composers really use mathematical ideas in their work? Is music an art or a science? Was Bach a mathematician? If you have ever pondered some of these questions, then this book is for you.

The connections between math and music from a structural perspective are plentiful. Both use a specialized form of notation to communicate their ideas. Each subject has its own logical structure and set of axioms finely tuned over centuries of study. Students in high school geometry learn the axiomatic technique of Euclid to write their first proofs. Students in music theory learn rules of four-part vocal writing, such as to avoid parallel fifths and octaves, in order to better understand harmony. Mathematicians use numbers as the invariant building blocks of their theory as musicians use pitch as the common denominator of their creations. Just as the number 3 has the same abstract meaning to mathematicians everywhere, the concert pitch A440 used to tune modern orchestras is a global standard.

At first glance, mathematics is often considered a "hard" science, while music is deemed to belong to the humanities. However, in addition to the many structural traits shared by each field, there are also many aesthetic and artistic links between the two disciplines. For example, both fields have produced great child prodigies, such as Mozart and Gauss. Parents play Bach and Beethoven to their babies and young children in order to foster their brain development and analytic skills. Many mathematicians are outstanding musicians, while many musicians, in particular composers, possess sharp mathematical minds. Scholars often speak of the "beauty" and "purity" of mathematics, although the same lofty descriptions could equally apply to music. While music has the obvious capacity to move the spirit, great mathematical discoveries and insights are often accompanied with an overwhelming sense of elation. The great, present-day mathematician Andrew Wiles wept on camera while speaking about his incredible proof of Fermat's Last Theorem.

Composers and musicians, whether they are cognizant of the fact or not, use mathematical concepts in their creations. Bach was a master at using symmetry to develop the themes in his fugues and create wonderfully rich counterpoint. Bell ringers in towers throughout England have been using permutations, symmetry, and group theory in change ringing to announce important events since the early 1600s. Symmetry and invariance play an important role in the 12-tone method of composition. Magic squares have been used by the composer Peter Maxwell Davies as an architectural blueprint for some of his pieces. The modern composer Iannis Xenakis used computers and probability theory to create his "stochastic music."

These are the kinds of topics and issues discussed in this book. The goal isn't just to point out the connections; the goal is to dive deep into their essence while learning the necessary mathematics to truly understand them. For example, the history and theory behind how we tune our instruments, and why we tune them the way we do, are long and involved. The simple ratios glorified by the Pythagoreans gave way to the just tunings of the Renaissance and the tempered systems of the Baroque period. These past systems used rational numbers to determine the frequency ratios of all musical intervals. Then came the compromise of equal temperament, the tuning system we use today. Its defining trait is the subdivision of the octave into 12 equal parts, a task that requires the use of an *irrational* number, namely, the 12th root of 2. But to really appreciate this evolution and comprehend it at a meaningful level requires understanding the difference between rational and irrational numbers. We need to be *sure* that the key frequency ratio given by the $\sqrt[12]{2}$ is actually irrational. This means learning some deep mathematics, carefully working through well-known proofs, and forming your own rigorous mathematical arguments.

One of the most impressive and gratifying aspects of mathematics is its ability to get to the heart of a problem; be able to describe its essential features, and develop the machinery to approach it; and then maybe, if you're lucky, actually solve it. This is the beauty and struggle inherent in mathematics. This book tries to capture some of that beauty using music as the motivating element.

As we proceed, we will develop the mathematics (or in some cases simply recall it) needed for understanding music at a fundamental level. Toward that end, this book has been written as a comprehensive textbook, with detailed explanations, worked-out sample problems, and exercises at the end of nearly every section. Altogether, there are over 200 exercises in the text intended to support the belief shared by many mathematicians that the best way to learn mathematics is to *do* mathematics. The same credo certainly applies to music. For those less inclined in the musical arena, Chapter 2 gives a thorough introduction to basic music theory, focusing on notation, scales, intervals, the circle of fifths, and the piano keyboard, among other topics. The ability to read music at a somewhat comfortable level is critical to understanding the musical examples presented in the latter half of the text.

The first half of the book explores rhythm (Chapter 1), music theory (Chapter 2), the science of sound (Chapter 3), and tuning and temperament (Chapter 4). The mathematics involved can be as simple as multiplying fractions or finding the least common multiple, or as complicated as using continued fractions to approximate irrational numbers. The musical examples in the first two chapters are taken from a variety of genres, including jazz and popular music, not just the Western classical tradition. The second half of the book, beginning with Chapter 5, focuses on the ways composers have used mathematical concepts to create meaningful music. We will learn about symmetry in music (e.g., Bach) in Chapter 5 and study the mathematical structure behind change ringing and 12-tone music in Chapters 6 and 7, respectively. The book ends with a chapter exploring the compositional techniques of three modern, mathematical composers and some illustrative examples of their work. Throughout our investigations into these topics, the goal is to attain a deeper level of comprehension, to pose insightful questions, and, above all, to enjoy the learning process as we investigate the fascinating connections between math and music.

Chapter 1

Rhythm

Music is the pleasure the human soul experiences from counting without being aware that it is counting. — Gottfried Wilhelm von Leibniz

Life is about rhythm. We vibrate, our hearts are pumping blood, we are a rhythm machine, that's what we are. — Mickey Hart

Rhythm is one of the most vital and distinctive features in music. The rhythmic flavor of a piece, its pulse, tempo, and character, defines the style of music almost instantly. A waltz is in three and a march in two. Pop and folk music are usually in four. Salsa and other Latin music are immediately recognizable by their trademark syncopated beats. (A rhythmic pattern is *syncopated* when beats that are normally not stressed become accented, while the expected beats are deemphasized.) The rhythms of hip hop and house music invite us to rise up and dance. The famous four-note motif "da-da-da-dum," which dramatically opens Beethoven's 5th Symphony, is world renowned.

In this chapter, we discuss the importance of rhythm in experiencing and understanding music. After explaining the basic notation, we investigate some of the key mathematical principles underlying the structure of rhythm. For example, the different durations of notes (whole, half, quarter, eighth, etc.) form a decreasing geometric sequence with ratio $r = 1/2$. When dots are added to a note to increase its length, a geometric series is being constructed. Time signatures are introduced and several contrasting musical examples are presented to highlight how rhythm strongly influences the "feel" of a piece of music. Polyrhythmic music is also discussed, and the underlying mathematical principle of the least common multiple is highlighted. Musical examples are presented from a variety of sources, including rock (e.g., Billy Joel, R.E.M., The National), jazz (Paul Desmond), folk (*Silent Night*), and classical (e.g., Bernstein, Chopin, Stravinsky, Tchaikovsky, and Verdi).

1.1 Musical Notation and a Geometric Property

One of the simplest mathematical concepts, of great importance to musicians, is counting. Listening to and playing music are inherently connected with the ability to count. To be a successful musician or singer, one needs to be able to count well. Most musicians have an intuitive pulse running through their heads as they play. Some will count quietly to themselves while waiting for the precise moment to begin playing; others internalize the pulse as a natural component of their playing. Counting is also significant in determining the length a given note should be held.

1.1.1 Duration: Geometric sequences

In most written music, there is a standard notation used to indicate the duration or length of a given note. This notation originated with European classical music and is now common to music all over the world. There is also notation for resting (not playing). Musical notes are written on the *staff*, five equally spaced horizontal lines, where the notes are played sequentially from left to right. From a mathematical perspective, the horizontal lines used as the standard format to describe music are akin to plotting time t on the horizontal axis of a graph to visualize the motion of a function such as $y = \sin(t)$. In this chapter, we focus exclusively on the duration of the note or rest. The location of the note on the musical staff, which defines its pitch, will be discussed in Chapter 2.

The key concept is the existence of a rhythmic pulse or *beat*, which is followed by all players. The standard symbols and terminology used for musical notes are shown in Figure 1.1.

FIGURE 1.1. Some basic symbols and terminology used for musical notation. In the rightmost figure, the line connecting the two notes is called a *beam*.

The number or fraction of beats a note is held depends on whether the note head is shaded or not, if it has a stem, and if there are any flags. There is also the possibility of adding a dot to a note to increase its length. Table 1.1 shows the different types of notes, their names, and their durations. Here we assume that a quarter note has a length that represents one beat.

Symbol:	𝅝	𝅗𝅥	𝅘𝅥	𝅘𝅥𝅮	𝅘𝅥𝅯	𝅘𝅥𝅰
Name of note:	whole	half	quarter	eighth	sixteenth	thirty-second
Number of beats:	4	2	1	1/2	1/4	1/8

TABLE 1.1. The different types of notes and their durations, assuming that a quarter note equals one beat.

Notice the interesting mathematical pattern in the durations of the notes in Table 1.1. Each successive note is half the length of the previous note. The list of durations of the different notes,

$$4,\ 2,\ 1,\ 1/2,\ 1/4,\ 1/8,\ 1/16,\ \dots,$$

is a nice example of a *geometric sequence* with ratio $r = 1/2$.

Definition 1.1.1. *A geometric sequence is a list of numbers where each successive number is obtained by multiplying the previous number by a common nonzero ratio r. If a_n represents the nth term in the sequence, then the next term is given by $a_{n+1} = r \cdot a_n$.*

The salient feature of a geometric sequence is that there exists a common *ratio* between successive numbers. This is in contrast to an *arithmetic sequence*, where there is a common *difference* between successive numbers in the list. The use of the term "geometric" refers to the fact that any number in a geometric sequence is the *geometric mean* of its two adjacent neighbors (assuming that the terms in the sequence are all positive). The geometric mean of the numbers a and b is defined to be $\sqrt{ab}$. For instance, 2 is the geometric mean of 4 and 1 because $2 = \sqrt{4 \cdot 1}$, and 1/4 is the geometric mean of 1/2 and 1/8 because $1/4 = \sqrt{\frac{1}{2} \cdot \frac{1}{8}}$. The geometric idea behind the geometric mean is that if a and b represent the lengths of the sides of a rectangle, then the geometric mean $\sqrt{ab}$ gives the length of the side of the square with equivalent area to that of the rectangle.

Example 1.1.2. *The sequence* 1, 3, 9, 27, 81, ... *is a geometric sequence with common ratio* $r = 3$. *Each term, except for the first, is the geometric mean of its adjacent neighbors. For example,* $3 = \sqrt{1 \cdot 9}$ *and* $9 = \sqrt{3 \cdot 27}$. *The sequence* 6, -4, 8/3, $-16/9$, ... *is also a geometric sequence, but here the common ratio is negative, with* $r = -2/3$. *The sequence* 2, 4, 6, 8, 10, ... *of even numbers is arithmetic, not geometric. The sequence* 1, 2, 4, 7, 13, 24, ... *is not geometric because there is no common ratio that works between all pairs of successive numbers in the list (e.g.,* $4/2 \neq 7/4$). □

A piece of music is typically divided up into *measures*, where each measure is separated by a vertical bar called a *bar line*. The number of beats per measure can vary between works, or can even change multiple times during a piece. This is notated using a *time signature*, an important musical concept we discuss in Section 1.2. The most commonly used time signature is $\frac{4}{4}$, called *common time* and denoted by $\mathbf{C}$, which has four beats per measure with the quarter note receiving one beat. A simple excerpt of music in common time is shown in Figure 1.2. The excerpt contains six measures, each with exactly four beats of music. The location of the notes on the staff is not important at this point; focus on the type of note (sixteenth, eighth, quarter, half, or whole), and the fact that each measure contains a total of four beats.

In Table 1.2, for a particular type of note, we list the number that is needed to fill one measure in common time. We also list what fraction of the measure is filled by a given note. Not surprisingly, two geometric sequences are apparent. The third row is a geometric sequence with common ratio $r = 2$, while the fourth row is another sequence with $r = 1/2$. Notice that the terms in the second row correspond exactly with the fractions in the fourth row, revealing the meaning behind the different names of the notes.

As mentioned above, there is also notation for silence, denoted by *rests*. The notation to indicate that a musician should rest for a specific number or fraction of beats is indicated in Table 1.3. Notice the similarities between the structure and names of the rests in Table 1.3 and the notes in Table 1.1.

FIGURE 1.2. A simple excerpt of music in $\frac{4}{4}$ time containing six measures, each of which has four beats with the quarter note equal to one beat. A single beam connecting two stems produces two eighth notes (e.g., measures 2 and 6), while a double beam between stems creates sixteenth notes (e.g., measure 5).

Symbol:	o	𝅗𝅥	♩	♪	♬	♬
Name of note:	whole	half	quarter	eighth	sixteenth	thirty-second
Number required to fill a four-beat measure:	1	2	4	8	16	32
Fraction of time held in a four-beat measure:	1	1/2	1/4	1/8	1/16	1/32

TABLE 1.2. The different types of notes and their durations in terms of a four-beat measure where the quarter note equals one beat.

Symbol:	▬	▬	𝄽	𝄾	𝄿	𝄿
Name of rest:	whole	half	quarter	eighth	sixteenth	thirty-second
Number of beats:	4	2	1	1/2	1/4	1/8

TABLE 1.3. The different types of rests and their durations, assuming that a quarter note equals one beat. A whole rest and half rest are each denoted by a small solid rectangle, except that a whole rest lies directly *below* the fourth line of the staff, while a half rest lies directly *above* the third line of the staff. The first two measures of the tenor part in Figure 1.3 contain whole rests, while the third measure begins with a half rest.

1.1.2 Dots: Geometric series

One common technique for extending the duration of a note is to add a dot after it. This has the effect of increasing the length the note is held by one-half its original value. For example, a dotted half note in $\frac{4}{4}$ time will get two beats for the half note plus one extra beat for the dot, giving a total of three beats. We can also have notes with two, three, or even four dots added. This illustrates the mathematical concept of *iteration*, where the same process is applied repeatedly. A note with two dots will be lengthened by half the original value and then lengthened again by half of the new value. Each additional dot lengthens the note by half the value of the previous dot. Thus, a double-dotted half note in $\frac{4}{4}$ time will get two beats for the half note, one beat for the first dot, and an additional half a beat for the second dot, yielding a total of $3\frac{1}{2}$ beats. It is also possible to add dots to rests to increase their length. The same rule applies: each dot increases the length of the rest by one-half its original value. Table 1.4 demonstrates the result of adding more and more dots to increase the duration of a half note in $\frac{4}{4}$ time.

Although it is unlikely to run across music that uses more than four dots, there are some examples with triple-dotted and quadruple-dotted notes. One such case occurs in the *Dies irae* (second movement) of Giuseppe Verdi's famous *Messa da Requiem* (1874). The vocal and orchestral parts of the *Rex Tremendae* section of this movement frequently contain triple- and quadruple-dotted notes (see Figures 1.3 and 1.4).

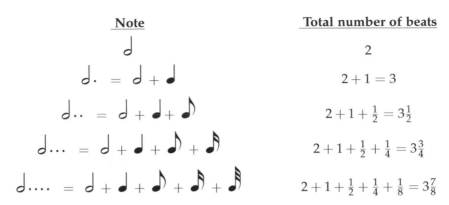

Note	Total number of beats
♩	2
♩. = ♩ + ♩	$2 + 1 = 3$
♩.. = ♩ + ♩ + ♪	$2 + 1 + \frac{1}{2} = 3\frac{1}{2}$
♩... = ♩ + ♩ + ♪ + ♪	$2 + 1 + \frac{1}{2} + \frac{1}{4} = 3\frac{3}{4}$
♩.... = ♩ + ♩ + ♪ + ♪ + ♪	$2 + 1 + \frac{1}{2} + \frac{1}{4} + \frac{1}{8} = 3\frac{7}{8}$

TABLE 1.4. The effect of repeatedly adding dots to a half note. The number of beats is given assuming that a quarter note equals one beat.

Another way to increase the length of a note is to connect it to an adjacent note using a *tie*. A tie is a slightly curved line that connects two consecutive notes of the same pitch (located in the same position on the staff). The tie instructs the musician to hold the given pitch for the length of each note combined. For example, a tie between a quarter note and a half note should be held for three beats (assuming that the quarter note receives one beat). A tie between a half note and a dotted half note will get five beats. The tie between the whole notes in the bass part between measures 324 and 325 in Figure 1.3 instructs the singer to hold that particular note for eight beats. Ties are often used to connect notes between adjacent measures.

From a mathematical perspective, each sum in the right-hand column of Table 1.4 is an example of a *geometric series*. A geometric series is just a sum of the terms in a geometric

FIGURE 1.3. Double- and triple-dotted notes abound for the tenors and bases in measures 322–329 of the *Dies irae* from Verdi's *Requiem*.

FIGURE 1.4. Triple- and quadruple-dotted notes in measures 356 and 358 of the *Dies irae* from Verdi's *Requiem*.

sequence. In general, a series is a sum, while a sequence is a list. For example, suppose that we have a geometric sequence with a common ratio $r = 2$, a first term of $1/4$, and seven terms. Then, the corresponding geometric series is

$$\frac{1}{4} + \frac{1}{2} + 1 + 2 + 4 + 8 + 16 \ = \ 31\tfrac{3}{4}. \tag{1.1}$$

Definition 1.1.3. *A geometric series is a sum of the terms in a geometric sequence. It has the form*

$$S_n \ = \ a_0 + a_0 r + a_0 r^2 + \cdots + a_0 r^{n-1}, \tag{1.2}$$

where r is the common ratio and a_0 denotes the first term.

By examining the exponents, we note that there are n terms being summed in the series in Equation (1.2).

For musicians, it is usually simple enough to just add the terms in a geometric series together to determine how long a note with dots should be played. In contrast, mathematicians, who enjoy finding formulas, can use a simplified expression to calculate the sum of a geometric series. This is particularly useful when there are a large number of terms, say, $n = 100$ or $n = 500{,}000$.

To derive the formula for the sum of a geometric series, we use a nice trick that highlights the underlying geometric principle. Beginning with the sum

$$S_n \ = \ a_0 + a_0 r + a_0 r^2 + \cdots + a_0 r^{n-1}, \tag{1.3}$$

multiply both sides by the common ratio r. Making sure to multiply every term on the right-hand side of Equation (1.3) by r, we obtain

$$r S_n \ = \ a_0 r + a_0 r^2 + a_0 r^3 + \cdots + a_0 r^n. \tag{1.4}$$

Notice that nearly all of the terms on the right-hand side of Equation (1.4) also appear on the right-hand side of Equation (1.3). If we subtract Equation (1.4) from Equation (1.3), these terms cancel out (even the terms represented by the $\cdots$ are canceled). We then obtain

$$S_n - rS_n \;=\; a_0 - a_0 r^n, \tag{1.5}$$

an expression that is easily solved for the sum S_n. Factoring out the S_n on the left-hand side of Equation (1.5) and dividing both sides by $1 - r$ gives an explicit formula for the sum of the first n terms in a geometric series:

$$S_n \;=\; \frac{a_0 - a_0 r^n}{1 - r}. \tag{1.6}$$

Returning to the example in Equation (1.1), we have $r = 2$, $a_0 = 1/4$, and $n = 7$. Using Formula (1.6), the sum is given by

$$S_7 \;=\; \frac{\frac{1}{4} - \frac{1}{4} \cdot 2^7}{1 - 2} \;=\; \frac{\frac{1}{4} - 32}{-1} \;=\; 31\tfrac{3}{4},$$

as expected.

Infinite geometric series

As a slight but enjoyable mathematical detour, let us consider what happens if we sum all of the terms in an *infinite* geometric series. While this may not have any specific musical applications, working with infinity (∞) is a common mathematical endeavor.

At first, summing an infinite geometric series may seem like an impossible task. After all, how can we sum an infinite set of numbers if we never stop adding? Formula (1.6) provides a way around this apparent conundrum. The key observation is that if the common ratio r satisfies $|r| < 1$, then the quantity r^n becomes smaller and smaller (in absolute value) as n gets larger and larger. For instance, if $r = 1/2$, then the sequence r^n is the familiar

$$1/2, \; 1/4, \; 1/8, \; 1/16, \; 1/32, \; 1/64, \; 1/128, \; 1/256, \; \ldots,$$

a sequence of values getting closer and closer to zero.

The relevant mathematical concept here is called a *limit*, abbreviated lim. Ignoring the rigorous mathematical details, we have

$$\lim_{n \to \infty} r^n \;=\; 0 \quad \text{provided that } |r| < 1.$$

This implies that the quantity $a_0 r^n$, found in the numerator of Formula (1.6), will vanish "in the limit." This yields a simple formula for the sum of an infinite geometric series with common ratio r and first term a_0:

$$S_\infty \;=\; \frac{a_0}{1 - r}. \tag{1.7}$$

It should be stressed that Formula (1.7) is valid *only* when the absolute value of the ratio r is less than 1. If this is not the case, then we say the infinite series *diverges* (does not limit on a particular numerical value).

Example 1.1.4. *Find the sum of the given infinite geometric series.*

   a. $3 + 1 + 1/3 + 1/9 + \cdots$

   b. $3 - 1 + 1/3 - 1/9 + - \cdots$

Solution: For **a.**, we have $a_0 = 3$ and $r = 1/3$. Using Formula (1.7), the sum of the infinite series is $S_\infty = 3/(1 - 1/3) = 9/2$. For the series in **b.**, we have $a_0 = 3$ and $r = -1/3$. This gives a sum of $S_\infty = 3/(1 + 1/3) = 9/4$.   □

Exercises for Section 1.1

1. Write out the first six terms of a geometric sequence with common ratio $r = 3$ and whose first term is $a_0 = 2/9$. Check that each term, other than a_0 or a_5, is the geometric mean of its adjacent neighbors.

2. Any three consecutive terms in a geometric sequence with common ratio r can be written as a, ar, ar^2. Compute the geometric mean of a and ar^2 and check that it is equal to the middle term ar. Assume that a and r are each positive. Conclude that any term of a geometric sequence (other than the first or last) is equivalent to the geometric mean of its adjacent neighbors.

3. Suppose that the length of a quarter note corresponds to one beat. How many beats does a triple-dotted quarter note get? How many beats does a double-dotted half note get?

4. Suppose that the length of a quarter note corresponds to one beat. How many beats does a double-dotted eighth rest get? How many beats does a double-dotted sixteenth rest get?

5. How many beats should the sopranos, altos, and tenors hold their opening notes in measure 356 of the *Dies irae* from Verdi's *Requiem* (see Figure 1.4)?

6. Write out the first eight terms of the geometric sequence starting with $a_0 = 1600$ and ratio $r = 1/4$. If these terms are added together, what is the sum of the resulting geometric series? If there were an infinite number of terms, what would the sum of the infinite geometric series be? Give all your answers as fractions.

7. Write out the first eight terms of the infinite geometric series with $a_0 = 250$ and $r = -1/5$. What is the sum of the infinite series?

8. Find the sum of the infinite geometric series $9 - 6 + 4 - 8/3 + 16/9 - + \cdots$.

9. Consider a geometric series with first term $a_0 = 1$ and ratio $r = 2$. Give a general argument (not just examples) showing that the sum of the first n terms is equal to one less than the next term in the series.

10. The following excerpt is in common time, which means that the quarter note is equal to one beat and there are four beats in each measure. However, the bar lines, indicating where each measure ends, are missing. Copy the excerpt onto staff paper and insert the bar lines correctly into the excerpt.

1.2   Time Signatures

Recall that written music is typically divided up into measures, with each measure separated from its neighbor by a bar line. One of the key rhythmic features in a piece of music is the number of beats per measure. There are many different styles and flavors of music that vary based on the number of beats in a measure, or how these beats are subdivided within the measure. Some modern classical works, as well as certain styles of folk music, often alter the number of beats in consecutive measures.

In written music, the *time signature* is used to indicate which type of note governs the beat in a measure and how many of those notes are needed to fill one measure. The time signature typically appears at the very start of the music, before any notes occur, although for more complicated music, it can change frequently within a piece, even from measure to measure. Another common term used for time signature is *meter*.

A time signature consists of two numbers, kind of like a fraction. We have already encountered one example, $\frac{4}{4}$ (common time $\mathbf{C}$), which indicates that four quarter notes are required to fill one measure of music. The time signature $\frac{9}{8}$ means that nine eighth notes fill one measure, while $\frac{3}{2}$ indicates that there are three half notes per measure. The bottom number of a time signature indicates which type of note represents the *principal beat* in a measure, while the top number describes how many of these notes are required to fill one measure. The bottom number can be 1, 2, 4, 8, or 16, referring to a whole, half, quarter, eighth, or sixteenth note, respectively. The principle beat may differ from what the listener or musician actually feels. For example, a piece in $\frac{9}{8}$, with nine eighth notes per measure, may actually sound as if it has three "beats" per measure with the eighth notes grouped into three sets of three. When applicable, we will distinguish the principal beat of a measure, the *microbeat* (indicated by the bottom number of the time signature), from the main pulse felt by a listener, the *macrobeat* [Gordon, 2012]. Other examples of these types of meter, often referred to as *compound meters*, are discussed in Section 1.2.1.

There are numerous ways to fill a measure of music in any given time signature. The only rule is that the total numerical value of the sum of the notes and rests agrees with that determined by the time signature. For example, a full measure of music in $\frac{6}{4}$ (six quarter-note beats per measure) can consist of three half notes, a whole note and two quarter notes, 12 eighth notes, or one quarter note and 20 sixteenth notes.

Example 1.2.1. *How many eighth notes are required to fill a measure in $\frac{5}{2}$ time? How many sixteenth notes are needed?*

Solution: In $\frac{5}{2}$ time, there are five half notes required to fill a measure. Since the length of a half note is equivalent to two quarter notes, and a quarter note is equivalent to two eighth notes, there are four eighth notes in a half note. Hence, we need $5 \cdot 4 = 20$ eighth notes to fill a measure in $\frac{5}{2}$ time. A sixteenth note has half the length of an eighth note, so we need eight sixteenth notes to equal one half-note principal beat. This means that there are $5 \cdot 8 = 40$ sixteenth notes required to completely fill a measure.   □

Example 1.2.2. *Interpret the time signature in the following musical excerpt and then insert the bar lines correctly into the music.*

Solution: The piece is in $\frac{5}{8}$ time, so there are the equivalent of five eighth notes per measure. The first measure starts with three eighth notes (three principal beats) followed by a quarter note (two beats, since one quarter note is equivalent to two eighth notes). The second measure starts with a dotted quarter rest that spans three principal beats because the dot adds half the value of a quarter rest ($2 + 1 = 3$). We then find four sixteenth notes, which is equivalent to two beats since a sixteenth note has half the length of an eighth note ($4 \cdot 1/2 = 2$). Continuing in this fashion, we place the remaining bar lines. One tricky spot is the tied note between the dotted quarter and quarter notes. This note is actually held for five principal beats ($3 + 2 = 5$), but the bar line is located in the middle of the tie. The full solution is shown in the following figure.   □

The first beat of a measure, called the *downbeat*, provides an important landmark that helps keep the musicians in a group together. The conductor of an orchestra is expected to clearly demarcate the downbeat of each measure. In a typical piece of music, a listener is usually able to discern the downbeat of each measure, making it easier to follow the music. It is important to realize that a piece of music does *not* have to begin with a complete measure. Many pieces begin with one or more notes, often called a *pickup*, before the downbeat of the first complete measure. Such an example is shown in Figure 1.8.

1.2.1 Musical examples

One of the best ways to become familiar with different time signatures is to listen carefully to contrasting examples. Common time is likely the most familiar time signature. Most popular music, in any genre, is in $\frac{4}{4}$ time, as this is one of the "easiest" rhythmic styles to understand. The time signature $\frac{3}{4}$, with three quarter notes per measure, is particularly well suited to dances such as the waltz. The "oom-pah-pah oom-pah-pah" rhythmic pattern, where the downbeat of each measure is stressed over the other beats, helps the dancers move along at a lively pace. One popular song in $\frac{3}{4}$ time is Billy Joel's classic *Piano Man* (1973). Although this meter is more common in the country/western genre, Joel's use of $\frac{3}{4}$ meter helps distinguish the song, creating a fun sing-along tune.

Music in $\frac{2}{4}$ time usually moves along at a quick pace because there are only two quarter-note beats per measure. This is the typical meter for marches and fast dance music, such

FIGURE 1.5. Franz Gruber's *Silent Night* (text by Joseph Mohn), demonstrating $\frac{6}{8}$ meter, where there are six eighth notes per measure.

as Tchaikovsky's *Russian Dance* from the *Nutcracker Suite* (1892). John Philip Sousa's *Stars and Stripes Forever* (1897) is a well-known march in *cut time* (denoted by ¢), which is equivalent to a $\frac{2}{2}$ meter. Latin music also features styles in $\frac{2}{4}$ meter, such as *merengue*, a popular kind of dance music originating in the Dominican Republic which features a fast rhythmic two-count.

Another frequently used time signature is $\frac{6}{8}$, with six eighth notes per measure. A good example of a piece in $\frac{6}{8}$ time is Franz Gruber's holiday classic *Silent Night* (1818; see Figure 1.5). For each musical example in this chapter, focus on the type of note or rest (quarter, dotted eighth, sixteenth, etc.), rather than where the note appears on the staff. Notice that each measure in Figure 1.5 has the equivalent of six eighth notes.

Although it is possible to count along with this music in three, it is most definitely not a waltz. Music in $\frac{6}{8}$ time often has a swaying, singsongy feel to it, and the typical measure is usually divided into two parts rather than six. Owing to this division, $\frac{6}{8}$ time is an example of a *compound meter* or *compound time*, since the main beat felt by the listener can be subdivided into three equal parts. One could count a piece in $\frac{6}{8}$ by counting two beats per measure, or by counting in six such as "*one*-two-three, *two*-two-three," with accents on the first and fourth eighth-note pulses. The English folk song *Greensleeves* (c. 1580) and R.E.M.'s *Everybody Hurts* (1992) are some other well-known examples in $\frac{6}{8}$ time.

Some composers juxtapose the compound $\frac{6}{8}$ time with the standard $\frac{3}{4}$ meter since it is easy to bounce back and forth between the two time signatures. A famous example that repeatedly alternates between $\frac{6}{8}$ and $\frac{3}{4}$ is *America* from Leonard Bernstein's *West Side Story* (1957, lyrics by Stephen Sondheim; see Figure 1.6). Each odd-numbered measure is in $\frac{6}{8}$ and features two groupings of three eighth notes, while the even-numbered measures are in $\frac{3}{4}$ and contain three quarter notes (rhythmically equivalent to three groups of two eighth notes). Even though the time signature changes in consecutive measures, the eighth-note pulse is constant throughout.

Notice the > marking above certain notes in the excerpt, instructing the singer, as well as the accompanying orchestra, to stress or accent the given note. The dots above each eighth note in the odd-numbered measures are *staccato* markings, instructing the musician to shorten the length of the note, while the bars above each quarter note in the even-numbered measures are *tenuto* markings, indicating that the note should be held for its full value. All of these markings serve to highlight the two versus three pattern in

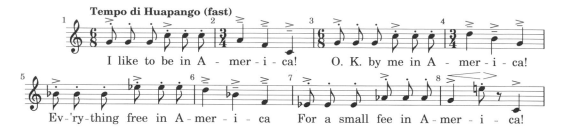

FIGURE 1.6. *America*, from Bernstein's *West Side Story*, displaying a distinctive rhythmic pattern that alternates between $\frac{6}{8}$ (two macrobeats per measure, each subdivided into three parts) and $\frac{3}{4}$ (three quarter-note beats, but the eighth-note pulse from the previous measure persists).

consecutive measures. Although the eighth-note pulse remains constant in each measure, the odd measures feel in two while the even measures are in three. Bernstein's musical depiction of the commutative property $(2 \cdot 3 = 3 \cdot 2)$ creates a memorable rhythmic pattern with a distinctive Latin flavor.

Other compound meters are $\frac{9}{8}$ and $\frac{12}{8}$ time, which have three and four macrobeats per measure, respectively, each subdivided into three eighth-note pulses. Dance and folk music from around the world employ these types of more complicated meters. The lively dance music of Ireland and Scotland, known as *jigs*, is often written using compound meters such as $\frac{6}{8}$ and $\frac{9}{8}$. The *Karşilama* folk dance of Northwest Turkey is in $\frac{9}{8}$ time. Some odd meters such as $\frac{5}{8}, \frac{7}{8}, \frac{7}{16}$, and $\frac{11}{16}$ can be found in the music of the southern Balkans (Bulgaria, Macedonia, and Greece). The popular Greek dance *Kalamatianós*, typically performed with the group arranged in a circle, is in $\frac{7}{8}$ time. All of these meters can be subdivided in different ways to create interesting and complex rhythmic patterns.

The second movement of Tchaikovsky's *Symphony No. 6, "Pathétique"* (1893), is an early illustrative example of a piece written in $\frac{5}{4}$ time. The entire movement has five quarter-note beats per measure, a very unusual meter for its day. The five beats of each measure can be grouped into a $2 + 3$ pattern, giving the music its dance-like character (see Figure 1.7). The third beat of the first and third measures shown (the cello part) features an eighth-note *triplet*, where three eighth notes are being compressed into the space normally occupied by two. For these triplets, the musician must play three equally spaced notes over one beat. Note the nice rhythmic symmetry of these measures: $1 + 1 + 3 + 1 + 1$, with the triplet placed in the center of the measure.

FIGURE 1.7. The opening melody (cellos) of the second movement of Tchaikovsky's *Symphony No. 6, "Pathétique,"* a movement written in $\frac{5}{4}$ meter.

Tchaikovsky's phrasing in each measure is also important. The curves above the music in Figure 1.7 are called *slurs*. The notes beneath a slur should be connected with as little gap as possible between consecutive notes. Each measure in the excerpt has a slur connecting the first two beats, and a slur or dotted half note over the last three beats. Here, the phrasing and rhythmic structure help delineate the $2 + 3$ pattern. The ♩ = 144 marking at the start of the excerpt is a tempo direction, giving the specific length of a quarter note. The value displayed is the number of quarter-note beats per minute, so in this case, 144 quarter notes is equivalent to 60 seconds, a quick tempo. The direction ♩ = 60 means one beat every second (a slow tempo), while ♩ = 120 implies two beats per second (twice as fast). Gustav Holst's *Mars, The Bringer of War*, from his famous seven-movement orchestral work *The Planets* (1914–1916), is another well-known example of a movement entirely composed in $\frac{5}{4}$ time.

From jazz, a famous example in $\frac{5}{4}$ meter is Paul Desmond's classic *Take Five* (1959), performed by the Dave Brubeck Quartet on their album *Time Out*. As the name indicates, this catchy tune also has five beats per measure and is offered as a contrast to the previous classical pieces in five. Unlike Tchaikovsky's *Pathétique*, here the subdivision of each measure is a $3 + 2$ grouping (see Figure 1.8). Notice that the dotted half notes occur at the start of certain measures, rather than the end. Underneath the groovy melody is a recurring rhythmic pattern played by the piano and drums which further accentuates the $3 + 2$ subdivision. Brubeck was inspired to produce jazz with alternative time signatures after hearing some Turkish music in $\frac{9}{8}$. The tune *Blue Rondo à la Turk*, on the same album, pays homage to its Turkish roots in name and through its $\frac{9}{8}$ meter.

FIGURE 1.8. The opening bars of Paul Desmond's famous *Take Five*, a jazz standard in $\frac{5}{4}$ time. The first four notes are a pickup, an incomplete measure.

An example in $\frac{7}{4}$ time from rock music is Peter Gabriel's *Solsbury Hill* (1977). This reflective song was written by Gabriel based on a spiritual experience he had atop Solsbury Hill in Somerset, England, just after leaving the band Genesis. Although there are seven quarter-note beats to a measure, the music combines so well with the underlying meter that it is hard to notice the unique time signature without counting along. Each measure is subdivided into a $3 + 4$ pattern. An apt use of a quarter-note triplet, three equally spaced notes sung over two quarter-note beats, occurs on the words "boom boom boom" in the chorus section. Pink Floyd's *Money* (1973), notable for its use of money-related sound effects at the start of the song, is another good example of a rock tune in $\frac{7}{4}$ time. As with *Solsbury Hill*, each of the seven-beat measures is subdivided into $3 + 4$.

1.2.2 Rhythmic repetition

Certain rhythmic patterns are so distinctive that they help define the character of a piece or even a whole musical style. These patterns are repeated continuously throughout a work, serving as a definitive foundation that connects musicians and listeners. A recurring rhythmic pattern in a piece of music is known as an *ostinato*. Sometimes a rhythmic

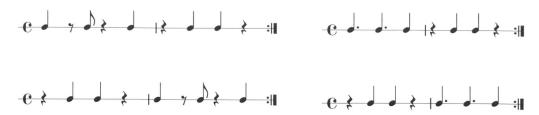

FIGURE 1.9. The 3-2 (*top*) and 2-3 (*bottom*) clave rhythmic patterns common in salsa music. Although notated differently, the left- and right-hand figures are rhythmically identical. The colon in front of the bar line at the end of the second measure is called a *repeat sign*, indicating that the two-measure pattern should be repeated.

ostinato is so memorable that it becomes central to an entire genre of music. Godfried Toussaint refers to such key rhythmic patterns as *timelines*.[1]

One important example of a timeline, which is instantly recognizable, comes from salsa music. The underlying rhythmic foundation for salsa is the *clave*, a unifying two-measure pattern shown in Figure 1.9. The word "clave" in Spanish means *key* or *keystone*. There is a "forward" 3-2 clave, which consists of three notes in the first measure and two in the second, or a "reverse" 2-3 clave, with two notes opening, followed by three in the second measure. This timeline is also known as *son clave* because of its Afro-Cuban origins and use in Cuban *son* music. The key clave rhythmic pattern is often highlighted in salsa music by the *claves*, two small wooden sticks that are struck together to produce a sharp, percussive sound.

One of the hallmarks of an interesting rhythmic pattern is syncopation, the notion of shifting notes off the primary beats in an unexpected manner. Try clapping (or better yet, find some claves) the 3-2 clave pattern, and observe how the notes fall with respect to the quarter-note beats of the measure. There is a note on the downbeat of the first measure (expected) and then again on the second half of the second beat (unexpected). There is no note on the third beat, but one occurs on the fourth beat (unexpected). In the second measure, regularity is thwarted again as notes are sounded on the second and third beats only.

If we divide the opening measure of the 3-2 clave pattern into eighth notes (writing it in $\frac{8}{8}$), then the notes fall on the first, fourth, and seventh pulses. Put another way, the first and second notes occupy a length of three eighth-note pulses (whether they actually sound for that duration is not important; they take up 3/8 of the measure until the next note is sounded), while the last note covers two eighth-note pulses. This $3 + 3 + 2$ subdivision of the measure is a syncopated pattern that is particularly noteworthy. In fact, it is common in music from all over the world, including Central Africa, Cuba, the Middle East, and the United States. It is featured prominently on the banjo for bluegrass music and was popular in early rock and roll (e.g., the bass line in Elvis Presley's version of *Hound Dog* [1956]). It is also the basis for the famous Charleston dance rhythm. The Grateful Dead used the full 3-2 clave pattern to underly their rendition of the popular song *Iko Iko*. A more recent example of the $3 + 3 + 2$ pattern is the song *Clocks* (2002) by British alternative rockers Coldplay, where the constant repetition of eighth notes is accented by the piano and drums to stress a $3 + 3 + 2$ subdivision in each measure.

---

[1]See Chapter 3 of Toussaint, *Geometry of Musical Rhythm*.

FIGURE 1.10. The string parts in the opening eight bars of *Les Augures Printaniers: Danses des Adolescentes* from Stravinsky's *Rite of Spring*, featuring unexpected accents (indicated by >) to liven up the repeating eighth-note pattern.

While the $3 + 3 + 2$ subdivision of a measure is fairly common, composers have also taken rhythmic repetition in different directions by providing accents in unexpected places. One of the most famous examples occurs in Igor Stravinsky's masterpiece *Le Sacre Du Printemps* (The Rite of Spring; 1913). This awe-inspiring musical work is actually a ballet, written in two parts. Stravinsky strove to create a musical depiction of raw and primitive life. The music is full of unconventional forms, ideas, and instrumentation. The meter often varies quickly within the work. For instance, at one point it changes from $\frac{5}{4}$ to $\frac{7}{4}$ to $\frac{6}{4}$ in consecutive measures.

Early in the piece, at the opening of the movement titled *Les Augures Printaniers: Danses des Adolescentes* (The Augurs of Spring: Dances of the Young Girls), the string section plays 32 consecutive staccato (short) eighth notes (see Figure 1.10). However, what makes the rhythm interesting is the seemingly random inclusion of accents by the strings and French horns located on the following eighth notes of the group: 10, 12, 18, 21, 25, 30. The gaps between successive accents, which are 2, 6, 3, 4, and 5, respectively, never repeat. The irregular placement of accents creates a dramatic and jarring percussive quality that keeps the listener on the edge of their seat. Two of the accents occur on downbeats, at the start of a measure, but there is no basic, intuitive pattern present.

After four measures, another group of repeated eighth notes occurs, this time for 10 consecutive measures (40 eighth notes) with accents on eighth notes 6, 8, and 14. Instead

of including accents in the last half of this group of eighth notes, Stravinsky begins inserting snippets of calls from other instruments such as the oboe and trombone. Later in the
section we find a third grouping of eighth notes which lasts for 35 consecutive measures.
Here the accents repeat the original structure, falling on the same numbered eighth notes
as in the first grouping. As more instruments begin to chime in with their response to
the recurring eighth-note pulse, Stravinsky continues the accents in the strings. However,
the accents become a bit less random and unexpected, as they feed off the calls of the
other instruments and the listener becomes more accustomed to their presence. Stravinsky's concept of misplaced accents over a repeating rhythmic pattern was quite novel and
striking for its time.

Exercises for Section 1.2

1. Using the numbers $1, 2, 3, 4, 5$, and 6, indicate the location of each eighth-note beat
   (the microbeats) in each measure of *Silent Night* in Figure 1.5.

2. How many quarter notes are required to fill up a measure in $\frac{7}{2}$ time? How many
   eighth notes are needed to fill up the same measure?

3. In $\frac{2}{4}$ time, what fraction of a measure is filled up by a double-dotted eighth note? How
   does your answer change if the time signature is $\frac{5}{8}$?

4. In $\frac{7}{8}$ time, a measure begins with a dotted quarter rest. How many eighth-note beats
   remain in the measure? What if the measure began with a double-dotted quarter rest?

5. The following measures of music are incomplete. In each case, the last note of the
   measure has been removed. Give the name of the note that completes the measure
   (e.g., quarter note, dotted eighth note, etc.).

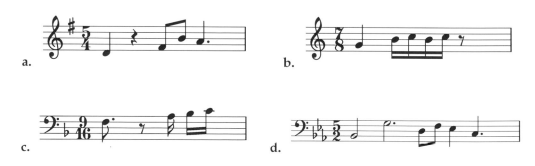

6. In each of the following excerpts of music, the time signature has been omitted. Assuming that an eighth note equals one beat, provide the correct time signature at the
   start of each excerpt.

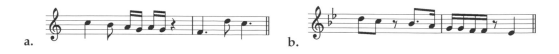

7. In each of the following excerpts of music, the time signature has been omitted. Assuming that a quarter note equals one beat, provide the correct time signature at the start of each excerpt.

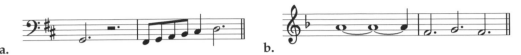

a.   b.

8. The following excerpts of music have different time signatures, but no bar lines. Insert the bar lines correctly into each excerpt.

a.

b.

9. Owen Blue is a wind player for an alternative "mathematical" symphony orchestra. While playing in $\frac{3}{2}$ time, he stumbles across a quarter note that is dotted infinitely often. (The composer has come up with some clever notation for this, so he is not actually looking at an infinite number of dots!) How long should Owen hold this particular note? Explain.

10. What is the time signature for the famous theme song to the TV show and movie series *Mission: Impossible*? Assume that a quarter note equals one beat.

11. Find a piece of music other than those described in the text which uses the $3 + 3 + 2$ subdivision of a measure.

1.3 Polyrhythmic Music

In the music of some cultures different rhythms often represent different elements of a story, and thus it is desirable to play them concurrently. This is particularly common in African tribal music and in the classical musical traditions of India. Playing two different rhythms simultaneously is called a *polyrhythm*. It occurs whenever a rhythmic pattern of one length is superimposed over another of a different length, such as three against four. For the polyrhythms discussed in this section, each distinct part is equally subdivided over the measure. For instance, a band might play a piece in $\frac{3}{4}$ time, with the drummer and bass playing three quarter-note beats per measure as the guitar and piano mark out four dotted eighth notes per measure. This is considerably challenging for the musicians involved and is very striking for the listener. Sometimes a polyrhythm may last for just a few measures, but there are examples, some of which we discuss below, where it is persistent

FIGURE 1.11. The tabla, a pair of Indian hand drums. © Andreas Gabler/iStock/Thinkstock.

throughout an entire piece. Polyrhythms sometimes occur in ballets or theatrical scores, where different characters dance to different beats.

In Indian classical music, the *tabla* is the primary percussion instrument, consisting of a pair of small hand drums (see Figure 1.11). The tabla player uses one hand on each drum and must play very complicated rhythmic patterns. These include difficult polyrhythms such as playing 11 beats in one hand while tapping 12 in the other, a feat that requires years of practice. The intricate rhythmic patterns are a form of language; the tabla player communicates by mimicking a set of syllables such as Dha, Tin, Na, Tun, or Ge. Syllables combine to form words, phrases, and sentences just as they would in any language.

Some basic polyrhythms

One of the simplest polyrhythms is the three-against-two pattern (see Figure 1.12). The top line of music features three quarter-note beats per measure, while the bottom line contains two dotted quarter-note beats. If we subdivide this measure into six eighth-note microbeats, as shown in the last two measures of Figure 1.12, then the top voice plays on beats one, three, and five, while the bottom plays on beats one and four. This is somewhat different from the triplets described in Section 1.2.1. A typical triplet consists of three notes played over a time interval of length two. In contrast, a polyrhythm occurs whenever both patterns are being played *simultaneously*.

Try tapping out the three-against-two polyrhythm using one hand (say, the right) for three beats and the other (the left) for two. The phrase "hot cup of tea" is a useful aid in mastering this pattern. Both hands play on "hot" at the start of the measure. The right hand also accents the words "cup" and "tea," while the left hand only plays on "hot" and "of."

FIGURE 1.12. The three-against-two polyrhythm, where the top voice plays three equally spaced notes per measure while the bottom plays two. The last two measures show the same polyrhythm in $\frac{6}{8}$ time, demonstrating the precise location of each note.

FIGURE 1.13. The four-against-three polyrhythm, where the top voice plays four equally spaced notes per measure while the bottom plays three. The last measure shows the same polyrhythm in $\frac{12}{16}$ time, demonstrating the precise location of each note.

Figure 1.13 shows a four-against-three polyrhythm. To determine the precise location of each note, and to compare the two contrasting parts, we must subdivide the measure into 12 sixteenth-note pulses. This is shown in the last measure of Figure 1.13, where the time signature changes to $\frac{12}{16}$ (12 beats per measure, with the sixteenth note equal to one beat). Observe that 3/4 and 12/16 are equivalent fractions. In this case, the top voice (four equally spaced pulses per measure) accents beats 1, 4, 7, and 10, while the bottom part (three equally spaced pulses) plays on beats 1, 5, and 9. To help play this polyrhythm, one possible accompanying phrase is "pass the golden butter," with the two hands sharing the beat on the word "pass."

1.3.1   The least common multiple

In the two previous examples, three-against-two and four-against-three, we subdivided the measure into 6 and 12 parts, respectively, in order to determine the precise location of each rhythm and their relation to each other. The relevant mathematical concept is the *least common multiple*. For example, to play a five-against-three polyrhythm, we require 15 subdivisions of the measure, while a six-against-eight pattern requires 24.

Throughout the book, we will use the symbol $\mathbb{Z}$ to denote the set of *integers* and $\mathbb{N}$ to represent the *natural numbers*. Specifically,

$$\mathbb{Z} = \{0, \pm 1, \pm 2, \pm 3, \pm 4, \ldots\} \quad \text{and}$$
$$\mathbb{N} = \{1, 2, 3, 4, \ldots\}.$$

Definition 1.3.1. *Suppose that m and n are two natural numbers. The* multiples of m and n *are, respectively,*

$$m, 2m, 3m, 4m, 5m, \ldots \quad and$$
$$n, 2n, 3n, 4n, 5n, \ldots .$$

A common multiple *is a number shared by each list, and the* least common multiple*, denoted by* LCM(m, n), *is the smallest of the common multiples.*

More succinctly, the least common multiple of two positive integers is the smallest integer that each number divides into evenly. In a polyrhythm featuring an m-against-n pattern, the least common multiple of m and n gives the smallest number of subdivisions required to break up the measure evenly and locate the precise placement of each note.

For example, if $m = 4$ and $n = 10$, then the multiples of m are $4, 8, 12, 16, 20, 24, 28, \ldots$ while the multiples of n are $10, 20, 30, 40, 50, \ldots$. Thus, the common multiples of m and n

are $20, 40, 60, 80, \ldots$ and the least common multiple is 20, that is, $\text{LCM}(4, 10) = 20$. Notice that the least common multiple of m and n divides evenly into any common multiple of m and n.

One of the simplest ways to find a multiple of two numbers is to just multiply them together. For instance, 120 is a common multiple of 6 and 20, and 132 is a common multiple of 11 and 12. The number mn is always a common multiple of m and n, for it is the nth number in the list of multiples of m, and the mth element in the list for n. However, the product mn is *not* always the least common multiple, as indicated by the example $\text{LCM}(4, 10) = 20$. The difference between this case and one such as $\text{LCM}(5, 12) = 60$ is that the numbers 4 and 10 have a common factor.

Definition 1.3.2. *The greatest common factor of two natural numbers m and n is called the* greatest common divisor, *denoted as* $\text{GCD}(m, n)$*. If two numbers m and n have no common factor other than 1, then* $\text{GCD}(m, n) = 1$*. In this case, m and n are called* relatively prime numbers.

For example, we have $\text{GCD}(4, 10) = 2$ and $\text{GCD}(15, 30) = 15$. The numbers 8 and 35 are relatively prime since they have no common factor other than 1. Recall that a natural number is a *prime number* if it has no factors other than itself and 1. The idea behind the terminology for two relatively prime numbers is that they are "prime" with respect to each other. For instance, the numbers 10 and 27 are not prime, but they have no common factor other than 1, so they are relatively prime. To avoid any confusion, we will assume that the number 1 is not prime.

Notice that 3 and 4 are relatively prime, and that $\text{LCM}(3, 4) = 3 \cdot 4 = 12$. Similarly, 10 and 27 are relatively prime, and $\text{LCM}(10, 27) = 10 \cdot 27 = 270$. Working through a few more examples, it becomes clear that when two numbers do not have a common factor, their least common multiple is simply the product of the two numbers.

Computing the least common multiple

To find a formula for the least common multiple when the two integers are not relatively prime, we consider some illustrative examples. Quick, what is the least common multiple of 15 and 18? The answer is 90. How could we obtain this answer promptly, without having to write down the multiples of each number? One way is to first compute their greatest common divisor, which is 3, and then write each number as a multiple of 3: $15 = 5 \cdot 3$ and $18 = 6 \cdot 3$. Now 15 and 18 are not relatively prime, but their divisors 5 and 6 are, so we have $\text{LCM}(5, 6) = 5 \cdot 6 = 30$. To find the least common multiple of 15 and 18, we just multiply this result by the greatest common divisor, $3 \cdot 30 = 90$, and presto, we have our answer.

Here is another example. To compute $\text{LCM}(12, 28)$, we first find that $\text{GCD}(12, 28) = 4$. Then, since $12 = 3 \cdot 4$ and $28 = 7 \cdot 4$, we find that $\text{LCM}(12, 28) = 3 \cdot 7 \cdot 4 = 84$. We can confirm this result by listing the first few multiples of 28, which are $28, 56, 84, \ldots$. Since 28 and 56 are not divisible by 12, but 84 is, we have that $\text{LCM}(12, 28) = 84$.

In each of the previous examples, note that $\text{LCM}(m, n) \cdot \text{GCD}(m, n) = mn$. For the first ($m = 15, n = 18$), we have $90 \cdot 3 = 270 = 15 \cdot 18$, while in the second ($m = 12, n = 28$), we have $84 \cdot 4 = 336 = 12 \cdot 28$. Put another way, the least common multiple of two numbers is found by taking the product of the two numbers and then dividing by their greatest common divisor. In essence, the known common multiple of mn is reduced by the factor that is common to both numbers.

While this sounds logical, checking a few examples is not the same as providing a general argument that works for *any* case. One of the strengths of mathematics is the ability to generalize from a specific example to an arbitrary one. This is typically done using variables (instead of numbers) and by presenting a logical and thorough argument based on some definitions and accepted concepts. Such an argument is known as a *proof*. Mathematicians specialize in writing clear and insightful proofs. It is the key step in turning a convincing hypothesis into an accepted theorem. As an example, we now provide a proof that definitively establishes the connection between the least common multiple of two numbers and their greatest common divisor.

Theorem 1.3.3. *Suppose that m and n are two positive integers. Then their least common multiple is found by*

$$\text{LCM}(m,n) = \frac{mn}{\text{GCD}(m,n)}. \tag{1.8}$$

The following proof relies on the fact that the least common multiple $\text{LCM}(m,n)$ always divides evenly into any common multiple of m and n. While this property can be shown rigorously, we accept it as true for the sake of brevity.

Proof: Let d denote the greatest common divisor of m and n, so $d = \text{GCD}(m,n)$. We then can write $m = j \cdot d$ and $n = k \cdot d$, where j and k are some positive integers satisfying $j \leq m$ and $k \leq n$. Note that j and k must be relatively prime; otherwise, we could pull out a common factor from each of them, and d would no longer be the *greatest* common divisor.

We claim that the number $j \cdot k \cdot d = jkd$ is the least common multiple of m and n. It is certainly a common multiple since it is equivalent to $k \cdot m$ and also $j \cdot n$, multiples of m and n, respectively. To see that it is the *least* common multiple, let $l = \text{LCM}(m,n)$. Since the least common multiple divides evenly into any common multiple, there exists a positive integer c such that $l \cdot c = jkd$. We would like to show that $c = 1$.

Since l is a common multiple of m and n, there exist positive integers a and b such that $l = a \cdot m = b \cdot n$. We then have

$$lc = jkd \implies amc = jkd \implies ajdc = jkd \implies ac = k,$$

and

$$lc = jkd \implies bnc = jkd \implies bkdc = jkd \implies bc = j.$$

The equations $ac = k$ and $bc = j$ imply that c is a common factor of both j and k. But since j and k are relatively prime, their only common factor is 1. Hence, it must be the case that $c = 1$. This shows that the least common multiple of m and n is in fact $l = jkd$.

To show that Equation (1.8) holds, we verify that it produces the correct result, namely, that $\text{LCM}(m,n) = jkd$. Substituting $m = j \cdot d$ and $n = k \cdot d$ into Equation (1.8), we find that

$$\text{LCM}(m,n) = \frac{jd \cdot kd}{d} = jkd,$$

as desired. This completes the proof.   □

Remark: Mathematics is full of definitions, theorems, and proofs. Understanding the underlying arguments is a key feature of the subject. Although the proof above is somewhat abstract and may be hard to follow, substituting numbers in for the variables may help to clarify the argument. The proof illustrates the power of mathematics. Using only the definitions of the relevant concepts, as well as a few facts, a logical and thorough argument has been constructed to verify that Equation (1.8) is valid in *all* cases.

Corollary 1.3.4. *If m and n are relatively prime, then their least common multiple is the product of the two numbers, that is,* $\text{LCM}(m,n) = mn$.

Proof: This follows directly from Theorem 1.3.3. If m and n are relatively prime, then $\text{GCD}(m,n) = 1$ by definition. Substituting this into Equation (1.8) quickly shows that $\text{LCM}(m,n) = mn$. $\qquad\square$

Example 1.3.5. *Find the least common multiple of* 12 *and* 44 *and check that Formula (1.8) yields the same result.*

Solution: We start by listing the multiples of 44, checking along the way for the first multiple that is divisible by 12. The first three multiples of 44 are $44, 88$, and 132. Both 44 and 88 are not multiples of 12 because 12 does not divide into either number evenly. However, since $132 = 11 \cdot 12$, we see that $\text{LCM}(12, 44) = 132$.

Alternatively, we note that $\text{GCD}(12, 44) = 4$. Plugging this value into Formula (1.8), we have

$$\text{LCM}(12, 44) = \frac{12 \cdot 44}{\text{GCD}(12, 44)} = \frac{528}{4} = 132,$$

as expected. $\qquad\square$

1.3.2 Musical examples

Most polyrhythms consist of numbers that are relatively prime. Patterns where one number is a multiple of the other are not considered polyrhythms. For example, in common time, a four-against-two pattern is simply four quarter notes against two half notes. Since two quarter notes equally subdivide one half note, this is not a particularly complicated rhythm. When m is a multiple of n, an m-against-n rhythmic pattern can always be equally subdivided, so the result becomes one straightforward rhythm, rather than a juxtaposition of two contrasting pulses. Hence, the truly interesting and complicated polyrhythms occur when the numbers involved are relatively prime.

There are musical examples where the numbers in question are not multiples yet still have a common factor greater than 1. However, these examples can be reduced to repeated cases where the numbers involved are relatively prime. For instance, a six-against-four polyrhythm is really equivalent to two consecutive three-against-two polyrhythms.

Chopin's piano music

The Polish composer and musician Frédéric Chopin (1810–1849) was a virtuoso pianist and prolific writer of piano music. Many of his compositions are artistically dazzling as well as technically challenging, in part due to the abundance of polyrhythms. A quick perusal of his collection of 20 nocturnes uncovers numerous examples of difficult polyrhythms between the right and left hands. For example, the final eight measures of Chopin's *Nocturne in C-sharp Minor* (1830; published posthumously [Chopin, 1998]) feature four different polyrhythms: 11-against-4, 13-against-4, 18-against-4, and a staggering

FIGURE 1.14. Some difficult polyrhythms in measures 9 and 10 of Chopin's *Nocturne in B Major*, op. 9, no. 3. Bar 9 features two five-against-three polyrhythms, while the second half of measure 10 shows a seven-against-three pattern.

35-against-4. To precisely locate the notes of a 35-against-4 polyrhythm would require subdividing the rhythm into 140 microbeats!

Chopin's *Nocturne in B Major*, op. 9, no. 3 (1830–1831) also contains many polyrhythms throughout the piece. Figure 1.14 shows three polyrhythms in measures 9 and 10. In the first measure of the excerpt, the right hand plays five equally spaced notes against the left hand's three. This polyrhythm occurs twice in a row. Then, in the second half of the following measure, the right hand plays seven against three in the left. The same piece also contains polyrhythms of 8-against-3, 14-against-3, and 11-against-4.

Most of Chopin's piano polyrhythms feature the higher-numbered part in the right hand, with the left-hand part keeping time in agreement with the primary beat. However, there are some examples where the left hand has a chance to shine with the more challenging part of the pattern. Figure 1.15 shows one such instance, featuring two polyrhythms from measures 79 and 80 of Chopin's *Barcarolle*, op. 60 (1845–1846). In the first measure of the excerpt, there is a six-against-seven polyrhythm, where the left hand plays seven and the right hand is in six (although not equally spaced because of the dotted eighth note). The second measure shows a six-against-five pattern, with the left hand again playing the odd number of beats. In each case, the left hand has the tougher subdivision.

It is worth mentioning that from a performance perspective, most of Chopin's polyrhythms are not expected to be played with strict mathematical accuracy. While the composer certainly indicates precise polyrhythms in his piano music, it is understood that a pianist is free to interpret these rhythms more lyrically than mathematically.

FIGURE 1.15. Two polyrhythms from Chopin's *Barcarolle*, op. 60, where the left hand has the more challenging rhythm.

FIGURE 1.16. Measure 3 of Chopin's *Nocturne in B-flat Minor*, op. 9, no. 1.

Example 1.3.6. *Figure 1.16 shows the third measure of Chopin's "Nocturne in B-flat Minor," op. 9, no. 1 (1830–1831). Identify the specific polyrhythm in the measure, and give the minimal number of parts needed to subdivide the measure in order to determine the precise location of each note.*

Solution: The right hand is playing 22 equally spaced notes (count them) against 12 in the left hand. Thus, we have a 22-against-12 polyrhythm. Although this could be divided into two 11-against-6 polyrhythms, the long connected beam in the right-hand part indicates that the 22 notes should be grouped as a single unit. The time signature is $\frac{6}{4}$, so there are six quarter-note beats per measure, or 12 eighth notes. Thus, the left hand maintains a steady eighth-note pulse while the right hand strives to subdivide the measure into 22 equal pieces. To determine the exact location of each note within the polyrhythm, we must subdivide the measure into 132 parts, since $\text{LCM}(12, 22) = 132$. Observe that $\text{LCM}(6, 11) = 66$, so if we had viewed the measure as two 11-against-6 polyrhythms, we would still require $2 \cdot 66 = 132$ subdivisions for the whole measure. $\square$

The Rite of Spring

One of the most memorable uses of polyrhythm in Western classical music coccurs in Stravinsky's *The Rite of Spring*, discussed earlier in Section 1.2.2. Polyrhythms occur in the movement *Cortège du Sage* (Procession of the Oldest and Wisest One) and the measures leading up to it (near the end of the first part). At first the tubas carry the melody (yes, the *tubas*) in four-bar phrases of $\frac{4}{4}$ time. The bass drum plays along but, just to make it interesting, accompanies the tubas quietly in three, hinting at the ensuing polyrhythmic chaos about to commence. Other instruments, such as the French horns, appear to be playing in four but repeat the same phrase every three measures. To add even greater rhythmic complexity, some instruments, such as the oboes, are accenting the second and fourth beats of every measure.

Then, the complexity becomes overwhelming when the time signature shifts to $\frac{6}{4}$ and the brass section enters in full force (rehearsal number 70). The dynamic markings throughout the orchestra are *ff* and *fff*, which means loud and very loud, respectively. At this point it is difficult to describe all the intricacies of Stravinsky's musical mayhem, but it can be compared to two competing orchestras trying to establish their own rhythmic will over each other. For example, the trumpets announce their parts with a quarter note followed by a dotted half note pattern (four beats total), three times every two measures. This is akin to a three-against-two polyrhythm. For further drama, the trumpets split into two groups, with one group playing the four-beat pattern at the start of the measure, while the other group, with an assist from the trombones, waits until the third beat of the measure to enter with the same pattern. A similar offset occurs in the percussion section, where the bass drum and tam-tam are playing three-against-four polyrhythms, but staggered from each other, while the timpani is accenting a four-against-six pattern. All of this occurs on top of the already-existing rhythmic structure carried over by the tubas, strings, and French horns. The end result is a wildly fantastic collage of competing rhythmic ideas.

In addition to the rhythmic innovations present in the work, *The Rite of Spring* also features complex harmonies, striking dissonance, and polytonality (multiple tonal structures occurring simultaneously). The work premiered as a ballet in Paris in 1913, and the jarring rhythms accompanied with the exotic, sexual dance led to a riot in the theater! Challenging to perform and thrilling to hear, *The Rite of Spring* is one of the great orchestral masterpieces of the twentieth century.

Poème Symphonique

Musical tempos such as ♩ = 96 (96 quarter-note beats per minute) can be measured and heard using a time-keeping device called a *metronome* (see Figure 1.17). The sound omitted by a metronome is a simple "tick-tock," back and forth. An accurate metronome is an indispensable tool for a musician, not only to determine the correct tempo, but also for accurately staying within that tempo.

Figure 1.17. An old-school metronome, a musical time-keeping device, and the only instrument used in Ligeti's polyrhythmic *Poème Symphonique*. © shipfactory/iStock/Thinkstock.

Given that its sole purpose is to determine a steady pulse, it is rare to find a piece of music that incorporates the metronome as an instrument. Rarer still is the musical work that *only* features a metronome for instrumentation. Such is the case for the piece *Poème Symphonique* by the Hungarian composer György Ligeti, written in 1962. The work requires 100 metronomes, all of which are set to *arbitrary* tempos. The metronomes begin together (after a brief silence), quickly become out of sync, and then one by one gradually come to a stop after several minutes. Old-school metronomes, which need to be repeatedly wound to work, typically stop after several minutes, so these are the "instrument" of choice for an actual performance. The piece sounds a bit like rain beginning to fall on a copper roof, although near the end, when only a few metronomes remain beating, the small polyrhythms are quite enjoyable from a mathematical perspective. The single beating metronome that closes the piece is delightfully simplistic.

Poème Symphonique is the ultimate cerebral experiment involving polyrhythms. Each metronome counts out its own distinct pulse, so in essence, there are 100 different rhythms occurring simultaneously (assuming that no two metronomes have the same tempo). It is a fun challenge to calculate when several metronomes with slightly different tempos will next beat together as one (see the last two exercises of this section). Incidentally, Ligeti was not the first to use the metronome as a musical instrument. The opening of Ravel's *L'Heure Espagnole* (1907–1909) and the third movement of Villa-Lobos's *Suite Sugestiva* (1929) both incorporate the metronome.

Fake Empire

We close this section with a fabulous example of a polyrhythmic rock song from the indie rock band The National. Written by Matthew Berninger and Bryce Dessner, *Fake Empire* (2008) features a four-against-three polyrhythm throughout the entire piece, perhaps the first rock tune to ever accomplish such a feat.[2] The piano part is shown in Figure 1.18. Both the right hand of the piano and the guitar play in four, while the left hand of the piano, vocals, and drums play in three! Dessner, a classically trained guitarist who is also a member of the avant-garde ensemble Bang on a Can, strives to expose his rock audiences to some of the more interesting forms from the modern classical world of music. In addition to the polyrhythm, Australian composer Padma Newsome, who was asked to write horn parts for the end of the song, provides a minimalist feel in the style of Steve Reich to close the piece. We will discuss some of Reich's music in Section 8.2. Although Dessner describes the piano part and its definitive polyrhythm as "*Chopsticks* simple,"[3] ultimately the polyrhythm is what distinguishes this remarkable song.

FIGURE 1.18. The primary piano part of The National's polyrhythmic hit *Fake Empire*. The right hand plays in four while the left hand remains in three for the entire piece.

Exercises for Section 1.3

1. Which of the following pairs of numbers are relatively prime?

   a. $5, 6$ **b.** $4, 18$ **c.** $11, 55$ **d.** $16, 81$ **e.** $18, 81$

   f. $2^m, 3^n$, where m and n are natural numbers

   g. p, q, where p and q are distinct prime numbers

   h. $p, 15p$, where p is a prime number ($p > 1$)

2. Find the least common multiple of 7 and 11, that is, find $\text{LCM}(7, 11)$. Find the greatest common divisor of 7 and 11, that is, find $\text{GCD}(7, 11)$.

---

[2]Other notable examples include *First Tube* by Phish, *Let Down* by Radiohead, and Frank Zappa's *The Black Page*.

[3]Richardson, National's Brand of Intelligent Art-Rock.

3. What is GCD(12, 20)? Use Formula (1.8) to find LCM(12, 20). Check your answer by listing the first few multiples of 12 and 20.

4. What is GCD(24, 30)? Use Formula (1.8) to find LCM(24, 30). Check your answer by listing the first few multiples of 24 and 30.

5. Find GCD(15, 100) and LCM(15, 100).

6. Vivian, an eager young percussionist, has joined a contemporary music ensemble. While practicing a new piece, she comes across a measure of music with a challenging polyrhythm of 12-against-9. How many parts should Vivian subdivide the measure into in order to determine the precise location of each note in the polyrhythm?

7. Figure 1.19 shows measure 14 of Chopin's *Nocturne in B-flat Minor*, op. 9, no. 1. Identify the specific polyrhythm in the second half of the measure and give the minimal number of subdivisions needed to determine the precise location of each note in the polyrhythm.

FIGURE 1.19. Measure 14 of Chopin's *Nocturne in B-flat Minor*, op. 9, no. 1.

8. Figure 1.20 shows measure 73 of Chopin's *Nocturne in B-flat Minor*, op. 9, no. 1. There are four polyrhythms in the measure. Identify each polyrhythm. Give the minimal number of subdivisions needed to determine the precise location of each note in the final polyrhythm of the measure.

FIGURE 1.20. Measure 73 of Chopin's *Nocturne in B-flat Minor*, op. 9, no. 1.

9. What property must two integers m and n have so that $\text{LCM}(m, n) = \dfrac{mn}{2}$? Give an example.

10. What property must two integers m and n have so that $\text{LCM}(m, n) = m$? Give an example.

11. Define $\text{LCM}(m, n, p)$ to be the least common multiple of the natural numbers m, n, and p. This is the smallest common multiple of all three numbers. Find $\text{LCM}(4, 6, 10)$, $\text{LCM}(4, 6, 11)$, and $\text{LCM}(4, 6, 12)$.

12. Define $\text{GCD}(m, n, p)$ to be the greatest common divisor of the natural numbers m, n, and p. This is the largest possible number that divides evenly into each number. Find $\text{GCD}(4, 6, 10)$, $\text{GCD}(10, 15, 50)$, and $\text{GCD}(7, 11, 15)$.

13. Find three natural numbers m, n, p, all greater than 1, such that $\text{LCM}(m, n, p) = mnp$.

14. Show by way of an example that $\text{LCM}(m, n, p)$ does *not* necessarily equal $\dfrac{mnp}{\text{GCD}(m, n, p)}$.

15. Find a formula for $\text{LCM}(m, n, p)$ in terms of $\text{LCM}(m, n)$ and p. Simplify your formula so that mnp is in the numerator. *Hint:* The initial formula begins as the least common multiple of a least common multiple.

16. Four metronomes are set ticking at the tempos of $\quarternote = 60$, $\quarternote = 70$, $\quarternote = 80$, and $\quarternote = 90$. Assuming that they all begin together, after how many seconds will they come together again at the same time?

17. Three metronomes are set ticking at the tempos of $\quarternote = 65$, $\quarternote = 68$, and $\quarternote = 75$. Assuming that they all begin together, after how many seconds will they come together again at the same time?

1.4   A Connection with Sanskrit Poetry

Manjul Bhargava is a renowned mathematician, winner of the prestigious Fields Medal in 2014. He is also a master tabla player. According to Bhargava, Indian classical music is inherently mathematical. In addition to some explicit connections, one of which is descibed here in detail, Bhargava also finds an artistic link between the two fields. He writes that "mathematical and musical ideas can frequently bounce off one another, ... because the creative processes involved are quite similar and they complement each other well."[4] The notion that mathematical concepts can influence musicians, particularly composers, will be explored in the second half of this book.

    One interesting connection between math and poetry involves the subject of *number theory* and a famous sequence of numbers first described by the Indian scholar Hemachandra (1089–1172). The Indian scholars Gopala (c. 1135) and Hemachandra were interested in analyzing the rhythmic structure in Sanskrit poetry. The rhythmic phrases were typically divided into long two-beat pulses, denoted by L, and short one-beat pulses, denoted by S. For example, a phrase six beats long could be subdivided as L L L, S S S S L, or S L S L.

    One of the properties that Hemachandra studied was the number of different ways to subdivide n beats into long and short pulses. For example, when $n = 3$, there are only three ways to subdivide three beats: S S S, S L, or L S. Any other S-L pattern will be either too short or too long. For $n = 4$, the number of possible subdivisions increases to five: S S S S, S L S, S S L, L S S, or L L.

---

[4] AMS Member Newsletter, Three Questions for Manjul Bhargava, p. 2.

There are eight ways to subdivide five beats, and it is instructive to assemble these eight different methods using the results for $n = 3$ and $n = 4$. Each pattern must end with either an L (two beats) or an S (one beat). If the pattern ends with an L, then we have three beats preceding it to make a total of five. This means we can append an L to each of the sequences found for $n = 3$:

$$S\,S\,S\,L, \quad S\,L\,L, \quad \text{or} \quad L\,S\,L.$$

These are the only possibilities that end with an L. Likewise, to list the subdivisions ending in an S, we take each pattern from the case $n = 4$ and append an S, giving the following five examples:

$$S\,S\,S\,S\,S, \quad S\,L\,S\,S, \quad S\,S\,L\,S, \quad L\,S\,S\,S, \quad \text{or} \quad L\,L\,S.$$

These are the only ways of subdividing five beats and ending with an S. Putting the two results together, the total number of ways to subdivide five beats is $3 + 5 = 8$. Note that we could have performed the same count by placing an S in front of the five sequences for $n = 4$ and an L in front of the three sequences for $n = 3$.

This method of counting is *recursive*. The number of subdivisions for the next value of n can be obtained by summing the previous two results. In other words, if H_n represents the number of ways to subdivide an n-beat pattern into S-L sequences, then we have just argued that $H_5 = H_4 + H_3$. More generally, we have the relationship

$$\boxed{H_n \ = \ H_{n-1} + H_{n-2}.} \tag{1.9}$$

Simply put, to create an S-L sequence of length n, append an S to every pattern of length $n - 1$ and append an L to every sequence of length $n - 2$. Since there are precisely H_{n-1} ways to do the former and H_{n-2} different ways to accomplish the latter, the total number of ways to create an S-L sequence of length n is their sum, $H_{n-1} + H_{n-2}$. We know that all possible n-beat patterns have been counted because each one ends with either an S or an L. This confirms that Equation (1.9) is valid for any $n \geq 3$.

Equation (1.9) is an example of a *recursive formula*. It is not an explicit relationship written in terms of n. Rather, to obtain the next term in the sequence, the nth term, we sum the previous two terms. For example, $H_6 = H_5 + H_4 = 8 + 5 = 13$, and $H_7 = H_6 + H_5 = 13 + 8 = 21$. One of the benefits of having this formula is that it provides a way to compute the total number of n-beat patterns without having to painstakingly write them all down. We know that there are 21 ways to subdivide seven beats into S-L patterns because $H_7 = 21$.

The Hemachandra–Fibonacci numbers

Note that $H_1 = 1$ since there is only one way to subdivide one beat, namely, just S. We have $H_2 = 2$ because two beats can be subdivided as either S S or L. Using Equation (1.9) repeatedly, we can generate a very famous sequence of numbers,

$$1, \ 2, \ 3, \ 5, \ 8, \ 13, \ 21, \ 34, \ 55, \ 89, \ \ldots, \tag{1.10}$$

which we will call the *Hemachandra numbers* and denote as H_n. The subscript n in H_n is called an *index*, since it indicates the location of the number within the list. Thus, $H_4 = 5$, $H_8 = 34$, and $H_{200} = H_{199} + H_{198}$.

In the West, the sequence (1.10) is more commonly known as the *Fibonacci sequence*, or the *Fibonacci series*, although calling the numbers a series incorrectly implies that they should be summed. The name Fibonacci was a pseudonym for the Italian mathematician Leonardo Pisano (c. 1170–1250), who wrote about the numbers in his important text *Liber Abaci* (Book on Computation; 1202), a text that moved Europe away from Roman numerals to the Hindu-Arabic system commonly used around the world today. Based on his solution to a riddle about rabbit breeding, Fibonacci presented the numbers as

$$1, 1, 2, 3, 5, 8, 13, 21, 34, 55, 89, \ldots, \tag{1.11}$$

a sequence we will refer to as the *Fibonacci numbers*. These numbers are identical to Hemachandra's except that they begin 1, 1 rather than 1, 2. The same fundamental recursive relationship generates both sequences; *the next number in the list is found by summing the previous two.* Given that Hemachandra, and likely other Indian mathematicians before him, studied the numbers earlier than Fibonacci, it only seems fair to include his name when referring to this famous sequence. We have established the following connection between ancient Indian poetry and mathematics.

Theorem 1.4.1. *The number of ways to subdivide n beats into a pattern of short (one beat) and long (two beats) pulses is given by the nth Hemachandra number or the $(n + 1)$th Fibonacci number.*

Example 1.4.2. *Find H_{11} and H_{12}. How many ways can 12 beats be subdivided into S-L patterns?*

Solution: From the list of Hemachandra numbers given in sequence (1.10), we see that $H_9 = 55$ and $H_{10} = 89$. Using the recursive formula, this implies that $H_{11} = H_{10} + H_9 = 89 + 55 = 144$ and $H_{12} = H_{11} + H_{10} = 144 + 89 = 233$. It follows from Theorem 1.4.1 that there are 233 different ways to subdivide 12 beats into S-L patterns. □

Exercises for Section 1.4

1. Find H_{13} and H_{14}. How many ways can 14 beats be subdivided into S-L patterns?

2. A common practice for mathematicians is to play around with numbers and look for any resulting patterns. In this exercise we will find an interesting identity involving the Hemachandra numbers.

   a. Compute each of the following sums: $H_1 + H_3$, $H_1 + H_3 + H_5$, $H_1 + H_3 + H_5 + H_7$, and $H_1 + H_3 + H_5 + H_7 + H_9$.

   b. Notice that each sum is very close to another Hemachandra number. Rewrite each sum in terms of the closest number in the sequence. For example, $H_1 + H_3 = H_4 - 1$.

   c. Using your results from part **b.**, find a formula for summing the first n odd Hemachandra numbers. In other words, find a formula for

   $$H_1 + H_3 + H_5 + H_7 + \cdots + H_{2n-1}$$

   (no proof necessary).[5] The index in your answer will depend on n (e.g., H_{n-3}).

   ---
   [5]A rigorous proof can be given using the Principle of Mathematical Induction.

3. Applying the technique used in the previous problem, find formulas for the following expressions of Hemachandra numbers. A typical answer will contain n in some way (e.g., $H_{n-3} + 2$ or $7n - 1$). A mathematical proof is not expected, just a conjectured formula.

     a. $H_1 + H_2 + H_3 + H_4 + \cdots + H_n$.

     b. $H_n \cdot H_{n+2} - H_{n+1}^2$.

     c. $H_1^2 + H_2^2 + H_3^2 + H_4^2 + \cdots + H_n^2$.

4. Edouard Lucas was a nineteenth-century French mathematician who studied and generalized the Hemachandra–Fibonacci numbers. One sequence of numbers he created uses the same recursive formula as Equation (1.9) but begins with the numbers 1 and 3. These numbers are called the *Lucas numbers* and are denoted L_n. Thus, $L_1 = 1$, $L_2 = 3$, and $L_n = L_{n-1} + L_{n-2}$ for each $n \geq 3$.

     a. Write out the first 12 Lucas numbers.

     b. Which sequence seems to be growing faster, Hemachandra or Lucas?

     c. Find a simple formula relating the Hemachandra H_n, Fibonacci F_n, and Lucas L_n numbers (no proof required).

5. Suppose we wanted to subdivide n beats, but instead of just using S (one beat) or L (two beats), we are also allowed to use a longer three-beat pulse, denoted by T. Thus, the patterns S L T S T and T L T L are two different ways to subdivide 10 beats. Let D_n be the number of ways to subdivide n beats into patterns using S, L, or T.

     a. What are the values of D_1, D_2, and D_3?

     b. Write down all the ways of subdividing four beats. What is D_4?

     c. Following the argument used in the text, what is the recursive formula that can be used to find D_n from the previous values? Use your formula to write out the first 10 terms of the sequence, that is, write out $D_1, D_2, \ldots, D_{10}$.

     d. How many ways can 10 beats be subdivided into patterns using S, L, or T?

References for Chapter 1

AMS Member Newsletter: Spring 2008, Three Questions for Manjul Bhargava, Professor of Mathematics at Princeton University, American Mathematical Society. `http://www.ams.org/membership/individual/membnewsltr-spring2008.pdf`.

Bernstein, L.: 1959, *West Side Story*, G. Schirmer, Inc., distributed by Hal Leonard Corporation. Lyrics by Stephen Sondheim.

Chopin, F.: 1989, *Fantasy in F Minor, Barcarolle, Berceuse and Other Works for Solo Piano*, Dover Publications, Inc., New York. Edited by Carl Mikuli.

Chopin, F.: 1998, *Nocturnes and Polonaises*, Dover Publications, Inc., Mineola, New York. Edited by Carl Mikuli.

Desmond, P.: 1960, *Take Five*, Desmond Music Company.

Fallows, D.: 2007–2014, Metronome (i), *Grove Music Online*, Oxford University Press.

Ferguson, K.: December 2004, Mastering Odd, Complex Time Signatures and Rhythms, *Guitar Nine*. http://www.guitar9.com/.

Garland, T. H. and Kahn, C. V.: 1995, *Math and Music: Harmonious Connections*, Dale Seymour Publications.

Gordon, E. E.: 2012, *Learning Sequences in Music: A Contemporary Music Learning Theory*, GIA Publications, Inc.

Hendrickson, S.: November 2006, Musical Traditions: Odd Meters, 7/8 Anyone?, *Victory Review*.

Kamien, R. (ed.): 1984, *The Norton Scores: An Anthology for Listening*, 4th edition, W. W. Norton & Company.

Richardson, D.: June 21, 2007, The National's Brand of Intelligent Art-Rock Quietly Hooks Listeners In, *San Francisco Chronicle*.

Schultz, S.: December 8, 2003, Bhargava Strikes Balance among Many Interests, *Princeton Weekly Bulletin* **93**(12).

Stravinsky, I.: 1989, *The Rite of Spring*, Dover Publications, Inc., New York.

Tchaikovsky, P. I.: 1993, *Symphony No. 6 in B Minor, "Pathétique."* New Edition of the Complete Works, Series II: Orchestral Works, vol. 39b, Schott Music. Edited by Thomas Kohlhase.

Toussaint, G. T.: 2013, *The Geometry of Musical Rhythm: What Makes a "Good" Rhythm Good?*, CRC Press.

Verdi, G.: 1965, *Messa da Requiem*, Edition Peters, no. 4251. Edited by Kurt Soldan.

Chapter 2

Introduction to Music Theory

The most important thing in our world is emotional connections—to people who are far away and to people who are near; humanity is our community, and music binds it together. Music addresses people's internal imagination and memory through sound; it helps to understand things from the inside. That's what music is for.

— Yo-Yo Ma[1]

This chapter introduces the key concepts from standard music theory which are necessary for understanding the examples and concepts in the text. The ability to read music, to comprehend the notes on the page, is fundamentally important. Becoming proficient at reading music and understanding its underlying theory is inherently a mathematical process. Here, some mathematical interpretations are explicitly given (e.g., the concept of invariance), while others are implied (e.g., learning musical intervals involves counting). The primary goal is to provide enough background, worked examples, and instructive exercises to teach the fundamental concepts of music theory. It is not assumed that the reader knows how to read music or is familiar with elementary music theory.

We begin with the staff and the naming of notes in different clefs, focusing mostly on the treble and bass clefs. To facilitate comprehension, the piano keyboard is presented as an important learning tool. Scales, intervals, accidentals, and chords are all easier to understand using a piano keyboard. Moreover, using an actual piano stresses the importance of hearing the notes on the page when learning to read and understand music. A detailed description of the standard scales (chromatic, whole-tone, major, and minor) follows with an emphasis on half steps and whole steps. Intervals and chords are discussed with specific musical examples. The useful circle of fifths is presented with a detailed explanation of how and why it works; the fact that 7 and 12 are relatively prime is crucial, but so is the structure of the piano keyboard and the way we tune our instruments. Key signatures are explained along with demonstrations of their usefulness when transposing to a different key. The chapter closes with a brief history of the evolution of polyphony in Western classical music, focusing on tonality and intervals. This helps develop a deeper musical appreciation, as well as provide some context for future material in the book (e.g., different tuning systems and modern music).

---

[1]Dyer, With Ear for Diversity, Cellist Yo-Yo Ma.

2.1   Musical Notation

Much like mathematics, music has developed its own highly specialized form of notation that is broadly accepted around the world. Although it can seem daunting at first, especially for those who have never learned to read music, with time and practice the ability to follow and understand musical notation can feel as natural as reading words on a page.

Musical notes are situated on a set of five equally spaced horizontal lines called a *staff*. The notes may be placed on the lines or in the spaces between lines. Lower notes are located at the bottom of the staff, while higher notes are found at the top (see Figure 2.1). The staff may be extended using *ledger lines*, small horizontal line segments equally spaced below or above the staff, which allow for even lower or higher notes, respectively.

Higher Notes

Lower Notes

FIGURE 2.1. The musical staff. Notes at the bottom of the staff are lower in pitch than those at the top.

Notes in Western music are named using a seven-letter alphabet:

A   B   C   D   E   F   G   A   B   C   D   E   F   G $\cdots$.

The process is *cyclic*, meaning that G is always followed by A, after which the list repeats again. The two notes above named "A" are an *octave* apart, a vitally important musical interval. Notes with the same letter name are all related by some number of octaves. Two notes that are an octave apart will sound the "same." We will explore why this is the case in Section 3.4.3 when we discuss the overtone series. When an orchestra is tuning to concert A440, the various instruments are all playing "A," but in different octaves.

2.1.1   The common clefs

There are a wide range of pitches available on most musical instruments, and the gaps grow even wider when comparing different instruments. To accommodate this variety, different *clefs* are used on the musical staff, depending on the natural range of the instrument or voice (see Figure 2.2). The *treble clef*, or *G clef*, is typically used for higher-pitched instruments such as the violin, trumpet, flute, oboe, clarinet, and piccolo and for soprano or alto voices. It is also the common clef for the right-hand part in a piano piece. The *bass clef* or *F clef* is usually reserved for lower-pitched instruments such as the trombone, tuba, cello, bassoon, timpani, and, of course, the bass (instrumental or vocal). It is the common clef for the left-hand part on the piano. In addition, there are two other clefs, the *alto clef* and the *tenor clef*, which are typically reserved for a few specialized instruments. For example, the viola part is usually written in the alto clef, while some bassoon and trombone parts utilize the tenor clef. There are always exceptions to these categorizations, so a good musician can move comfortably from one clef to another. While other clefs exist, these four are the ones typically found in Western music. Of the four, the treble and bass clefs are by far the most common.

FIGURE 2.2. The various clefs used to indicate different notes on the staff. From left to right are shown the treble, bass, alto, and tenor clefs.

The key to reading music in different clefs is to determine one particular note on the staff and then use the cyclic musical alphabet to find the other notes. All clefs use the same seven letters, but the location of those particular notes varies from clef to clef. Climbing up one position on the staff, from line to space or vice versa, corresponds to moving up one letter of the musical alphabet.

For example, the final spiral of the ornamental treble clef always encircles a G on the second line of the staff from the bottom. Thus, the space above this G corresponds to an A, and the line above this G corresponds to a B (the middle of the staff). Similarly, the space below the G represents an F and the line below G gives an E (the bottom line of the staff). Continuing in this fashion, we can label each note on the different lines and spaces of the staff for the treble clef (see Figure 2.3). Notice that the notes on the spaces of the treble clef spell out the word "FACE." A popular phrase to help recall the notes on the lines of the treble clef is "Every Good Boy Deserves Fudge," where the first letter of each word gives the names of the notes.

FIGURE 2.3. The notes on the staff for the treble or G clef. In this case, the note G is located on the line encircled by the spiral of the clef. As the notes move up the staff, the letter names move up the musical alphabet cyclically.

The bass clef, or F clef, always has its two dots surrounding the line corresponding to the note F. Consequently, the note on the space below this line (two spaces from the top) is an E, while the note on the top space of the staff is a G. The full set of notes for the bass clef is shown in Figure 2.4. Common phrases for the notes on the lines and spaces of the treble clef are "Good Boys Deserve Fudge Always," and "All Cows Eat Grass," respectively.

As described above, ledger lines may be used to place notes higher or lower than those given on the staff. These lines are small, horizontal segments that should be equally spaced at a distance equivalent to the space on the staff. The names of the notes on ledger lines and in the spaces between follow the same cyclic pattern of the musical alphabet.

FIGURE 2.4. The notes on the staff for the bass or F clef. Here the note F is located on the line between the two dots of the clef.

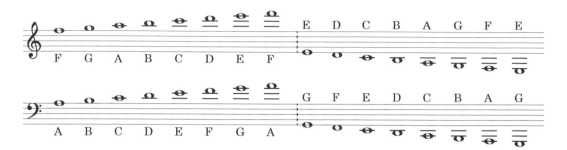

FIGURE 2.5. Ledger lines can be used to extend the staff above or below, in either the treble or bass clefs. Notes on and between ledger lines are named by following the musical alphabet cyclically.

For instance, in the treble clef, the note on the first ledger line above the top of the staff is an A, the note in the space above the first ledger line is a B, and the note on the ledger line above that is a C. A similar pattern occurs for the notes below the staff and with the bass clef (see Figure 2.5).

As can be seen in Figure 2.5, the note names below the staff for the treble clef match those above the staff for the bass clef. The corresponding pitches are actually identical. The C that is on the first ledger line *below* the staff of the treble clef is the same note (identical pitch) as the C on the first ledger line *above* the staff of the bass clef. This C is called *middle C* and corresponds to the "middle C" close to the center of a piano keyboard. On many pianos this note is directly below the first letter of the brand name. For singers, a middle C is relatively low in the voice range for an alto or soprano, while it is at the higher end of the spectrum for a tenor or bass.

Figure 2.6 indicates the location of middle C in each of the four clefs. To reiterate, playing this note in all four cases will produce the exact same pitch (or same frequency). Notice that the symbols for the alto and tenor clefs are symmetric with respect to the line on the staff containing middle C. For this reason, each clef is an example of a *C clef*. For the alto clef, middle C occurs on the middle line of the staff, while it is on the second line from the top for the tenor clef. The other notes on the staff for these two clefs are then determined in the same fashion as before, applying the musical alphabet in a cyclic fashion. Thus, the note names on the lines of the alto clef, moving from bottom to top, are F A C E G, while those on the lines of the tenor clef are D F A C E. The letters between those listed provide the names of the notes in the spaces. By cycling through the musical alphabet, the names of the notes on and between ledger lines are found in a similar manner to those for the treble and bass clefs.

FIGURE 2.6. The location of middle C for the treble, bass, alto, and tenor clefs (positioned from left to right, respectively).

The piano staff

Piano music combines the treble clef (right hand) and bass clef (left hand) into one staff for easier reading. This is called the *piano staff* or *grand staff* and is denoted by a large curly bracket connecting the two staves. In order to read piano music successfully, pianists must learn to be comfortable reading each clef simultaneously. In Figure 2.7 we show the piano staff along with the names of all the notes on the lines of each staff. Middle C connects the two staves, as it represents the same pitch on either staff.

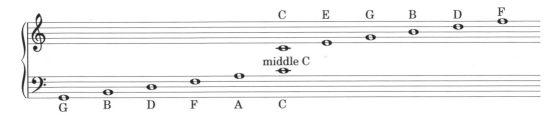

FIGURE 2.7. Climbing up the lines of the piano staff. The right hand reads the treble clef, while the left hand follows the bass clef. The two clefs are connected by their common note, middle C.

2.1.2   The piano keyboard

When learning to read and understand music, the piano keyboard is an extremely useful tool. Even for those who have little experience playing the piano, visualizing the notes on the piano keyboard facilitates the comprehension of music theory. For the exercises and examples in this chapter, it is highly recommended that the corresponding notes on the piano be played in order to make both an aural and visual connection with the music.

The modern piano consists of 88 keys that are divided into just over seven octaves. Each octave consists of 12 notes, containing seven white keys and five black keys. The seven white keys correspond to the seven notes of the musical alphabet, A B C D E F G. The black keys represent notes in between some of these notes, as indicated in Figure 2.8. These notes require an additional symbol called an *accidental*. A *sharp* (♯) raises the note up a step, while a *flat* (♭) lowers the note down a step. Thus, the black key between C and D can be named either C♯ or D♭. Describing the same note with two different names is known as *enharmonic equivalence*. For example, the notes B♭ and A♯ are enharmonically equivalent because they represent the same note between A and B. In addition to the dual naming, the other tricky feature of the piano keyboard is that there are only five black keys in an octave, as compared with seven white keys. There is no black key between the notes B and C, nor is there one between the notes E and F. The reason for this will become apparent when we study major scales in Section 2.2.

Figure 2.8 shows the names of the keys of a section of the piano keyboard. Notice the asymmetry of the keyboard, with the black keys grouped in subsets of two and three. C is the white note just below the black key in the group of two, while F is the white note just below the black key in the group of three. The asymmetry helps pianists quickly find the correct notes.

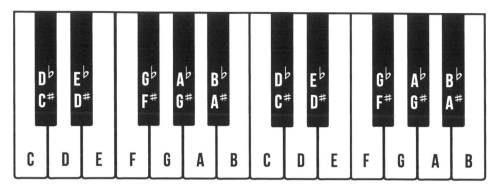

FIGURE 2.8. A section of the piano keyboard beginning on C. Note the different but equivalent names of the black keys. If we start on any particular note, the next time we encounter that same note we will have traveled an octave.

Half steps and accidentals

The notion of distance between two musical notes, called an *interval*, is critically important in music. We have already learned one important interval, the octave, which is the distance between a note and the next appearance of that note in the musical alphabet. The basic building block in Western music is the *half step*, which corresponds to the interval between two consecutive keys on the piano. For example, the interval between A (white key) and A♯ (the adjacent, higher black key) is a half step, as is the distance between D♯ (black key) and E (white key). Moving from B to C is also a half step, even though there is no black key between the two notes. The octave is divided into 12 equal half steps, a fact that may seem obvious by counting the number of keys between one note and its next appearance on the piano. However, the basic idea of a 12-note equally subdivided octave took many centuries to establish as an accepted rule in our methods of tuning.

To indicate a ♯ or ♭ on the staff, the accidental is placed *before* the note it is intended to modify. Even though we say "G-sharp" and write G♯, the sharp symbol is placed before the note head on the staff (see Figure 2.9). The same rule applies for flats. Also, the accidental is always placed on the *same* line or space as the note head; otherwise, it would appear to be altering a different note. By convention, accidentals attached to a note are assumed to last for the remainder of a measure. Thus, if a D♭ occurs at the start of a measure, every remaining D in the measure is also considered to be a D♭, even if there is no ♭ symbol attached to the remaining Ds. This only applies to notes located on the same line or space as the note receiving the initial accidental. A D in a different octave still remains a D.

FIGURE 2.9. Using sharps and flats to modify a note. The notes in each measure are identical, that is, they are enharmonically equivalent. The accidental is always placed before the note and on the same line or space as the note head.

In addition to the ♯ and ♭ symbols, there are three other types of accidentals used to modify a note. The most important of these is the natural sign ♮, which is used to cancel a previous accidental (of any type). This is important when two notes with the same letter name occur in the same measure. For example, suppose we had a measure of music that began with C♯ and finished with a C. To make sure the last note is a C, we must "cancel" the sharp from the earlier C by writing C♮. The technical rule is that natural signs are only required when the conflicting notes (e.g., a C and C♯) occur within the same measure. However, natural signs are frequently used to remind a musician of the note being played.

Figure 2.10 demonstrates some typical situations where a natural sign is needed. In the first two measures, the natural sign ♮ is used to change the A♯ to an A and the E♭ to an E, respectively. In the third measure, notice that the A♭ on the third beat does not require an accidental because the flat from the start of the measure carries through. The natural sign on the fourth beat of the third measure converts the A♭ to an A. In the final measure, the natural sign is not technically required but is included as a reminder that the flat on the D in the previous measure does not persist into the next measure.

FIGURE 2.10. Some music demonstrating the use of the natural sign (♮).

The other two kinds of accidentals are the double sharp x, used to raise a note by two half steps, and the double flat ♭♭, used to lower a note by two half steps. Thus, G^x and B♭♭ are both equivalent to A. The note E^x is the same as F♯, while C♭♭ is another name for B♭. Double sharps and flats are far less common than the other accidentals. For convenience, the five accidentals are summarized in Table 2.1.

Symbol	Name	Meaning
♯	Sharp	Raises the note by a half step
♭	Flat	Lowers the note by a half step
♮	Natural sign	Cancels any previous accidental
x	Double sharp	Raises the note by two half steps
♭♭	Double flat	Lowers the note by two half steps

TABLE 2.1. The five different types of accidentals and their meaning.

Exercises for Section 2.1

1. Provide the letter name (including accidentals) of each note in the following examples.

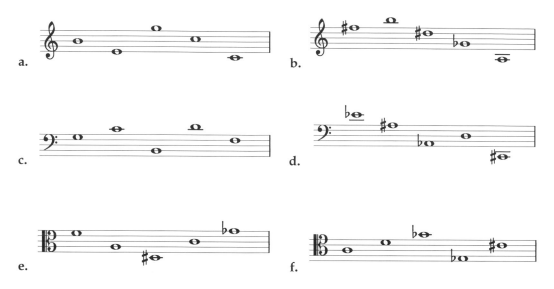

a.

b.

c.

d.

e.

f.

2. Copy the following excerpt onto staff paper and then write the name of each note (including accidentals) under each note.

3. Copy the following excerpt onto staff paper and then write the name of each note (including accidentals) under each note.

4. Take a good look at the piano keyboard in Figure 2.8. Now try to answer the following questions without using the keyboard. There is more than one possible answer to each question.

    a. Name a note that is enharmonically equivalent to A♭.

    b. Name a note that is enharmonically equivalent to D♯.

    c. Name a note that is enharmonically equivalent to F×.

    d. Name a note that is enharmonically equivalent to C♭.

      e. Name a note that is enharmonically equivalent to F$\flat\flat$.

      f. Name a note that is a half step above A.

      g. Name a note that is a half step below G.

      h. Name a note that is two half steps above E.

      i. Name a note that is three half steps below G.

5. Place the notes shown on the piano keyboard in their correct locations on the staff using the treble clef. Observe the location of middle C. Include accidentals where necessary.

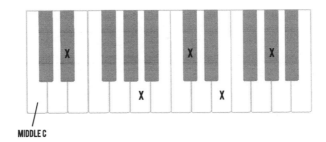

6. Place the notes shown on the piano keyboard in their correct locations on the staff using the bass clef. Observe the location of middle C. Include accidentals where necessary.

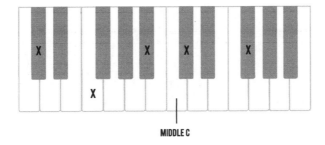

2.2 Scales

In this section we learn about the most common and important scales used in Western music. A solid grasp of the different scales helps lay the foundation for understanding the structure of music and the rationale for harmony.

    A *scale* is a collection of ordered notes which follows some prescribed pattern. The term originates from the Italian word *scala*, which means ladder. Scales typically begin and end an octave apart, and the pattern followed by a particular scale remains the same regardless of the opening note. For instance, the G major scale and E$\flat$ major scale follow the same precise interval pattern between successive notes. While the names of the notes differ, one scale is just a shifted version of the other.

2.2.1   Chromatic scale

The simplest scale is constructed by climbing up the piano keyboard one half step at a time. This 12-note scale is called the *chromatic scale*, and the note it begins on is basically irrelevant. The interval between any two successive notes is always a half step. Since there are 12 half steps in an octave, there are 12 notes in a chromatic scale. We exclude the top note from the count because it has the same name as the bottom note. A chromatic scale is easy enough to play on the piano by walking your fingers up the keyboard one key at a time. The scale should sound like a gradual rise in pitch over the span of an octave.

An ascending chromatic scale beginning and ending on C is shown in Figure 2.11. Here we have utilized the piano staff to show the scale in both the treble and bass clefs. As the scale is traversed, the right- and left-hand parts are always an octave apart. Notice the use of accidentals (sharps) and that each ♯ sign is placed on the line or space directly in front of the note head. Also, there are two notes, B and E, which do not have any accidentals. This is because B♯ = C and E♯ = F (enharmonic equivalence).

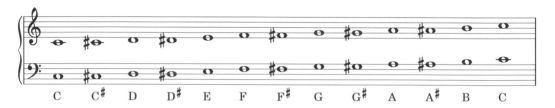

FIGURE 2.11. An ascending chromatic scale beginning on C. The scale has 12 notes (we consider the bottom and top notes to be equivalent), with consecutive notes always separated by a half step.

Regardless of the note a chromatic scale starts on, the total set of notes forming the scale is always the same. It is the 12 notes of the piano keyboard. If we begin a chromatic scale on a new note, say, F, then the order of the notes is still preserved (F♯ still follows F, G follows F♯, etc.). While the starting and ending notes change, the full set of notes making up the scale is unchanged. This illustrates an important mathematical property called *invariance*. Starting a chromatic scale on a different note has no effect on the overall set of pitches; they remain invariant under any shift. In essence, there is only one chromatic scale.

In general, if an object remains the same after applying a mathematical transformation, then it is considered *invariant* with respect to that operation. For instance, a square is invariant under a 90° rotation about its center, or under a reflection about a vertical line through its center. After applying each of these operations, the square is in the same position as it started. On the other hand, a rectangle with a width different than its height is not invariant under a 90° rotation, although it is invariant under a rotation of 180°. Shapes that are invariant under a transformation such as a rotation or reflection have symmetry. This will be explored further when we discuss examples of symmetry in music in Section 5.1.

One way to visualize the invariance of the chromatic scale is to place the 12 notes consecutively and equally spaced around a circle, just like the hours on a clock. Locate C at the top (12:00), C♯ at 1:00, D at 2:00, and so forth, until reaching B at 11:00 (see

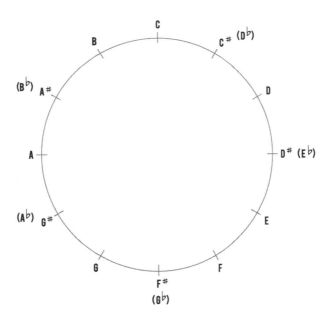

FIGURE 2.12. A musical clock demonstrating the 12 notes of the chromatic scale and their invariance under rotation. A rotation of the clock by a multiple of 30° changes the starting note of the scale but does not alter the total set of pitches.

Figure 2.12). Beginning at the top of the clock, an ascending chromatic scale opening on C moves clockwise around the circle until returning to the starting point. The descending scale is found by traversing the circle in a counterclockwise manner. Each chromatic scale is created by simply rotating the clock some multiple of 30° ($1/12 \cdot 360 = 30$), so that the new starting note appears at the top of the clock. For example, a 30° counterclockwise rotation of the circle will slide each note over one position in the scale (C♯ becomes the first note, D becomes the second, etc.), while the first note (C) cycles to the end of the list.

Example 2.2.1. *Using the piano staff, write a descending chromatic scale beginning and ending on the note G. Use flats for accidentals.*

Solution: The solution is shown in Figure 2.13. Here we choose to use flats rather than sharps for the descending scale; otherwise, we would have required five natural signs (on the notes F, D, C, A, and G), muddying the musical notation.   □

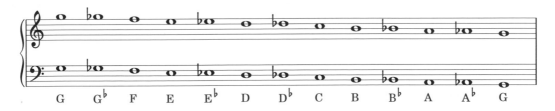

FIGURE 2.13. A descending chromatic scale beginning on G using flats and the piano staff.

FIGURE 2.14. The opening measures of Monk's chromatic melody in *Blue Monk*.

The great jazz pianist and composer Thelonious Monk (1917–1982) used the chromatic scale, or at least parts of it, in very creative ways. Monk tended to prefer harmonic structure that moved in half steps. For example, instead of using D–G–C (ii–V–I), a common jazz chord progression to finish a musical phrase, Monk preferred the chromatic sequence D–Db–C. He also wrote some popular jazz tunes that featured chromatic melodies. One famous example is *Blue Monk*, shown in Figure 2.14.

2.2.2   Whole-tone scale

A *whole step* is equivalent to two half steps. On the piano keyboard, this is equivalent to skipping a single note between two keys. For example, the interval between C and D, or from F♯ to G♯, is equal to a whole step. Moving from Eb to F, or from B to C♯, is also equivalent to a whole step.

The *whole-tone scale* is a six-note scale consisting of only whole steps between successive notes. Since a chromatic scale contains 12 notes and is made up solely of half steps, it follows that a whole-tone scale will contain half as many notes, which is six. Figure 2.15 displays two different whole-tone scales in the treble clef, one starting on C and the other starting on C♯. Each scale contains half of the possible notes from the chromatic scale, with no overlap between the two. Since the same interval is used between consecutive notes, a whole-tone scale that begins on D or E or F♯ will consist of the same set of notes as the whole-tone scale starting on C. In analogy with the musical clock created for the chromatic scale, a whole-tone clock will contain only six notes but will have the same invariance properties as the chromatic version. There are two such clocks, one with the pitches for the whole-tone scale starting on C, and the other for the scale beginning on C♯. In terms of the total set of pitches, there are only two different whole-tone scales.

FIGURE 2.15. The two different whole-tone scales, one beginning on C and the other on C♯. In each case, the interval between consecutive notes is always equal to a whole step (two half steps). Since a whole-tone scale is invariant, shifting the starting note in either example yields a scale with the same total set of notes as the original.

Example 2.2.2. *Write an ascending whole-tone scale starting on G using the bass clef.*

Solution: By referring to the piano keyboard, or by shifting the second whole-tone scale in Figure 2.15, the notes of the whole-tone scale we seek are G, A, B, C♯, D♯, F, and G. The scale, written in the bass clef, is shown in Figure 2.16.   □

G   A   B   C#   D#   F   G

FIGURE 2.16. A whole-tone scale starting on G in the bass clef.

It is worth playing the two whole-tone scales in Figure 2.15 on the piano. The resulting music may sound a little hollow and unfamiliar. Surprisingly, even though a whole-tone scale is perfectly symmetric and consistent with its interval structure, it does not produce an aesthetically appealing sound. The absence of certain notes gives the scale its odd feel. The reason for this will become apparent when we discuss the overtone series in Section 3.4.3.

Due to its odd nature, the whole-tone scale took a long time to find its way into Western music. It is featured in the works of some nineteenth-century Russian composers, including Glinka and Dargomïzhsky, as well as the later French impressionists, such as Claude Debussy. The whole-tone scale is used extensively by Debussy in his piano prelude *Voiles* (1910). It is also prominently used in jazz harmony by composers such as John Coltrane and Wayne Shorter. Thelonious Monk and the legendary jazz pianist Art Tatum incorporated the whole-tone scale into their compositions and solos, often to increase the tension in a melodic phrase. The opening bars of Monk's *Ruby My Dear* (1945) feature overlapping, descending whole-tone scales.

A slick use of the whole-tone scale appears in the bridge (B section) of the jazz classic for saxophone *Harlem Nocturne*, music written by Earle Hagen (1939). In measures 7–8 and 15–16 of the bridge, the eighth-note melody alternates between two descending whole-tone scales, one starting on F and the other on B♭ (see Figure 2.17). The two scales are distinct and collectively encompass all 12 notes of the chromatic scale. To further distinguish these measures, the notes of the whole-tone scale are typically played short and straight, that is, they are not swung in the typical jazz style.

FIGURE 2.17. Bars 7–8 of the bridge in *Harlem Nocturne* feature two alternating, descending whole-tone scales, one starting on F, the other on B♭.

2.2.3 Major scales

The most important musical scale is arguably the *major scale*, a scale critical for understanding the fundamental principles of music theory. Unlike the previous two examples, the chromatic and whole-tone scales, the major scale is constructed using an asymmetric pattern of half and whole steps. It contains seven notes, excluding the final note.

The simplest major scale to understand and to play on the piano is C major. Beginning with the C key, playing the white keys consecutively forms the C major scale. The notes are simply

C D E F G A B C.

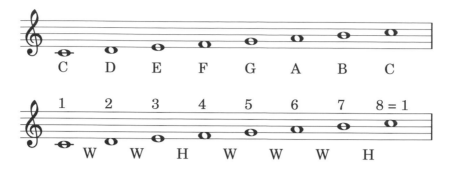

FIGURE 2.18. The C major scale. Notice the pattern of whole (W) and half (H) steps between consecutive notes. This same pattern holds for any major scale.

Figure 2.18 displays the C major scale in the treble clef, along with the numbered positions of each note. These numbers are known as the *scale degrees*. If we consider the successive intervals between each note of the scale, the pattern of half (H) and whole (W) steps is

$$\boxed{\text{W W H W W W H.}} \tag{2.1}$$

The sequence of whole and half steps in (2.1) is critical to understanding and writing major scales. Every major scale uses this same sequence between notes. Thus, a major scale consists of five whole steps and two half steps, with the half steps coming between scale degrees 3 and 4 and between 7 and 8. Notice that the pattern in (2.1) is asymmetric.

Constructing major scales

How do we build major scales other than C major? We will demonstrate by constructing a D major scale. It is important to have a picture of the piano keyboard for easy reference (see Figure 2.8).

To construct a major scale, first write down the names of the notes in the seven-letter musical alphabet consecutively, beginning and ending with the same note that identifies the scale. This is known as a *diatonic* scale spelling. Since we would like to build a D major scale, we have

D E F G A B C D.

Next we add accidentals to obtain the crucial H-W sequence shown in (2.1). The general rule is to use accidentals to modify the notes in the above list in order to avoid introducing new letter names. If the scale starts with a sharped note or requires a ♯, then use only sharps throughout. If the scale begins with a flatted note or requires a ♭, then stick to using just flats. Never mix flats and sharps when writing a major scale.

The first interval in our example is a whole step, so E is fine. But the next interval, from E to F, is a half step when it should be a whole step. Consequently, we must raise the F up a half step, replacing it with F♯:

D E F♯ G A B C D.

Now the first two intervals are whole steps, as required. The interval between F♯ and G is a half step, which is what we want, so we leave the G alone. Since the next two intervals, from G to A and from A to B, are each whole steps, no accidentals are required. But the interval between B and C is a half step, so we must raise the C up a half step, replacing it with C♯. This also serves to adjust the final interval of the scale from a whole step (incorrect) to a half step (correct). In sum, the notes of the D major scale are given by

$$D \quad E \quad F♯ \quad G \quad A \quad B \quad C♯ \quad D,$$

as shown in Figure 2.19.

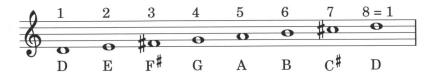

FIGURE 2.19. The D major scale in the treble clef.

There are a few important traits of the D major scale just created. First, notice that it contains two sharped notes. Even though it has the same H-W pattern as a C major scale, we are forced to use two sharps in order to obtain the required pattern. Hence, the collection of notes forming each scale are *different*. Second, observe that each note in the D major scale is precisely one whole step higher than the corresponding note in the C major scale. For instance, the fifth note of the D major scale is an A, which is one whole step above the fifth note of the C major scale (a G). This consistent gap between the two scales follows because the same H-W pattern is used in each case. Since the starting notes differ by a whole step, *all* corresponding notes of the two scales differ by a whole step. Thus, an alternative method of constructing the D major scale is to raise each note of the C major scale by a whole step.

This presents a rather interesting dichotomy. On the one hand, a C major scale and a D major scale are clearly different since they do not contain the same set of notes. On the other hand, the scales are closely related since one is obtained from the other by moving each note a whole step.

Example 2.2.3. *Construct an ascending A♭ major scale and write the corresponding notes using the bass clef.*

Solution: Since the name of the scale contains a flat, we must use flats only. We start with the musical alphabet beginning and ending on A, except that A is replaced by A♭:

$$A♭ \quad B \quad C \quad D \quad E \quad F \quad G \quad A♭.$$

Using the piano keyboard as a guide, we now adjust the notes above to match the required H-W pattern in (2.1).

The first interval, from A♭ to B, is too wide at three half steps. This means we must flat the second note, changing the B to a B♭. The interval between B♭ and C is a whole step, so the C remains unchanged. Next, the gap between C and D is too large, so we flatten the D to make the third interval the required length, a half step. Thus far, we have

$$A♭ \quad B♭ \quad C \quad D♭ \quad E \quad F \quad G \quad A♭.$$

Continuing along, the next interval, from D♭ to E, is also too large at three half steps. Thus, we must modify the E by adding a flat. The remaining three intervals are now correct, as the scale is completed with the intervals W W H. The final result is displayed in Figure 2.20, with the music notated in the bass clef. An A♭ major scale requires four flatted notes: A♭, B♭, D♭, and E♭.   □

FIGURE 2.20. The A♭ major scale, with its four flats, in the bass clef.

Musical keys

Thus far we have constructed three different major scales: C (no sharps or flats), D (two sharps), and A♭ (four flats). It turns out that every major scale requires a different number of sharps or flats. In fact, the number of sharps or flats *uniquely* determines the scale. This leads to the idea of a defining *key* in a musical work.

If a piece of music primarily features the notes from a specific major scale, say, A major, then we say that the piece is in the *key of A major*. If the music consistently has four flats, then it is likely in the key of A♭ major. There are certainly exceptions to this rule (e.g., many modern works of classical music have no key at all), but it is a useful assumption when first learning to play or analyze music. For example, knowing the key of the current piece allows jazz musicians to improvise since the notes of that key's scale will blend better with the accompanying instruments.

As an example, consider the famous "Ode to Joy" theme from Ludwig van Beethoven's popular 9th Symphony (1824; see Figure 2.21). What key is this excerpt in? Based on the prevalence of F♯s, we might guess the key of G major, since this turns out to be the only key containing one sharp (namely, F♯). However, notice that each measure except for the last begins on either a D, an F♯, or an A, and that the phrase finishes on a D. In the D major scale, D, F♯, and A correspond to scale degrees 1, 3, and 5, respectively. These three notes sound particularly pleasing when played together (known as a major chord). The prevalence of scale degrees 1, 3, and 5 of the D major scale throughout the excerpt suggests that it is in the key of D major.

FIGURE 2.21. The famous "Ode to Joy" theme from Beethoven's 9th Symphony, featuring scale degrees 1–5 of the D major scale.

2.2.4   Minor scales

As a general rule and rather gross oversimplification, music in a major key tends to sound happy and uplifting, whereas a piece or excerpt in a minor key feels solemn and reflective. Bobby McFarrin's *Don't Worry Be Happy* (1988), Beethoven's "Ode to Joy" theme, and the *Happy Birthday Song* are, not surprisingly, all written in a major key. In contrast, Alicia Keys' *Fallin'* (2001), off her debut album *Songs in A Minor*, is in a minor key, as is Beethoven's melancholy *Moonlight Sonata* (1801). Simon & Garfunkle's emotionally powerful ballad *The Sound of Silence* (1964), written just after the assassination of John F. Kennedy, is in the key of D minor. The intervals that arise from the minor scale tend to sound less comfortable than their major counterparts. For instance, vocal sports fans who enjoy heckling the opposing team during a game have mastered the use of the minor third interval.

There are certainly many exceptions to this characterization. Songs can be quite sad, even if they are based in a major key. One such example is R.E.M.'s sorrowful ballad *Everybody Hurts*, once voted "the saddest song of all time,"[2] which is actually in the key of D major. While some of the emotional contrast between major and minor is probably due to cultural conditioning, we offer some scientific explanations in Section 4.2.2.

There are two types of minor scales we will focus on: the *natural minor scale* and the *harmonic minor scale*. There is only one note that differs between the two scales. When the adjective "minor" is used, it typically refers to the natural minor scale.

Figure 2.22 demonstrates a natural C minor scale. Notice that three notes, the flatted notes E♭, A♭, and B♭, are different from the C major scale. The first five notes are identical to those of the major scale, except for the third flatted note. This small difference has a huge impact on the feel of the scale. The natural minor scale is created using the following sequence of half and whole steps:

$$W \quad H \quad W \quad W \quad H \quad W \quad W. \tag{2.2}$$

This same pattern is used for every natural minor scale. As with the major scale, there are two half steps and five whole steps, but here, the half steps occur between scale degrees

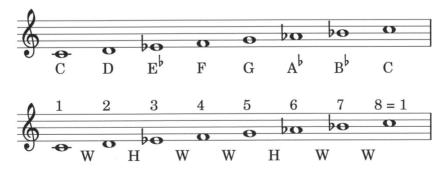

FIGURE 2.22. The natural C minor scale. Notice the pattern of whole (W) and half (H) steps between consecutive notes. This same pattern holds for any natural minor scale. Unlike C major, which has no sharps or flats, C minor requires three flats.

---

[2]McCormick, What Is the Happiest Song of All Time?

2 and 3 and between 5 and 6. Similar to the major scale, the defining sequence in (2.2) is asymmetric.

The harmonic minor scale is equivalent to the natural minor scale except for the seventh scale degree, which is raised up a half step. For example, in harmonic C minor, the B♭ near the end of the scale is adjusted to the note B (see Figure 2.23). This causes a jump of three half steps between the sixth and seventh scale degrees. It also changes the last interval to a half step. For reasons that will become apparent in the next section, raising the seventh scale degree to construct the harmonic minor scale allows for more intuitive melodies and harmonies to be created.

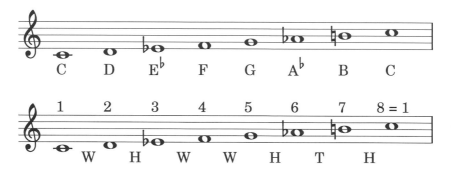

FIGURE 2.23. The harmonic C minor scale. The only difference from the natural C minor scale is that the seventh scale degree has been raised up a half step, resulting in a jump of three half steps (denoted by T) between scale degrees 6 and 7.

Example 2.2.4. *Construct the natural and harmonic versions of the B minor scale and write the notes in the treble clef.*

Solution: We adapt the approach used to construct a major scale, adding accidentals as needed to achieve the H-W sequence in (2.2). Beginning with the notes

$$B \quad C \quad D \quad E \quad F \quad G \quad A \quad B,$$

notice that the first interval is only a half step. Thus, we add a sharp to C to make this interval a whole step. The second interval, from C♯ to D, is now a half step, so we leave the D alone. The third interval is the correct length, but the half step between E and F is too short. Hence, we raise F up a half step, changing it to F♯. The new list of notes is

$$B \quad C♯ \quad D \quad E \quad F♯ \quad G \quad A \quad B.$$

The final three intervals are now H   W   W, as required, so we have completed the natural minor scale. To obtain the harmonic version, we raise the penultimate note, A, by a half step to A♯. The two scales are shown in Figure 2.24.   □

2.2.5   Why are there 12 major scales?

We close this section on scales with an analysis concerning the number of distinct major scales. As opposed to the chromatic (one) and the whole-tone scales (two), there are 12 different major scales, one for each note of the chromatic scale. The set of notes for a

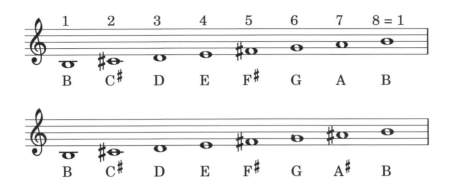

FIGURE 2.24. The natural (*top*) and harmonic (*bottom*) B minor scales.

particular major scale is unique to that scale. For instance, there is only one C major scale, with no sharps and flats, and it must begin on C. A scale containing the same sequence of notes, but starting on a note other than C, will not produce the correct H-W pattern.

For example, consider the notes of the C major scale, but beginning on D (see Figure 2.25):

$$D \quad E \quad F \quad G \quad A \quad B \quad C \quad D.$$

The sequence of half and whole steps for this list of notes is W H W W W H W, which is different than the required pattern for a major scale. Even though the set of notes is the same as the C major scale, this new scale will sound minor rather than major, because the opening five notes are identical to those of a D minor scale.

No matter which note we begin with, attempting to create another major scale using only the notes of the C major scale will never work. A different H-W pattern is always obtained. In fact, the new H-W sequence is equal to some number of *cyclic shifts* of the pattern (2.1) that defines the major scale. Let us introduce some notation and a definition to make this argument more precise.

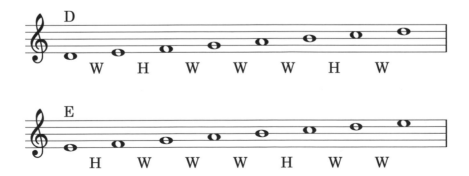

FIGURE 2.25. The notes of the C major scale starting on D (*top*) and E (*bottom*). Notice the H-W sequences obtained, each different from the defining pattern for a major scale.

Definition 2.2.5. *The cyclic shift σ is the function that takes a sequence of items and shifts each element one space to the left, with the first element moving to the end of the list. In other words,*

$$\sigma(1 \; 2 \; 3 \; \cdots \; n) = (2 \; 3 \; 4 \; \cdots \; n \; 1).$$

We will denote $\sigma^2 = \sigma \circ \sigma$ as the composition of two consecutive shifts. This has the effect of sliding each element two spaces to the left. Likewise, σ^3 is three consecutive shifts, σ^4 is four, and so on. If there are n items in the sequence, then the map σ^n will return the sequence to its original ordering, that is, σ^n is really the identity map (nothing moves). For example, on a 12-hour clock, σ^{12} returns the hours to their original positions. The inverse of σ, denoted as σ^{-1}, is a right cyclic shift, where each element shifts to the right one position and the last element cycles around to the front of the list. For a sequence with n items, σ^{-1} has the same effect as σ^{n-1}.

The H-W sequence for a C major scale starting on D, W H W W W H W, is the cyclic shift σ of the major scale pattern W W H W W W H (slide each letter to the left one position, and move the first letter to the end). Shifting twice by applying σ^2 yields the sequence H W W W H W W. This is the H-W sequence for the notes of a C major scale that starts on E:

E   F   G   A   B   C   D   E

(see Figure 2.25). Each cyclic shift σ^n always produces a new pattern until we return to the original H-W sequence with σ^7. Consequently, the C major scale is truly unique, and it must begin on C. Since every major scale follows the same pattern of half and whole steps, this argument can be applied to each scale. This explains why there are 12 different major scales.

Contrast this line of reasoning with the symmetric whole-tone scale, for which there are only two different versions. The scale beginning on C has the same set of notes as the one that begins on D, as well as the scale starting on E, and so forth. Applying the cyclic shift σ^n to the sequence W W W W W W always produces the same sequence, regardless of the value of n. The invariance of the pattern is what causes the whole-tone scale to be invariant under a shift in its starting note. We have arrived at the heart of the matter.

The uniqueness of the major scale, and the reason that there are 12 different versions, is a result of the fact that its defining H-W sequence, W W H W W W H, does not map back to itself under repeated iterations of the cyclic shift map σ^n, for $n \in \{1, 2, \ldots, 6\}$.

The beauty of this mathematical argument is that it also applies to other H-W sequences, such as those used for the natural and harmonic minor scales. The pattern for the natural minor scale, W H W W H W W, also never returns to itself under σ^n until $n = 7$. This means that there are 12 different natural minor scales. The same argument works for the harmonic minor scale as well.

What is so special about these defining H-W sequences? Are they specifically designed to avoid repetition under σ^n? Surprisingly, the answer is no. What is special about these sequences is that they contain two elements (H or W) and are seven units long. The number seven is prime, so it has no factors other than one and itself. To obtain repetition in a sequence, we need there to be smaller versions within itself, called *subsequences*, that repeat. To do this, the total length of the sequence must *not* be prime.

For example, the H-W sequence

$$\text{H W W H W W H W W,}$$

which is nine units long, will be sent back to itself after just three cyclic shifts (σ^3). Here the subsequence H W W, with a length of three, fits three times into the full sequence, a construction made possible since three is a factor of nine. It is not possible to construct such an example if the total length is a prime number. The only way to obtain invariance under the cyclic shift map is to choose just one symbol, either H (the chromatic scale) or W (the whole-tone scale). Involving both symbols in a sequence of prime length will never be invariant. Even an H-W sequence with symmetry, such as W W H W H W W (a palindrome), will not return to itself under σ^n until $n = 7$. Although this pattern is symmetric, it will still produce 12 different scales (see Exercise 8).

Returning to music theory, the scales obtained by shifting the order of the notes in a major scale are called *modes*. For example, starting the scale on the second scale degree produces the *Dorian* mode, while beginning with the third scale degree yields the *Phrygian* mode. The top scale in Figure 2.25 is D Dorian, while the bottom is E Phrygian. The use of modes is common in both jazz and modern classical music.

For review and easy reference, the five types of scales discussed in this section are summarized in Table 2.2.

Scale	# of Notes	H-W Sequence
Chromatic	12	All half steps
Whole-Tone	6	All whole steps
Major	7	W W H W W W H
Natural Minor	7	W H W W H W W
Harmonic Minor	7	W H W W H T H

TABLE 2.2. Five types of scales and their defining sequences of half (H) and whole (W) steps. The symbol T represents three half steps.

Exercises for Section 2.2

1. Take a good look at the piano keyboard in Figure 2.8. Now answer the following questions without using the keyboard. Note that because of enharmonic equivalence, there is more than one possible answer for questions **a–g**.

   a. Name a note that is a whole step above A.

   b. Name a note that is a half step below E$\flat$.

    c. Name a note that is a half step above B.

    d. Name a note that is a whole step below F♯.

    e. Name a note that is a whole step above F♯.

    f. Name a note that is a half step below F.

    g. Name a note that is a half step below C.

    h. How many half steps apart are B and D♭?

    i. How many half steps apart are C♯ and E?

2. Write an ascending chromatic scale (one octave) using sharps in both the treble clef and the bass clef beginning on F. Use whole notes and write the letter name below each note.

3. Write a descending chromatic scale (one octave) in the bass clef beginning on B♭. Use whole notes and write the letter name below each note.

4. Write an ascending whole-tone scale (one octave) using only flats in both the treble and bass clefs beginning on A♭. Use whole notes and write the letter name below each note.

5. Write a descending whole-tone scale (one octave) using only sharps in the treble clef beginning on F♯. Use whole notes and write the letter name below each note.

6. Make several copies of the blank piano keyboard shown in Figure 2.26. For each of the scales given below, mark the corresponding keys on the piano keyboard, using numbers for each scale degree. Assume that each scale is ascending and is one octave in length. A sample solution is shown on the right in Figure 2.26.

    a. G major   **b.** A major   **c.** B♭ major

    d. E♭ major   **e.** B major   **f.** E natural minor

    g. B♭ natural minor   **h.** F harmonic minor   **i.** C♯ harmonic minor

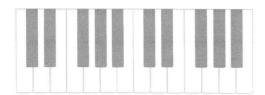 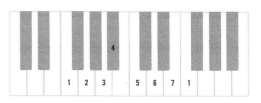

FIGURE 2.26. A blank piano keyboard for Exercise 6. To the right is a sample solution showing the corresponding keys of an F major scale.

7. Using your answers to the previous exercise, notate each of the following ascending scales in both the treble and bass clefs. Use whole notes and write the correct letter name below each note.

      a. G major   **b.** A major   **c.** B♭ major

      d. E♭ major   **e.** B major   **f.** E natural minor

      g. B♭ natural minor   **h.** F harmonic minor   **i.** C♯ harmonic minor

8. Consider the H-W sequence W W H W H W W. This pattern has a vertical symmetry since it is the same forward and backward (a palindrome). Beginning on C, write down the seven notes that follow using this particular H-W sequence (you should end on C). Denote this new scale as "Scale R." Next, using repeated applications of the cyclic shift map σ^n, explain why starting Scale R on a different note, while keeping the total set of notes the same, will always yield a *different* H-W sequence than the original. Conclude that this H-W sequence also produces 12 different scales, even though it is symmetric.

2.3   Intervals and Chords

In music theory, the distance between two notes is called an *interval*. We have already learned three important intervals that were vital in defining different scales: a half step, a whole step, and an octave. In this section we use the major and minor scales to describe some other important musical intervals.

Definition 2.3.1. *Two notes are an <u>nth interval</u> apart if they are n steps from each other on the musical staff. When counting the number of steps, be sure to include the location of* **both** *the starting and ending notes.*

For example, a G and a B are a *third* apart since there are three staff steps between them. In the treble clef, this is the second line from the bottom (G), the space above (A), and the middle line of the staff (B) (see Figure 2.27). Assuming that G is the lower note, a G and a D are a *fifth* apart, while the interval between G and A is a *second*. Each of these examples assumes that the notes are within the same octave; otherwise, the intervals would be a length greater than eight.

From a mathematical perspective, the definition of a musical interval is counterintuitive because it overcounts the actual distance by one unit. The notes G and A are next to

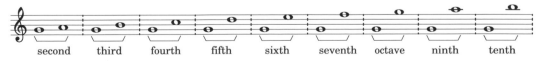

FIGURE 2.27. Some basic musical intervals. When counting the number of steps between two notes, always include the line or space of *both* the starting and ending notes.

each other on the piano or on the staff, so that should really be considered one step, not two. The notes A and C are two steps away on the piano keyboard, so that interval should really be a second, not a third. The musical version of measuring distance causes problems when combining intervals. For instance, we would expect that a second and a third combine to produce a fifth, but this is false! The interval from G to A (a second) plus the interval from A to C (a third) is equal to the interval from G to C (a fourth). According to music theory, the equation $2 + 3 = 4$ is valid! In general, we must always subtract 1 from the mathematical sum of two intervals in order to obtain the correct musically defined length.

2.3.1 Major and perfect intervals

In addition to the number of steps in a musical interval, there is also the designation of perfect, major, or minor, as well as augmented or diminished. The terms "perfect" and "major" are assigned to intervals corresponding to those found in the major scale. More specifically, if the major scale beginning on the bottom note contains the upper note as part of its scale, then the interval is perfect (in the case of the fourth or fifth) or it is major. The abbreviation for a major interval with n steps is Mn. Thus, M3 represents a major third and M6 means a major sixth. We use P4 and P5 to denote a perfect fourth and fifth, respectively. Some major and perfect intervals are shown in Figure 2.28, using middle C as the bottom note. Since the C major scale contains no sharps or flats, no accidentals are required to notate any of the intervals shown.

FIGURE 2.28. Some major (M) and perfect (P) intervals. In each case, the top note is a part of the C major scale.

Recall that every major scale consists of the same sequence of half (H) and whole (W) steps: W W H W W W H. Consequently, every major or perfect interval will contain the same number of half steps, regardless of the starting note. For example, the interval from a C up to an F, a perfect fourth, consists of five half steps. This can be determined by counting the number of keys on the piano keyboard between C and F. Here we do *not* include both the top and bottom keys in our count. Starting on a C, count one for C♯, two for D, three for D♯, four for E, and five upon reaching the F key. Alternatively, we could use the H-W pattern for the major scale. Since F corresponds to the fourth scale degree, we need to compute the number of half steps in the sequence W W H. A whole step equals two half steps, so the sum is $2 + 2 + 1 = 5$, as expected. Any other perfect fourth (e.g., an E up to an A, or a B♭ up to an E♭) will also have five half steps.

Table 2.3 lists all of the major and perfect intervals within the octave, including the number of half steps in each interval. Two identical notes are considered to be in *unison*. Note that a major second (two half steps) is identical to a whole step. Each scale degree in the major scale has a particular name, with the *tonic* (scale degree 1) and *dominant* (scale degree 5) being two of the most frequently used terms. One aspect of learning music theory is developing the ability to aurally recognize different musical intervals.

Scale Deg.	Interval	Half Steps	Name	Musical Example(s)
1	Uni.	0	Tonic	
2	M2	2	Supertonic	*Frère Jacques* *Happy Birthday to You*
3	M3	4	Mediant	*Oh, When the Saints* *Kumbaya*
4	P4	5	Subdominant	*Here Comes the Bride* *Oh Christmas Tree*
5	P5	7	Dominant	*Twinkle Twinkle Little Star* *My Favorite Things*
6	M6	9	Submediant	*My Bonnie Lies over the Ocean* *It Came upon a Midnight Clear*
7	M7	11	Leading tone	*Take on Me*
8 = 1	Oct.	12	Octave	*Somewhere over the Rainbow*

TABLE 2.3. The major (M) and perfect (P) intervals within the octave, including the number of half steps in each interval, and some musical examples whose melodies open with the given interval. "Uni." stands for unison. The names given to each scale degree are also listed.

To facilitate that goal, examples of pieces that open with a particular interval are also included in Table 2.3.

One of the most important intervals is the perfect fifth, a musical distance that we will repeatedly encounter throughout the book. It has seven half steps. Two notes a perfect fifth apart sound particularly nice when played together. In *Twinkle Twinkle Little Star*, the pitches for the first "twinkle" and the second "twinkle" are a perfect fifth apart.

2.3.2   Minor intervals and the tritone

If an interval is not perfect or major, we will consider it to be minor. The one exception is the interval that divides the octave in half, that is, an interval of six half steps, called a *tritone*. Some important minor intervals are shown in Figure 2.29, using middle C as the bottom note. Minor intervals are notated with a lowercase m. Observe that the minor third, sixth, and seventh shown in Figure 2.29 are each part of a C natural minor scale. However, some of the intervals in the C natural minor scale are actually major or perfect, namely, the major second (C to D), the perfect fourth (C to F), the perfect fifth (C to G), and the octave (C to C).

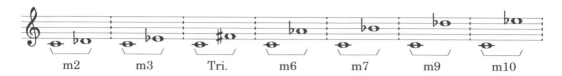

FIGURE 2.29. Some minor (m) intervals and the tritone (Tri.).

Table 2.4 lists all of the minor intervals within the octave, as well as the tritone. The number of half steps in each interval is provided, a sequence that complements the list for the major and perfect intervals shown in Table 2.3. Note that a minor second is equivalent to a half step, a simple interval that movie composer John Williams ominously converted into the terrifying two-note motif for the killer shark in *Jaws* (1975). Notice the contrast between the musical examples listed for the major and minor intervals in each table. It is recommended that a piano be used to aurally explore the differences between the major and minor intervals discussed in this section.

The terms "augmented" or "diminished" are used to raise or lower, respectively, an interval by a half step. The interval from C up to G♭ is technically a diminished fifth (it is a half step below the perfect fifth from C to G), but since it is equal to six half steps, we will refer to it as a tritone. Similarly, the interval from C up to F♯, which is the same two notes, is technically called an augmented fourth. Since the use of the terms "augmented" and "diminished" is largely one of music theory semantics, we will not include them in our discussion of intervals.

Notes	Interval	Half Steps	Musical Example(s)
C – D♭	m2	1	Theme from *Jaws*
C – E♭	m3	3	"Air-ball!" (the heckle interval) *Greensleeves*
C – F♯	Tri.	6	*Maria* Opening choir in *The Simpsons* theme
C – A♭	m6	8	Theme from *Love Story* *Go Down Moses*
C – B♭	m7	10	*There's a Place for Us*

TABLE 2.4. The tritone (Tri.) and the minor (m) intervals within the octave, including the number of half steps in each interval, and some musical examples.

Example 2.3.2. *Notate the musical interval in each of the following measures. Include major, minor, and perfect designations.*

Solution: There are two basic approaches to determining each interval. The first measure shows an interval between F and C, which is five steps on the staff. Since C is in the F major scale, the interval is a perfect fifth. Alternatively, we could use the piano keyboard to count the seven half steps between F and C and then use Table 2.3 to obtain a perfect fifth. Counting the number of half steps is an easier approach if the major scale is less familiar. The last three measures are all in the bass clef. The final measure shows the interval between D and F, which is a third. We know from Section 2.2.3 that the D major scale contains an F♯, so the interval must be a minor third. The complete solution is shown in Figure 2.30. □

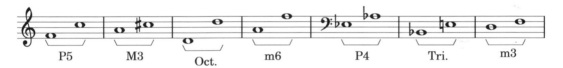

FIGURE 2.30. The solution to Example 2.3.2.

2.3.3 Chords

> *Sting makes me feel like a musical Neanderthal. When we get together, we always have the same argument. He insists that there are more than three chords, while I insist that there are not.*
>
> — Bruce Springstein[3]

Now that we have discussed major and minor scales and the basic intervals, we are ready to tackle some harmonic theory by discussing chords. A *chord* is a collection of three or more notes played simultaneously. When the chord contains exactly three notes, it is often called a *triad*. The more notes a chord contains, the more complicated (and interesting) it becomes. Just like scales and intervals, chords have major and minor designations, as well as descriptive terms such as augmented, diminished, and suspended. Once again, using a piano to play the different chords described here is an important aspect of learning music theory.

Major chords

The two most important chords are the *major* and *minor* triads. They are easiest to understand using the C major and minor scales. A C major chord or triad is simply the chord built with the first, third, and fifth scale degrees, namely, C, E, and G. These three

---

[3]"The Boss," paying tribute to his friend Sting, a 2014 Kennedy Center honoree for the arts. Zongker, Kennedy Center Honors Five Artists.

FIGURE 2.31. Some major chords in root position.

notes played together on the piano will produce a pleasing sound. Combining the notes on scale degrees 1, 3, and 5 of any major scale produces the major chord corresponding to the name of that scale. For example, D, F♯, and A form a D major chord and B♭, D, and F form a B♭ major chord. Some examples of major chords are shown in Figure 2.31. One of the reasons it is so important to master the major scales is the ease and flexibility obtained for building major chords.

Major chords are named according to the first scale degree (the tonic). The notes of a major triad may be given in a different order, such as 3 on the bottom, 5 in the middle, and 1 on the top, known as an *inversion*. We will primarily stick to *root position* with our chords, where scale degree 1 is on the bottom, 3 is in the middle, and 5 is on top, as demonstrated by each chord in Figure 2.31.

Minor chords

Minor chords are formed using the first, third, and fifth scale degrees of a minor scale (both the natural and harmonic versions give the same notes). In C minor, this is the notes C, E♭, and G, while in D minor, it is the notes D, F, and A (see Figure 2.32). Notice that the only difference between a C major and a C minor chord is the middle note, which is shifted down a half step from the major to the minor version. This half step has a surprisingly big impact on the feel of the chord, giving minor chords their solemn flavor.

Let us consider the musical intervals within major and minor chords. First, the interval between the outer notes in both versions is always a perfect fifth (the distance between the first and fifth scale degrees). Second, the interval between the bottom two notes in a major chord, from scale degree 1 to 3, is a major third. For a minor chord in root position, this interval is a minor third. On the other hand, the interval between the top two notes in a major chord, between scale degrees 3 and 5, is a *minor* third. For example, the interval between an E and a G, the top two notes of a C major chord in root position, is only three half steps, a minor third. Likewise, the interval between the top two notes in a minor chord is a *major* third. For instance, in a C minor chord, the interval between the top two notes, an E♭ and a G, is four half steps, which is a major third (see Figure 2.33).

FIGURE 2.32. Some minor chords in root position. Each chord, named after its root note and with an "m" for minor, is obtained from its major counterpart in Figure 2.31 by modifying the middle note down one half step.

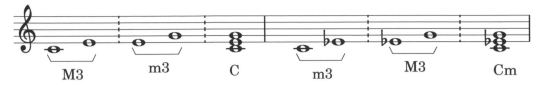

FIGURE 2.33. The structure of a major and minor chord. The major chord (*left*) consists of a major third on the bottom and a minor third on top. The order is flipped for a minor chord (*right*).

In sum, the major and minor chords subdivide the seven half steps of a perfect fifth in two different ways. A major chord is built with a major third as the foundation with a minor third on top ($7 = 4 + 3$), while a minor chord consists of a minor third beneath a major third ($7 = 3 + 4$). Mathematically speaking, this is a simple application of the commutative property. From a musical perspective, it is a contrast fundamental to the theory of harmony.

Augmented, diminished, and suspended chords

There are many other types of chords found in various styles of music. Some of these are triads that can be obtained by raising or lowering pitches from the major and minor triads. More complex chords can be obtained by adding one or more notes to existing triads. We explore each of these types below.

Raising the fifth up a half step in a major triad forms an *augmented chord*. A C augmented chord contains the notes C, E, and G♯ and is often denoted as C+ in jazz. This chord consists of two major thirds. Lowering the fifth by a half step in a minor triad forms a *diminished chord*. A C diminished chord has the notes C, E♭, and G♭ and is notated as C°. A diminished chord contains two minor thirds.

One fun chord is constructed by playing the first, fourth, and fifth scale degrees of a major scale together. This chord is called a *suspended chord* because the tension between the fourth and fifth scale degrees wants to resolve. The fourth is "suspended" until it can resolve to the third scale degree, the middle of the major triad. A C suspended chord, denoted as C^sus, consists of the notes C, F, and G. A great use of a suspended chord in a pop song is The Who's *Pinball Wizard*, from their rock opera *Tommy* (1969). A suspended chord is used repeatedly to generate excitement at the opening of the song and then features prominently in the chord progression underlying each verse. Some examples of augmented, diminished, and suspended chords are shown in Figure 2.34.

FIGURE 2.34. Some examples of augmented (+), diminished (○), and suspended (sus) chords.

Seventh chords

Seventh chords are obtained by adding the seventh scale degree to the previously described triads. The additional note can be either a minor or major seventh away from the bottom note. Seventh chords are particularly important for the theory of harmony. For example, a G *dominant seventh chord*, denoted G7, wants to move toward a C major triad. A G7 chord contains the notes G, B, D, and F. Notice that the F does not belong to the G major scale (the seventh scale degree of G major is F♯, the leading tone), although it does belong to the C major scale, the key where the chord is heading. The tritone between B and F, which are the seventh and fourth scale degrees in C major, respectively, wants to close to the major third between C and E. In musical parlance, we say that the G7 chord *resolves* to C major. All four notes of the G dominant seventh chord are contained in the C major scale.

In general, a seventh chord features a top note that is a half step below the leading tone. So, a C7 chord has a top note of B♭, and an E7 chord has a top note of D. A seventh chord that contains the leading tone of the scale of the tonic is called a *major seventh chord* and is denoted as Cmaj7. The notes C, E, G, and B make up a Cmaj7 chord, while G, B, D, and F♯ are the pitches in a Gmaj7 chord. The Beatles song *Something* (1969) features a nice sequence of seventh chords during its main verse, following a Cmaj7 chord with a C7 chord, which then resolves to F major.

Seventh chords are very common in jazz, as they add some additional spice and color to the harmony. In addition to dominant and major seventh chords, jazz music features minor seventh chords (e.g., Cm7, which adds a minor seventh above a minor triad), minor-major seventh chords (e.g., Cm-maj7, which adds a major seventh above a minor triad), diminished seventh chords (e.g., C°7, which adds a major sixth above a diminished triad), and augmented seventh chords (e.g., C7♯5, which adds a minor seventh above an augmented triad). These examples, as well as a few others, are all demonstrated in Figure 2.35. Notice that the diminished seventh chord C°7, which can function as a passing chord between two minor triads or as an alternative to the dominant seventh chord, contains only minor thirds between adjacent notes.

FIGURE 2.35. Some examples of different seventh chords with C as the root.

The jazz standard *Harlem Nocturne*, mentioned earlier for its use of the whole-tone scale, features a number of minor-major chords, such as a G minor chord with a major seventh on top. This is the chord Gm-maj7, with the notes G, B♭, D, and F♯, featuring a minor third as the bottom interval and a major third for the top. This is the perfect harmonic background for the sultry opening saxophone melody.

Finally, we mention that some classical composers have superimposed chords to combine multiple keys, a technique called *polytonality*. One of the first composers to attempt this was Stravinsky in *The Rite of Spring* (see Figure 1.10). The section with repeating staccato eighth notes and jarring accents features an E major chord (E, G♯, and B) in the basses and cellos beneath an E♭ dominant seventh chord (E♭, G, B♭, and D♭) in the violas and violins. The simultaneous combination of two different major keys a half step apart paved the way for future composers to discard tonality altogether.

Chord progressions

A sequence of chords is known as a *chord progression*. When notating progressions, it is common to use roman numerals and label the chords by the scale degree of the root. Capitalization is used for major chords and lowercase for minor chords. For example, using only the notes from the C major scale, the triads built on the notes C, F, and G are all major chords (C, E, and G; F, A, and C; G, B, and D). These are notated I, IV, and V, respectively. The triads beginning on D, E, and A are all minor chords (D, F, and A; E, G, and B; A, C, and E) and are notated as ii, iii, and vi, respectively. The triad beginning on B in the C major scale, which is the notes B, D, and F, is a diminished triad, typically notated as vii°.

For the purposes of understanding harmony, the most important chord progression moves from the dominant major chord (V; with or without the seventh) to the tonic (I). In the key of C major, this is the progression from a G major triad to a C major chord. In the key of A major, it is the progression from E to A. The V–I progression is the most common way of ending a phrase or piece of music.

Another well-known progression, nearly universal in pop and folk music, is the chord sequence I–IV–V, or some variation thereof. It can be found at the start of several classic rock songs such as *Louie Louie, La Bamba,* and *Summer Nights.* The related sequence I–IV–I–V–IV–I defines the standard blues progression. It is typically spread out over 12 measures, commonly known as the *12-bar blues,* as follows:

Measure:	1	2	3	4	5	6	7	8	9	10	11	12
Chord:	I	I	I	I	IV	IV	I	I	V	IV	I	I

Sometimes the second measure features the IV chord as opposed to the I chord (e.g., Monk's *Blue Monk*; see Figure 2.14). In jazz, the sequence ii–V or ii–V–I is quite common, with the ii chord usually appearing as a minor seventh chord. A musician with a solid knowledge of chords and these progressions will be able to play and appreciate a wide variety of musical styles and examples.

Exercises for Section 2.3

1. After studying Tables 2.3 and 2.4 carefully, answer the following questions without referring back to either table. The piano keyboard is a helpful (and allowable) tool.

   a. How many half steps are there in a major third?

   b. How many half steps are there in a minor seventh?

   c. What interval consists of seven half steps?

   d. What note is a perfect fourth above D♭?

   e. What note is a tritone above G?

   f. What interval is sometimes called the "heckle interval"?

   g. Starting on E, go up a major sixth and then down a minor third. What note have you arrived at?

   h. Starting on C, go up three perfect fifths and down one octave. What note have you arrived at?

2. Notate the musical interval in each measure below, including major, minor, and perfect designations.

3. Notate all the intervals (include major, minor, and perfect designations) between successive notes in the musical excerpt below. Be sure to indicate the intervals between measures as well.

4. Since a tritone consists of six half steps, two consecutive tritones fill an octave (e.g., the tritone from C to F♯ combined with the tritone from F♯ to C). Similarly, there are six major seconds or 12 minor seconds in an octave. What other intervals can be combined repeatedly with themselves to form an octave? State how many of each are needed to fill an octave.

5. Construct the major and minor triads using E♭ and F♯ as the bottom notes. For each of the four chords, list the names of the notes and place them in root position on the staff in the treble clef. Use diatonic spellings for the chords, that is, use the names of the notes that arise from the respective major and minor scales.

6. Construct augmented and diminished triads using F and A as the bottom notes. For each of the four chords, list the names of the notes and place them in root position on the staff in the treble clef. Use diatonic spellings for the chords, that is, use the letter names corresponding to scale degrees 1, 3, and 5.

7. What is the middle note in a B♭ suspended triad (B♭ sus)? What note does it want to move toward?

8. What are the four notes in a B diminished seventh chord (B°7)? What four notes make up an F diminished seventh chord (F°7)? What do you notice about your two answers? Thinking in terms of the total set of notes, how many distinct diminished seventh chords are there? List the notes in each chord.

2.4   Tonality, Key Signatures, and the Circle of Fifths

As explained in Section 2.2.5, there are 12 different major scales, one for each note of the chromatic scale. Due to enharmonic equivalence, some scales have more than one name. For example, an F♯ and a G♭ major scale have the same exact pitches, although the first scale has six sharps while the other has six flats. Each major scale is uniquely determined by the number of flats or sharps it contains. For example, G major has one sharp (F♯), and E♭ major contains three flats (B♭, E♭, and A♭). This arises from the existence of the 12 different scales, as well as the diatonic naming conventions used to describe the notes of each scale, that is, using all seven letters of the musical alphabet, plus accidentals. Recall that a piece or excerpt of music predominantly featuring the notes from a particular major scale is said to be in the *key* of that scale.

2.4.1 The critical tonic-dominant relationship

As previously discussed, one of the most important chord progressions in Western music is the dominant (V) to tonic (I) progression. In the key of C major, the dominant chord is constructed from scale degrees 1, 3, and 5 of a G major scale (the notes G, B, and D). The dominant chord, particularly when the minor seventh is added (in this case, an F), strongly wants to resolve to the tonic chord. The tonic-dominant relationship is further reinforced by the fact that the two keys determined by each pitch have all but one note in common.

For example, the C major and G major scales share the same set of notes except for the F in C major, which is modified to F♯ in G major (see Figure 2.36). Likewise, the keys of C major and F major are also musically close, as the B in C major and the B♭ in F major are the only notes not common to each. In this case, F is the tonic and C, a perfect fifth higher, is the dominant. The pitch that the two scales do not have in common is important. It is the leading tone of the scale built on the dominant (F♯ when comparing C and G, or B when comparing F and C). This note is crucial, as it naturally "leads" back to the tonic.

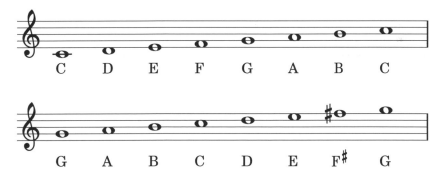

FIGURE 2.36. The C major and G major scales have the same set of notes, except the F in the key of C becomes an F♯ in the key of G.

There is nothing special about these particular examples. Since a major scale is always constructed with the same H-W sequence, we can translate the above examples to begin on any note we please. This leads to an important fact:

> Two major scales built on notes a perfect fifth apart will always share the same set of pitches except for one. The different note for the scale built on the dominant is the leading tone of that key.

Major Scale	# of Sharps	Sharp Sequence
C	0	
G	1	F♯
D	2	F♯, C♯
A	3	F♯, C♯, G♯
E	4	F♯, C♯, G♯, D♯
B	5	F♯, C♯, G♯, D♯, A♯
F♯	6	F♯, C♯, G♯, D♯, A♯, E♯
C♯	7	F♯, C♯, G♯, D♯, A♯, E♯, B♯

TABLE 2.5. The sharp keys and their sequence of sharps.

It follows that raising a major scale with zero or more sharps by a perfect fifth will add a sharp to the new key. For instance, since D is a perfect fifth above G, a D major scale will contain two sharps: F♯, which is inherited from the G major scale, and C♯, which is the leading tone to D (a half step below D). Continuing the process, A major (A is the dominant of D) will contain three sharps: F♯ and C♯ from the D major scale, as well as G♯, the leading tone to A. Each time we go up by a perfect fifth, a new sharp is added to the key of the dominant. This special note is a perfect fifth higher than the new note of the previous key (e.g., C♯ is a perfect fifth above F♯, and G♯ is a perfect fifth above C♯.)

Based on this observation, it makes sense to list the sharps for the sharp keys (those major scales containing only sharps) in the order in which they are generated by repeatedly moving from the tonic to the dominant. The list of sharp keys and their corresponding sequence of sharps is shown in Table 2.5. This is a standard list in music theory. Notice that each successive key and each additional sharp is a perfect fifth higher, respectively, than the previous one.

The flat keys

A similar argument can be used to construct the sequence of successive flats for the different flat keys. However, in this case we consider the effect of *lowering* the pitch by a perfect fifth. For example, when moving from the key of C down a perfect fifth to the key of F, the new note added is a B♭. This pitch results from *lowering* the leading tone of the original key (B in the key of C major) by a half step. Again, by translation, the argument can be applied repeatedly to all seven flat keys. Moving from F down a perfect fifth to the key of B♭ will yield two flats: the B♭ inherited from the key of F and the new flat E♭, which is a half step below E, the leading tone of F. Continuing in this fashion, we can determine the order in which the flats are added by repeatedly dropping the newest flat by a perfect fifth (see Table 2.6).

Major Scale	# of Flats	Flat Sequence
F	1	B♭
B♭	2	B♭, E♭
E♭	3	B♭, E♭, A♭
A♭	4	B♭, E♭, A♭, D♭
D♭	5	B♭, E♭, A♭, D♭, G♭
G♭	6	B♭, E♭, A♭, D♭, G♭, C♭
C♭	7	B♭, E♭, A♭, D♭, G♭, C♭, F♭

TABLE 2.6. The flat keys and their sequence of flats. Notice that each successive key and each additional flat is a perfect fifth *lower*, respectively, than the previous one.

2.4.2   Key signatures

The correct sequences of sharps and flats are reflected in an important musical symbol called a *key signature*. Writing music with a large number of accidentals, particularly in keys that have a significant number of sharps or flats, can become difficult to read. The top line of music in Figure 2.37 shows the opening four measures of *Twinkle Twinkle Little Star* in the key of F♯ major. Every note receives a sharp except for the two Bs at the start of the third measure.

A cleaner way of writing the same music is to use a key signature, as shown on the bottom line of music in Figure 2.37. Here the group of six sharps at the start of the line indicates which notes on the staff should be sharped *automatically*. These are the six sharps in the key of F♯ major. In other words, whenever an F, C, G, D, A, or E appears in the music, it should be raised up a half step. This applies to *all* versions of the given notes, regardless of the octave they lie in. Notice that the order of sharps shown in the key signature, F♯, C♯, G♯, D♯, A♯, and E♯, agrees with the order specified in Table 2.5. Each sharp in the list is a perfect fifth higher than its predecessor.

FIGURE 2.37. The first four measures of the lullaby *Twinkle Twinkle Little Star* in the key of F♯ major, written with the key signature (*bottom*) and without (*top*).

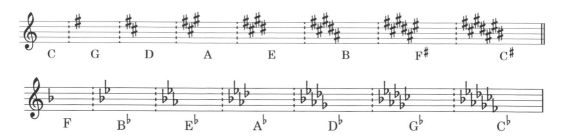

FIGURE 2.38. The key signatures for the sharp and flat keys.

Key signatures are a useful device for simplifying musical notation. They also immediately identify the key of a piece or musical excerpt. The key signature is placed at the start of every line of music and is assumed to apply to each measure in the given line. A natural sign is required to cancel an accidental listed in the key signature. For instance, if we are in the key of E♭ major (three flats: B♭, E♭, and A♭) and want to write the note B, we must include a natural sign ♮ in front of the B to cancel the automatic flat given in the key signature.

The key signatures for the sharp and flat keys are shown in Figure 2.38. Observe that the order of the accidentals on the staff in each key is in agreement with the order we derived by moving up a perfect fifth in the case of the sharp keys (compare with Table 2.5) and down a perfect fifth in the case of the flat keys (compare with Table 2.6). Accidentals in a key signature are always placed within the staff (no ledger lines).

Example 2.4.1. *Figure 2.39 shows a melody written in a major key, using a key signature. Indicate which notes receive accidentals and give the names of those notes.*

FIGURE 2.39. A sample melody written using a key signature.

Solution: According to the key signature, the excerpt is in the key of A major, with the three sharps F♯, C♯, and G♯. This means that *every* F, C, and G, regardless of the octave, should receive a sharp. The solution is shown in Figure 2.40, where the notes affected by the key signature are circled. Note the use of the natural sign in the fifth measure to change the C♯ to a C♮.   □

FIGURE 2.40. The melody from Figure 2.39 with the notes modified by the key signature identified.

2.4.3 The circle of fifths

The *circle of fifths* is a clever musical device that captures the crucial tonic-dominant relationship we have been discussing. It summarizes the number of sharps or flats in the major scales and can be used to determine the correct order in which accidentals are listed for a particular key. It is also useful for understanding the relationship between two different keys.

The circle of fifths is constructed in a manner similar to the musical clock we built for the chromatic scale, except instead of going clockwise around the circle by half steps, we go up by perfect fifths. The top of the circle begins with C. Moving clockwise, each ensuing notch on the 12-note clock is a perfect fifth above the previous note. This yields the sequence C, G, D, A, E, B, F♯ (at the bottom of the circle), and C♯. In the opposite direction, moving counterclockwise around the circle means lowering the pitch by a perfect fifth. Beginning at the top of the clock, this gives the sequence C, F, B♭, E♭, A♭, D♭, G♭ (at the bottom of the circle), and C♭ (see Figure 2.41). For reasons that will be discussed shortly, we do not include other enharmonic equivalents such as G♯, D♯, or F♭.

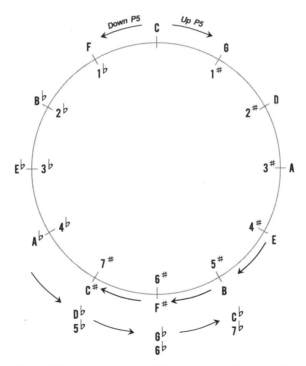

FIGURE 2.41. The circle of fifths, a useful tool for learning the number and order of sharps or flats in a given key.

As discussed in Sections 2.4.1 and 2.4.2, the advantage of moving around the circle by perfect fifths is that going clockwise adds a sharp to the next key, while traveling counterclockwise adds a flat. This gives a quick way to determine the number of flats or sharps in a given key, without having to construct the entire major scale. Moreover, the correct ordering of accidentals is also captured by the circle of fifths, since each successive accidental is either higher (sharps) or lower (flats) than its predecessor by a perfect fifth.

Beginning at the 11:00 position on the circle of fifths (the F) and traveling clockwise produces the sequence F, C, G, D, A, E, B. Adding a sharp to each of these notes produces the correct sequence of sharps for the sharp keys. Adding a *flat* to each of these notes and reading the list *backward* produces the correct sequence of flats.

One way to help remember or reconstruct the circle of fifths is to notice that every other note forms a whole-tone scale. For example, beginning on C and skipping every other note in the clockwise direction (the even-numbered hours) gives the notes of the whole-tone scale starting on C. Similarly, starting with G and skipping every other note (the odd-numbered hours) yields the notes of a whole-tone scale starting on G.

Each pitch of the 12-note chromatic scale is listed exactly once on the circle of fifths. From a mathematical perspective, this occurs because 7 and 12, the number of half steps in a perfect fifth and an octave, respectively, are relatively prime. Thus, according to Theorem 1.3.3, the least common multiple of 7 and 12 is their product, 84. Recall that the modern piano keyboard has 88 keys. If we begin on the lowest C on the piano and proceed up by perfect fifths, we will touch each note of the chromatic scale once before returning to a C, all the way at the top of the piano. Traveling 12 perfect fifths is equivalent to going seven octaves ($12 \cdot 7 = 7 \cdot 12 = 84$). Since the number of half steps in a perfect fifth and the number in an octave are relatively prime, we are guaranteed to hit all 12 notes of the chromatic scale before returning back to C. In contrast, a circle constructed by shifting up an interval that has a number of half steps not relatively prime to 12 will close up too early. For instance, if we begin on C and move up by consecutive minor thirds (three half steps), we would obtain the notes C, E♭, G♭, A, and then C. This is only a third of the notes of the chromatic scale.

Tonal proximity

Keys that are near each other on the circle of fifths will have more notes in common than those far away. Adjacent keys, such as C and G, or B and F♯, share all of the same notes except for one. Keys that are two steps away on the circle of fifths will share all but two notes. For instance, C major has a C and an F, while D major contains a C♯ and an F♯; otherwise, their notes are the same. Keys that are far away on the circle of fifths are not musically close. A C major scale and F♯ major scale, diametrically opposite on the circle, have only two notes in common, B and F (technically E♯ in the key of F♯).

This is a useful fact for understanding how music works. When composers change keys, to do so smoothly means keeping them musically close. The more notes in common between the two major scales, the less jarring the switch to the new key will seem. A well-practiced method of developing a musical theme in an alternative key is to shift from the key of the tonic to the key of the dominant, since they have all but one note in common. The circle of fifths provides an easy reference to determine the tonal proximity of two different keys.

Why is there no key of G♯?

The key of C♯ contains seven sharps, that is, all seven notes of the musical alphabet receive sharps. Rising up a perfect fifth, the next key would be G♯. In this case, the newest sharp would correspond to a perfect fifth above B♯ = C, which is the note G. However, due to the convention of using diatonic scale spellings, G and G♯ are not allowed to be used in the same scale. Thus, we are forced to introduce a double sharp, and write F♯ for G. This

FIGURE 2.42. A G♯ major scale written with diatonic scale spellings has eight sharps.

implies that the key of G♯ has eight sharps (see Figure 2.42). Since it is bizarre to have a seven-note scale contain eight sharps, G♯ is discarded in favor of the simpler A♭, a key that has four flats. The same argument can be applied to the flat keys. Moving a perfect fifth down from C♭, the key with seven flats, yields the key of F♭ major. Using a diatonic scale spelling would require eight flats, including a B♭♭. Writing this key in E major, with its four sharps, is considerably easier.

The relative major of a minor key

Now that we have a better grasp of the different major keys courtesy of the circle of fifths, we are ready to identify a simple connection between the natural minor and major scales. Recall from Section 2.2.4 that the key of C minor has three flats, and these are B♭, E♭, and A♭. An E♭ major scale also has three flats, and these are precisely the same flats found in the C minor scale. It follows that the two scales possess the exact same set of notes, just with a different starting note (a different tonic). In fact, taking a C minor scale and beginning on the third scale degree produces an E♭ major scale (see Figure 2.43). The key of C minor and the key of E♭ have the same exact key signature (three flats).

This connection between a natural minor scale and the corresponding major key, called the *relative major*, works for any minor scale. For example, a B natural minor scale, which has the two sharps F♯ and C♯ (see Figure 2.24), contains the same notes as a D major scale.

Recall the H-W sequence for a natural minor scale, W H W W H W W. Applying two cyclic shifts (σ^2) to this sequence gives the defining sequence for a major scale, W W H W W W H. Hence, shifting a minor scale to start on the third scale degree will always produce a major scale, and the number of sharps or flats for the initial minor scale and its

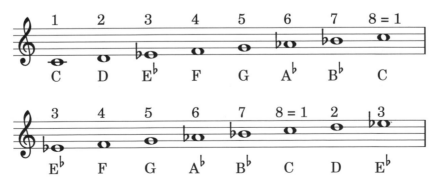

FIGURE 2.43. Taking the notes of the C minor scale and starting on the third scale degree yields an E♭ major scale.

relative major will be identical. The third scale degree of a natural minor scale is a minor third above the starting note. Thus, to find the key of the relative major, we go up a minor third from the tonic (e.g., E♭ is a minor third above C, and D is a minor third above B). This motivates the following definition.

Definition 2.4.2. *The relative major of a natural minor key has the same number of flats or sharps as its minor counterpart. The key of the relative major is found by going up a minor third from the tonic of the minor scale. Similarly, the relative minor of a major key is the minor key with the same number of flats or sharps as its major counterpart. The key of the relative minor is found by going down a minor third from the tonic of the major scale.*

For example, the relative minor of C major is A minor, since A is a minor third *below* C. Both C major and A minor have no sharps or flats. The relative major of G minor is B♭ major, since B♭ is a minor third *above* G. Each key has two flats, B♭ and E♭. Since a minor key and its relative major share the same set of notes, they have the same key signature. Hence, care must be taken when identifying a piece of music by its key signature, as the work may be in either a major or a minor key.

Just as with the major keys, enharmonic equivalence allows for certain minor keys to share the same notes but have different names. The relative major of E♭ minor is G♭ major because G♭ is a minor third above E♭. Thus, the key of E♭ minor has six flats. On the other hand, the relative major of D♯ minor is F♯ major, with six sharps. While the pitches of E♭ minor and D♯ minor are identical, the names of the notes are entirely different.

Johann Sebastian Bach's famous *Well-Tempered Clavier* consists of two books of 24 preludes and fugues, one for each of the 12 major keys and one for each of the 12 minor keys. Interestingly, in Book I (1722) Bach wrote the prelude in E♭ minor (six flats), but the fugue in D♯ minor (six sharps). Since these two pieces are played back to back, this was a rather peculiar shift of key signatures. In Book II (1742), Bach chose to keep this particular prelude and fugue in the same key, D♯ minor. Figure 2.44 connects the relative minor keys with their major counterparts in an "ancient" version of the circle of fifths.

2.4.4 Transposition

One of the main advantages of key signatures is that they allow music to be easily transferred into different keys. Writing a melody or piece of music in a different key is known as *transposing*.

Definition 2.4.3. *A transposition of a piece of music is rewriting it in a different key. Using key signatures, this is a vertical translation along the staff.*

While changing keys sounds complicated, the basic idea is quite simple. Each note is translated by the same amount, an interval determined by the number of steps between the old key and the new one. For example, to change a melody from C major to E♭ major, each note needs to be raised a minor third, the interval between C and E♭. This could take some work if it were done note by note. But using key signatures makes the job a relatively simple task because any additional accidentals are handled by the key signature. To accomplish the transposition, we add three flats to the key signature and then raise each note three steps on the staff (three because the interval is a third). Remember to count *both* the starting and ending locations when shifting notes on the staff.

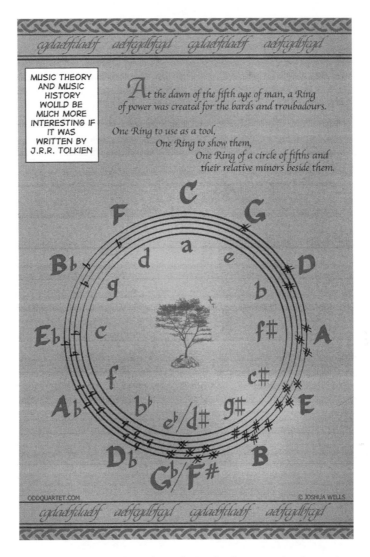

FIGURE 2.44. What if Frodo had taken music theory?

The process is demonstrated on the opening four bars of *Twinkle Twinkle Little Star* in Figure 2.45. The initial melody is shown on the top line of music, in the key of C major. Each note is then moved up three steps on the staff. Thus, notes on a line of the staff are translated up to the next highest line, while notes in a space are shifted up to the next space. There is no need to worry about adding any accidentals because the relationship between the notes of the melody and the tonic is preserved under a vertical translation. For example, the starting note C is the fist scale degree in C major. It moves up to an E♭ (no flat is required in front of the note because of the key signature), which is the first scale degree of E♭ major. The A in the second measure of the original melody is scale degree 6 in C major. This note shifts up to a C, which is scale degree 6 in E♭ major. A vertical shift moves the tonic, but it preserves the scale degree of each note in relation to the tonic.

FIGURE 2.45. The first four measures of *Twinkle Twinkle Little Star* written in the keys of C major, E♭ major, and B major.

The bottom line of Figure 2.45 shows the melody transposed into the key of B major. Since B is one step below a C, this is achieved by shifting each note of the original melody down one step and then changing the key signature to have five sharps. The only subtlety when transposing occurs if there are notes foreign to the scale of the original key (usually indicated by an accidental not in the key signature). In this case, it is best to calculate the number of half steps between the old key and the new one in order to determine the correct location of the new note. Still, the combination of a distance-preserving transformation (a vertical shift) and the use of key signatures makes transposition a straightforward task.

2.4.5   The evolution of polyphony

Condensing the history of Western classical music into a few pages is a rather daunting exercise, one that is sure to exclude many features and examples. Nevertheless, there are some key historical developments that provide an important context for our ensuing discussion on tuning and temperament in Chapter 4. Understanding some of this history helps clarify the evolution of the instruments we play and the music we listen to. It also strengthens our musical appreciation and breadth of knowledge. Our discussion will highlight the important changes in tonality which transpired over the centuries.[4]

The music of ancient civilizations such as the Egyptians and Babylonians is mostly lost to us. Although very few written examples of ancient Greek music exist, we do know that Pythagoras and his followers applied mathematical analysis to study the ratios of frequencies between different pitches. The *Pythagorean scale*, discussed in Section 4.1, is the precursor to our current equally-tempered major scale, different by only slight modifications.

---

[4]Much of the material for this section comes from Machlis, *Enjoyment of Music*.

One of the earliest organized attempts to notate music and preserve it for future generations was undertaken by the Catholic Church. This music is known as *Gregorian chant*, in reference to Pope Gregory the Great, who reigned from 590 to 604 and commanded that the singing of the monks of the Church be transcribed. The melodic texture is *monophonic*, consisting of voices in unison singing a single-line melody. There is no rhythmic structure to Gregorian chant, but rather a free-verse rhythm accompanying the Latin text. The melodic lines typically feature small steps between successive notes, with occasional jumps of larger intervals such as fourths and fifths.

Early polyphonic music

An important development in Western music occurred near the end of the Romanesque period (c. 850–1150), when multiple lines of music played *simultaneously* began to appear. This is known as *polyphonic* music, where multiple voices play or sing together on different lines of music ("poly" means *many*). This new type of music required rhythmic structure in order for multiple parts to stay together. Soon followed the need to put notes to the page, and the musical staff and accompanying notation were born. At this point, music began to shift away from primarily an oral tradition to a written one, allowing for future musicians (and composers) to benefit from the works of the past.

Moving forward to the thirteenth and fourteenth centuries, the *motet* emerged as one of the earliest forms of polyphonic music. Often a fragment of Gregorian chant was taken and arranged into a more precise rhythmic structure (usually with longer notes) that formed the structural blueprint for the piece. This became known as the *cantus firmus* (fixed melody). Other faster lines of countermelodies accompanied and contrasted the cantus firmus, creating more varied polyphonic music. The separate lines of music were thought to be independent in a harmonic sense (sometimes they even carried different lines of text), but at cadences (the end of a phrase), the intervals of an octave or a perfect fifth were commonly used. One of the great motet composers was the Frenchman Guillaume de Machaut (c. 1300–1377), who helped shift the motet from a religious to secular art form. The sound of a motet is fuller than Gregorian chant, but not as rich as the music of the following centuries.

During the Renaissance (c. 1450–1600), music started to embrace more secular ideals such as clarity and sensuality. The *a cappella* musical style (unaccompanied singing) became prominent, featuring a polyphonic trait known as *continuous imitation*. In this case, the vocal lines were much more dependent, with different parts imitating each other in close succession. This concept was the precursor to the more sophisticated *fugue* so artfully developed later by Johann Sebastian Bach. Some famous Renaissance composers include the Flemish Josquin des Prez (c. 1450–1521) and the deeply religious Catholic composer Giovanni Pierluigi da Palestrina (c. 1525–1594). While tonality, the idea of having a specific key, is not quite established in their music, the cadences (the measures leading up to a final chord) are suggestive of a dominant-to-tonic relationship.

Major/minor tonality

In the Baroque period (c. 1600–1750), a significant musical development occurred with the firm establishment of the system of major and minor keys. Composers began to write in a specific key with a definitive tonal center. One of the main drivers of this change was the technological improvements to musical instruments, particularly keyboard instruments such as the organ and harpsichord (an early version of the piano). Music evolved from primarily vocal works to those that required instrumental accompaniment, as well as pieces written for instruments only. This necessitated a more evolved harmonic structure, in order for the instrumental parts to properly support the voices. The need for a substantial and comprehensive music theory, with its emphasis on the major and minor keys discussed in this chapter, had arrived.

One of the greatest Baroque composers, perhaps one of the best of all time, was Johann Sebastian Bach (1685–1750). Bach was a prolific composer who wrote music for many instruments, including the organ, keyboard, violin, and cello, as well as the full orchestra. His *Well-Tempered Clavier*, the 48 preludes and fugues in each major and minor key, has been characterized as the "pianist's Old Testament." He also composed over 200 *cantatas*, choral works featuring four parts (soprano, alto, tenor, and bass), vocal soloists, and instrumental accompaniment, which were performed as part of a Protestant service every Sunday in Leipzig, Germany. Bach's extensive use of symmetry in his music, a topic we will explore in detail in Section 5.1, is a mathematical trait that underscores his brilliance as a composer.

After Bach, the era from 1750 to 1900 produced some of the most well-known names in classical music history: Mozart, Beethoven, Haydn, Chopin, Verdi, Brahms, Tchaikovsky, Dvořák, Puccini, and Debussy (to name a few). Building on the foundations laid by Bach and other Baroque composers, these artists produced some of the most memorable works in the classical repertoire. Certain forms, such as the symphony and opera, reached the pinnacle of their development under these gifted composers.

In terms of harmony and music theory, a deep understanding of consonance and dissonance was achieved during this period and used to great effect. The notion of "tension and release," as epitomized with the dominant (V; tension) to tonic (I; release) relationship, became the governing mantra. Figure 2.46 demonstrates this relationship in the opening four bars of Mozart's famous *Eine Kleine Nachtmusik* (A Little Night Music; 1787). The piece is in the key of G major, so the first two bars are clearly rooted in the tonic G (the notes are G, B, and D, forming a G major triad), while the next two measures feature the dominant D (the notes D, F♯, A, and C form a dominant D seventh chord). The fifth measure desperately wants to return to the tonic. The intimate connection between V and I is explored throughout *Nachtmusik*.

FIGURE 2.46. The opening bars of Mozart's *Eine Kleine Nachtmusik*, demonstrating the basic tonic (I) to dominant (V) relationship.

Moving beyond tonality

> *Tonality is not an eternal law of music, but simply a means toward the achievement of musical form.*
>
> — Arnold Schoenberg[5]

---

[5]Machlis, *Enjoyment of Music*, p. 471.

In 1865, Richard Wagner's groundbreaking opera *Tristan und Isolde* premiered in Munich. The work displayed a new and unconventional approach to tonality and harmonic structure and incorporated a wide range of orchestral sounds. Of particular note was the manner in which Wagner experimented with musical tension. Instead of resolving dissonant chords at a pace that most audiences of the day were accustomed to, the composer frequently surprised his listeners by delaying resolution or discarding it altogether. For instance, the opening chord of the piece, the "Tristan" chord, consists of the notes F, B, D♯, and G♯, a tense combination built over the dissonant tritone between F and B (see Figure 2.47). The ensuing chord (in measure 3) appears to be an E dominant seventh chord, except that the fifth begins as an A♯ (the highest pitch at the start of measure 3, a tritone above E), before rising to its expected home on B. Moreover, this dominant seventh chord remains hanging; it does not resolve to the expected tonic, presumably A minor. This type of harmonic progression, featuring unresolved tension and repeated use of dissonant intervals such as the tritone, is prominent throughout the piece. Wagner's seminal work is widely viewed to have laid the foundation for later composers to challenge and eventually discard the commonly held tenants of major-minor tonality.

FIGURE 2.47. The opening three measures of Wagner's opera *Tristan und Isolde*. The dissonant "Tristan" chord in the second measure is eventually followed by a dominant seventh chord that does not resolve.

At the start of the twentieth century, composers such as Bartók and Stravinsky wrote works using simultaneous multiple keys (polytonality), as well as complex rhythmic structures, such as polyrhythms (e.g., *The Rite of Spring*; see Section 1.3.2). Some composers began to ignore tonality altogether, writing pieces without any definitive key. So-called *atonal* music, first developed by Arnold Schoenberg (1874-1951) and his followers, abandoned the idea of tension and release, as there was no central key to organize around. This was a very radical departure, one that challenged composers to create new concepts to unify their works.

One important compositional technique, developed by Schoenberg in the 1920s, was the *12-tone method* of composing. In this rather democratic style of composition, instead of favoring one pitch as a tonal center, each pitch of the chromatic scale is given equal importance. A sequence of 12 notes in some prescribed order, called a *tone row*, is used to generate all the musical material for a work. Instead of a melody, the idea is to compose by following the order determined by the row. The composer is free to choose the rhythm, or in which octave a particular note in the row should be placed. Moreover, a particular tone row can be transposed, reflected horizontally, written backward, or some combination of these operations. This generation of new tone rows from the original one is an inherently mathematical process. In Chapter 7, we will study this groundbreaking method of composition in great detail.

By the middle of the twentieth century, composers were not only redefining and re-making tonality but experimenting with the actual concept of music and how it is created. Composers such as John Cage (1912–1992) explored different timbres by altering traditional instruments. His famous "prepared piano" music involves placing incongruous objects such as rubber, felt, coins, screws, and bolts on and between the strings of a piano to greatly alter its usual timbre. Cage also glorified *chance music*, where randomness ruled over structure and form. The piece *Imaginary Landscapes* (1951) involves 12 radios tuned to different stations all playing simultaneously.

In a different vein, mathematics and science became a source of inspiration for composers such as Milton Babbitt (1916–2011), György Ligeti (1923–2006; whose piece *Poème Symphonique* was discussed in Section 1.3.2), Iannis Xenakis (1922–2001), and Sir Peter Maxwell Davies (b. 1934). We will examine some of the mathematical methods and concepts used by these composers in Chapter 8.

It is important to realize that the music we love, in whatever genre or style, did not just magically appear one day. Our brief synopsis gives some indication to the remarkable evolution of music over the past 1500 years. The changes in the tonality, structure, form, concept, and purpose of music, changes often mirroring and reflecting the culture and artistic media of their times, have been significant milestones. The development of music has taken a long and varied path through human history, a path where, as we shall see, mathematics has played an important role.

Exercises for Section 2.4

1. After studying Figure 2.41 carefully, try to answer the following questions without referring back to the circle of fifths.

   a. How many sharps are in the key of E major? List them in the correct order.

   b. How many sharps are in the key of F♯ major? List them in the correct order.

   c. How many flats are in the key of D♭ major? List them in the correct order.

   d. Which major key has six flats?

   e. Which major key has five sharps?

   f. Which two keys are the closest musically to E♭ major?

   g. Which two keys are the closest musically to A major?

   h. Which key is the furthest musically from A major?

2. The following questions concern the relative major and minor keys.

   a. What key is the relative major of F minor?

   b. What key is the relative major of C♯ minor?

   c. What key is the relative minor of D♭ major?

   d. What key is the relative minor of A major?

   e. Which minor key has six flats?

   f. Which minor key has five sharps?

3. Each excerpt of music below uses a key signature. Indicate which notes receive accidentals and give the names of those notes.

a.

b.

4. The first four bars of Mozart's *Eine Kleine Nachtmusik*, in the key of G major, are shown in Figure 2.46. Using key signatures, transpose the first four bars into the keys of F major and D major.

5. Using key signatures, transpose the following excerpt into the keys of D♭ major and E major.

References for Chapter 2

Andrews, H. K.: 2007–2014, Whole-Tone Scale, *Grove Music Online*, Oxford University Press.

Dyer, R.: June 22, 2003, With Ear for Diversity, Cellist Yo-Yo Ma Has the World on a String Star Continues Tanglewood Tradition, *Boston Globe* p. N.6.

Groos, A. (ed.): 2011, *Richard Wagner: Tristan und Isolde*, Cambridge University Press.

Hagen, E.: 1940, *Harlem Nocturne*, Shapiro, Bernstein & Co., Inc., New York.

Kelley, R.: 2009, *Thelonious Monk: The Life and Times of an American Original*, Free Press.

Machlis, J.: 1984, *The Enjoyment of Music: An Introduction to Perceptive Listening*, 5th edition, W. W. Norton & Company.

Manoff, T.: 2001, *The Music Kit: Workbook*, 4th edition, W. W. Norton & Company.

McCormick, N.: February 1, 2012, What Is the Happiest Song of All Time?, *The Telegraph*. http://blogs.telegraph.co.uk/culture/neilmccormick/.

Meadows, E. S.: 2003, The Musical Language of Thelonious Monk, in *Bebop to Cool: Context, Ideology, and Musical Identity*, Greenwood Publishing Group. The African American Experience (online resource): http://testaae.greenwood.com/.

Monk, T.: 1962, *Blue Monk*, Thelonious Music Corp.

Zongker, B.: Dec. 9, 2014, Kennedy Center Honors Five Artists, *Telegram & Gazette* p. B.5.

Chapter 3

The Science of Sound

Suppose you are at an orchestra concert with your eyes closed, listening to a virtuoso violinist perform with an accompanying orchestra. When the violinist produces a particular note, say, middle C on the piano, how does your ear-brain system process that information? How do we hear that particular note and recognize that it is coming from that particular violin, even if all the other musicians are busily engaged producing their own sounds? Moreover, why do we enjoy certain combinations of notes more than others? As the violinist comes to the end of a phrase, why do we eagerly anticipate the music to resolve and then delight when that expectation is satisfied?

In this chapter we answer some of these questions by focusing on the mechanics of sound and how pitch is perceived. We first describe the incredible ear-brain system, emphasizing the key role played by the brain. Then we discuss some key attributes of sound such as loudness (measured in decibels) and pitch (determined by frequency). Since sound waves can be decomposed into a sum of sine waves, a review of the sine function and its properties is provided. The way we perceive pitch, the interesting concept of a residue pitch, and the importance of autocorrelation are investigated. We then describe the mechanics of a harmonic oscillator and a vibrating string, which motivates the all-important overtone series. The phenomenon of beats is developed using some lesser known but easily derived trigonometric identities. The chapter closes with a fun project involving a one-stringed instrument called a monochord. By discovering a few basic connections between the ratios of string lengths and the corresponding musical intervals, we construct the frequency ratios of the Pythagorean scale, a precursor to our modern major scale.

3.1   How We Hear

Sound is formed by changes in air pressure and the propagation of those changes. When an object vibrates, air molecules (primarily nitrogen and oxygen) push back and forth on each other and a wave forms, a *sound wave*. Air molecules are very sensitive to changes in pressure and can pass along information after colliding with their neighbors. These new molecules bump up against others, and the domino effect creates a propagating sound wave. If this mass of collisions occurs at a fast enough speed, we will hear a sound.

Interestingly, it is not the individual molecules that constitute the wave, but rather their collective interaction. This is similar to the movement of a water wave, or the famous *stadium wave* at a sporting event. Individual water molecules or sports fans do not flow with the wave, but their movement is the reason it exists. Their motion is transverse to the direction of the wave. While individual air molecules move at an average speed of 500 meters per second (m/s), the actual speed of sound through air at 20°C is 343 m/s, or 767.3 miles per hour.

Air is the medium that sound travels through. Without air, or another medium, there is no sound because there is no pressure to vary. Sound is absent in a vacuum, a point that seems lost on many creators of science fiction movies and television shows.

3.1.1   The magnificent ear-brain system

How do localized and fast-moving changes in air pressure become recognized sounds in our head? The answer lies with our truly phenomenal ear-brain system.

When sound reaches our outer ear, the oddly shaped pinna helps focus the pressure variations toward our middle ear. These variations pass down the ear canal (meatus) and vibrate our eardrum (tympanic membrane). Here the three tiny bones called the hammer (malleus), anvil (incus), and stirrup (stapes) work to propagate the vibrations through an oval window and into the perilymph fluid of the cochlea (see Figure 3.1).

The snail-shaped cochlea, which is a little over an inch when unrolled, filters the sound by frequency (the number of cycles a wave makes in a second). This is accomplished by

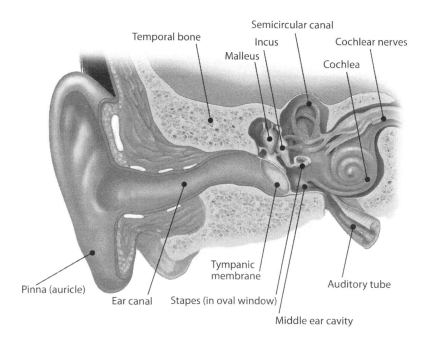

FIGURE 3.1. The human ear. Pressure variations reaching the pinna are converted into frequency data that immediately travel to the brain via the cochlear nerves. © Kocakayaali; Dreamstime.com—Human Ear Photo.

tiny rows of hairs on the remarkable *basilar membrane*. The fluid inside the cochlea helps the hairs on the basilar membrane transmit the frequency data to the brain via neural pathways. Here we have a physical-to-electrical transformation, as ion channels are engaged to transmit information from the molecules connecting the hair tips to nerve impulses heading to the brain. There are about 16,000 hair cells connected to the basilar membrane, and the hairs themselves are connected via tip links, with thickness approximately equal to 10 atoms. Amazingly, there is a *feedback effect* in real time, as the brain sends messages back to the basilar membrane to distinguish and amplify certain frequencies. The existence of this feedback loop is accepted; the understanding of how it actually works is unknown.

The role of the brain

As these auditory nerve impulses travel out of our inner ear, the brain takes over in some truly remarkable ways. Air pressure is a function of time and location. When an orchestra plays, each instrument creates its own pressure fluctuations, and the combination of these variations forms a sequence of changes in air pressure at a listener's ear. Even though there is only one combined impact at the ear, the brain is smart enough to instantly discern which instrument is playing what music.

In the past, an oscilloscope was used to record the pressure variations from a sound wave and create a *signal*, a graph of pressure versus time. The horizontal axis indicated the time, while the amount of pressure was plotted on the vertical axis. A graph of this kind represents a mathematical *function*, where each time t is mapped to the corresponding pressure fluctuation p occurring at that particular time. In mathematical terms, we write $p = p(t)$ and say that p is a function of t. Nowadays we can produce such graphs digitally using a standard computer with the appropriate software.

Given a complicated musical signal produced by a large group of instruments such as a symphony orchestra, it is virtually impossible to determine the individual pieces that combined to produce that particular graph. However, our brain can do this decomposition instantly and with little effort. Part of this is due to the fact that the brain is a learning organ. It can memorize a plethora of sounds and recall them instantly when required. The first time a child hears a trumpet, the characteristic sound is stored in the memory banks. When the trumpet plays again, the brain activates its memory and the child knows that a trumpet is sounding. This same learning process is what helps babies learn to speak, a fact that should discourage the use of frivolous "baby talk" to young children. Another reason our brains are capable of decomposing complicated sounds into their individual components is that the ear, through the basilar membrane, is able to do some of the initial frequency analysis before passing along the sound pressure information to the brain. The refined signal is easier for our brains to interpret than the raw pressure signal produced by the orchestra.

Another remarkable feature of the brain is the ability to adapt and adjust to a given situation. Sounds that seem incomprehensible to some, such as the chatter over an intercom or a heavy accent, are completely clear to others. In a foreign country, we naturally begin to pick up the local accent as we listen and learn to speak the native language. After living abroad for many years, a person's ability to communicate and understand a foreign tongue can come easily.

Finally, the brain is good at focusing on the sounds we care about, while eliminating extraneous "noise." It is easy enough to have a conversation at a party with the person next to you despite the fact that there are many other sounds competing for your attention (e.g., other conversations, music playing, televisions blaring). Some people can become so engaged in an activity that surrounding sounds are literally blocked out by the brain. Professional athletes develop this skill in order to ignore the jeers of opposing or disgruntled fans.

3.2   Attributes of Sound

When we hear a sound, certain properties are immediately familiar, such as loudness, pitch, timbre, and duration. These qualities are traits we perceive; they can also be measured in a physical sense with units. For example, a trumpet blasting a high C in a rock band could be described physically as a note with frequency 1045 Hz (cycles per second), intensity 80 dB (decibels), and length 2 seconds. The key attributes of sound, described in both physical and perceptual terms, are shown in Table 3.1. The notion of pitch is actually more complicated than it might initially seem. Most sounds do not have a clearly defined frequency, but rather a combination of frequencies, known as *partials* or *harmonics*. We will discuss this important aspect of pitch in the coming sections. The word "timbre" refers to the unique characteristics of a sound, the aspects that distinguish it from other sounds (e.g., the difference between a plucked string and a blown horn).

Perceptual	Physical	Units
Loudness	Intensity	dB (decibels)
Pitch	Frequency	Hz (cycles per second)
Duration	Length of time	s (seconds)
Timbre	Spectrum	

TABLE 3.1. The attributes of sound described in perceptual and physical terms.

3.2.1   Loudness and decibels

The human ear is amazingly sensitive to different degrees of loudness. Our ears can discern sounds ranging from the softest whisper in a quiet room to the loud jet engine on an airport runway. The power of sound, known as *sound intensity*, is measured on a logarithmic scale with a unit called a *decibel* (dB). Logarithms are useful when measuring quantities that have a huge range of values. For instance, the Richter scale, used to measure the amplitude of motion during an earthquake, is a logarithmic scale. Instead of measuring the quantity, it is the log of the quantity that is important. By using logarithms, values that are orders of magnitude apart become much closer together.

Sound intensity I is measured with respect to a particular reference value denoted by I_0. The formula to convert intensity into decibels is

$$\text{number of decibels} \ = \ 10 \log_{10} \left(\frac{I}{I_0} \right), \tag{3.1}$$

where I_0 is typically taken to be the *threshold of human hearing*, or $I_0 = 1 \times 10^{-12}$ watts/m^2.

The key aspect of understanding Formula (3.1) is that it converts a *multiplication* of intensity into a *sum* of decibels. Specifically, by substituting $I = d \cdot I_0$ into Formula (3.1), we deduce the following important relationship:

| Raising the sound intensity by a factor of d means adding $10 \log_{10}(d)$ decibels. | (3.2) |

For instance, increasing the intensity of a sound 100 times (which seems like a large amount) corresponds to a rise in volume of only 20 dB, since $10 \log_{10}(100) = 10 \cdot 2 = 20$. If we double the volume, then this corresponds to an approximate increase of just 3 dB, as $10 \log_{10}(2) \approx 3$.

Table 3.2 displays some common sounds and their approximate decibel equivalents. Sustained exposure to a sound over 85 dB can lead to hearing loss. It is important to understand that a decibel is a physically dimensionless unit, unlike a second or a meter. A decibel value is measured in reference to I_0 via Formula (3.1). The numbers shown in Table 3.2 are with respect to $I_0 = 1 \times 10^{-12}$ watts/m^2, the threshold of human hearing.

Sound	Decibels
Threshold of human hearing	0
Whisper	15
Mosquito buzz	40
Regular conversation	60
Jackhammer	100
Rock concert	120
Threshold of pain	130
Jet engine at 30 meters	150

TABLE 3.2. Some sounds and their approximate intensity measured in decibels.

Understanding logarithms

Logarithms are defined as the inverses of exponential functions. The base of the logarithm, which is b in the expression $\log_b(d)$, is the same as the base of the corresponding exponential function, so b^x and $\log_b(x)$ are inverse functions.

The simplest way to understand logs is to remember that the output of a logarithm is an *exponent*. For example, $\log_2(d) = x$ means that $2^x = d$, and $\log_{10}(d) = y$ means that $10^y = d$. In other words, when trying to find the log of the number d, raising the base of the logarithm to the value of the logarithm returns the number d. We have $\log_2(16) = 4$, since $2^4 = 16$, and $\log_{10}(1000) = 3$, because $10^3 = 1000$. In each of these examples, the output of the logarithm (the number we want to find) is an exponent. Note that $\log_b(1) = 0$ for any base b because $b^0 = 1$. This explains the 0 in the first line of Table 3.2; plugging $I = I_0$ into Formula (3.1) gives $10 \log_{10}(1) = 0$.

Example 3.2.1. *Find the value of each logarithm, if it exists.*

a. $\log_2(1/32)$ **b.** $\log_5(625)$ **c.** $\log_{10}(\sqrt{10})$ **d.** $\log_{10}(0)$

Solution: For part **a.**, $32 = 2^5$, so $1/32 = 2^{-5}$, and thus $\log_2(1/32) = -5$. In **b.**, $625 = 25^2 = 5^4$, so $\log_5(625) = 4$. For **c.**, the number $\sqrt{10}$ is equivalent to $10^{1/2}$, which is a power of the base of the logarithm. Thus, the solution is the value of the exponent, that is, $\log_{10}(\sqrt{10}) = 1/2$. Finally, the last logarithm does not exist because the equation $10^x = 0$ has no solution; the graph of the exponential function b^x always lies above the horizontal axis. □

The fact that the exponential function b^x and the logarithm to the base b, $\log_b(x)$, are inverses of each other implies the identities

$$b^{\log_b(x)} = x \quad \text{and} \quad \log_b(b^x) = x.$$

These formulas can be understood using the definition of the logarithm.

One of the useful properties of logarithms is that they turn the operation of multiplication into addition. This is captured by the formula

$$\log_b(xy) = \log_b(x) + \log_b(y),$$

which can be checked by raising both sides to the base b and applying the rule for exponents $b^{m+n} = b^m \cdot b^n$. Similarly, we have the properties

$$\log_b\left(\frac{x}{y}\right) = \log_b(x) - \log_b(y) \quad \text{and} \quad \log_b(x^m) = m\log_b(x).$$

Example 3.2.2. *Suppose that the volume of sound coming from a speaker is increased by 30 decibels. By what factor has the sound's intensity increased?*

Solution: Using Rule (3.2), we must solve the equation $30 = 10\log_{10}(d)$, which simplifies to $3 = \log_{10}(d)$. Raising both sides to the base 10 gives $10^3 = 10^{\log_{10}(d)} = d$. The intensity of the sound has increased by a factor of 1000. □

3.2.2 Frequency

Frequency is measured by the number of cycles a wave makes in a second. One cycle occurs when the wave has returned to its starting position and is about to repeat itself. Figure 3.2 shows a wave with three cycles in 1 second. The standard unit to measure frequency is a *hertz* (Hz), named after the German physicist Heinrich Hertz (1857–1894), the first scientist to discover how to transmit and receive radio waves. A value of 1 Hz means a rate of one cycle per second. Bigger frequencies, which means faster waves, correspond to higher notes, while smaller frequencies (slower waves) produce lower-sounding pitches. A singer moving up the scale is increasing the speed of the notes being sung. Some frequencies are too low (below 20 Hz) or too high (above 20,000 Hz) for the human ear to perceive.

One important musical frequency is 440 Hz, which is the frequency of the A just above middle C on the piano. This is the A on the second space of the staff in the treble clef, commonly referred to as A440. It is the standard note that orchestras and musical groups

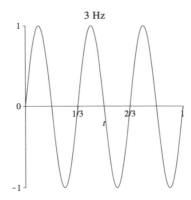

FIGURE 3.2. A wave with a frequency of three cycles per second.

around the world use as their baseline pitch for tuning. Table 3.3 shows the audible frequency range of some mammals, as well as the range for notes produced by different instruments and voices.

While the particular frequency of a note determines its pitch, the *ratio* of two frequencies is important for understanding the musical interval between two notes. We will explore the importance of these ratios and how different musical scales arise from different ratios in Chapter 4. One of the most important ratios in music is 2:1. We have already seen its importance in rhythm (see Section 1.1) through geometric sequences and series with a ratio of $r = 1/2$. Two notes whose frequencies are in a 2:1 ratio are an octave apart. For example, the pitch corresponding to 220 Hz is the A just below middle C on the piano, since $440 \div 220 = 2$.

Mammal or Instrument	Frequency Range (Hz)
Human Ear	20–20,000
Dog	50–46,000
Dolphin	1000–130,000
Bat	2000–110,000
Gerbil	100–60,000
Piano	27–4186
Violin	196–3520
Tuba	40–440
Soprano	262–1047
Bass (voice)	80–330

TABLE 3.3. The approximate frequency range heard by some mammals contrasted with the range of some instruments and voices.

The French priest Marin Mersenne (1588–1648), who made significant contributions to the fields of both mathematics and music, was the first scientist to accurately measure the frequency of a plucked string with an audible pitch (at 84 Hz). Mersenne was aware of the key connection between the ratio $2 : 1$ and the octave. Using a long enough string, he was able to count the number of cycles produced under vibration, and then by repeatedly folding it in half, he multiplied by the correct power of 2 to determine the frequency of the given note. His technique was described in his treatise *Harmonie Universelle* (1636). Mersenne is sometimes remembered as the "father of acoustics."

Exercises for Section 3.2

1. Compute each of the following logarithms, if they exist.

    a. $\log_{10}(1{,}000{,}000)$ **b.** $\log_{10}(0.00001)$ **c.** $\log_{10}(1)$ **d.** $\log_{10}(\text{googol})$

    e. $\log_2(1/64)$ **f.** $\log_3(243)$ **g.** $\log_4(1024)$ **h.** $\log_6(1/36)$

2. Suppose that the intensity of a sound is increased by a factor of d. In each of the following cases, give the number of decibels (approximate if necessary) by which the volume rises.

    a. $d = 4$ **b.** $d = 200$ **c.** $d = 1000$ **d.** $d = 1{,}000{,}000$

3. Suppose the music coming out of an amplifier is increased by 10 dB. By what factor has the intensity of the sound increased? What is the answer if the sound is increased by 15 dB?

4. If increasing the sound intensity by a factor of d raises the volume by 18 dB, how much will it change the volume if the intensity is increased by a factor of $2d$?

5. Recall that two pitches whose frequencies are in a $2 : 1$ ratio are an octave apart, and that A440 corresponds to the first A above middle C on the piano.

    a. What is the frequency of the A on the first ledger line above the staff in the treble clef?

    b. What is the frequency of the A on the top line of the bass clef?

    c. What is the frequency of the A on the bottom space of the bass clef?

    d. What is the frequency of the lowest note on the modern piano?

3.3   Sine Waves

Recall that pressure variations recorded from a sound wave form a signal, a graph of pressure versus time. To understand pitch and why some groups of notes sound more consonant than others, we need to understand the fundamental building block of signals: the *sinusoid*. It turns out that most vibrating objects oscillate in a particular way. Their motion may be described by a single sine wave (e.g., a tuning fork or pure tone) or, more likely, by a combination of sine waves with different frequencies. One of the big advances in the theory of acoustics was the realization that different signals can be decomposed into a combination of sine waves. The mathematical foundation for this discovery was established by Jean Baptiste Joseph Fourier (1768–1830), and the ensuing theory is now called Fourier analysis. Fourier realized the importance of using an infinite series of trigonometric functions to study a natural process, such as the motion of a wave or the vibrations of a string. It follows that understanding sine waves is vital to our discussion of pitch, music theory, and harmony. And yes, this means having a firm grasp of trigonometry!

3.3.1   The sine function

A tuning fork consists of two metal tines above a handle (see Figure 3.3) and is designed to produce a clear pure tone of a particular frequency, such as A440. The tines move air between them periodically, creating a signal that is essentially a perfect sine wave. If we were to trace the vibrations of a struck tuning fork on a moving piece of paper, the result would be a graph of a sine function with frequency equal to that of the tuning fork.

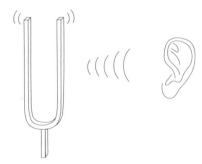

FIGURE 3.3. A vibrating tuning fork emits a simple sinusoidal sound wave toward our ear.

There are multiple definitions of the sine function $y = \sin t$, but one of the simplest to understand is that it represents the y-coordinate on the unit circle at an angle of t radians.

Definition 3.3.1. *The sine of t, denoted by* $\sin(t)$ *or just* $\sin t$, *is the y-coordinate of the point of intersection between the unit circle and a ray emanating from the origin at an angle of t radians (see Figure 3.4).*

It is important to remember that the input into the sine function $y = \sin t$ is an *angle t*, while the output is a *number y*.

Recall that the unit circle is the circle of radius $r = 1$ whose center lies at the origin $(0,0)$. Since $r = 1$, it follows that the largest possible y-value on the unit circle is 1, and the smallest possible value is -1. This implies that the output of the sine function (the range) is always between -1 and 1, that is, $-1 \leq y \leq 1$. Any point (x, y) on the unit circle satisfies the equation $x^2 + y^2 = 1$ by a simple application of the Pythagorean theorem.

The complementary trigonometric function to sine is the cosine function, $x = \cos t$, defined as the x-coordinate on the unit circle at the angle t (see Figure 3.4). Since the equation of the unit circle is $x^2 + y^2 = 1$, it follows quickly that

$$\cos^2(t) + \sin^2(t) = 1 \quad \text{for any angle } t,$$

a fundamental trig identity. The notation $\cos^2 t$ means $(\cos t)^2$.

Unit Circle

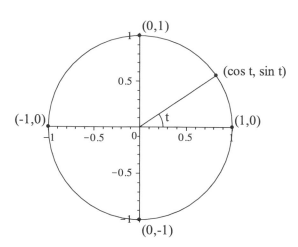

FIGURE 3.4. The unit circle along with a few key points. The values of $\cos t$ and $\sin t$ are defined as the x- and y-coordinates, respectively, of the point of intersection at angle t.

Angles in trigonometry are typically measured in *radians*. A radian is a physical unit of measurement for an angle, just as a foot represents a particular length, or a second measures a certain length of time.

Definition 3.3.2. *One <u>radian</u> is the angle formed by traveling one unit of length along the unit circle.*

Let us explore this definition further. Recall that the circumference of a circle is given by $C = 2\pi r$, where r is the length of the radius. Since the unit circle has a radius of length 1, its circumference is simply 2π. Therefore, going all the way around the unit circle, an angle of 360°, corresponds to an angle of 2π radians, which we denote as 2π rad. This is a fundamental relationship:

$$360° = 2\pi \text{ rad} \qquad \text{or} \qquad 180° = \pi \text{ rad.} \tag{3.3}$$

Thus, π radians is equivalent to an angle of 180°, and a right angle, 90°, is the same angle as $\pi/2$ radians. One radian is approximately 57.3°. It can be visualized by taking a piece of string of unit length (one radius) and placing it along the unit circle. The angle formed between the start and the end of the string is one radian. (Notice the etymology of the word "radian"; it comes from "radius.") Since radians correspond to an actual physical unit of measurement, unlike degrees, which basically rely on the fact that 360 has numerous factors, we will *always* assume that our angles are measured in radians. Moving forward, we will drop the "rad" suffix from our angles.

When measuring angles on the unit circle, the angle t is measured from the positive x-axis in the counterclockwise direction (see Figure 3.4). Thus, the point $(1,0)$ corresponds to an angle of 0, the point $(0,1)$ is located at $\pi/2$ radians, and the point $(-1,0)$ is found at π. Each time we traverse the entire circle in the counterclockwise direction, we add a multiple of 2π to the angle. Thus, the point $(1,0)$ also corresponds to the angle $2\pi, 4\pi, 6\pi$, and so on. Going in the clockwise direction decreases the value of the angle t. Hence, the point $(0,-1)$ corresponds to the angle $-\pi/2$, and the point $(-1,0)$ is located at the angle $-\pi$.

We are now ready to determine some values of the sine function and examine its graph.

Example 3.3.3. *Find the values of each of the following expressions:*

a. $\sin(0)$ **b.** $\sin(\pi/2)$ **c.** $\sin(7\pi)$ **d.** $\cos(7\pi)$ **e.** $\cos^2(\pi/75) + \sin^2(\pi/75)$.

Solution: For part **a.**, the angle 0 corresponds to the point $(1,0)$ on the unit circle. Thus, $\sin(0) = 0$, since the sine function returns the y-coordinate. In **b.**, a ray at an angle of $\pi/2$ radians ($90°$) meets the unit circle at the point $(0,1)$, so $\sin(\pi/2) = 1$. The angle 7π corresponds to going around the circle three and a half times, reaching the point $(-1,0)$. Hence, the answers to **c.** and **d.** are $\sin(7\pi) = 0$ and $\cos(7\pi) = -1$, respectively. Finally, the famous trig identity $\cos^2(t) + \sin^2(t) = 1$ holds for *any* angle t, including $t = \pi/75$, so the answer to **e.** is simply 1. $\square$

3.3.2 Graphing sinusoids

The graph of one cycle of the sine function $y = \sin t$ is shown in Figure 3.5. Let us follow the y-coordinate as the angle t moves counterclockwise around the unit circle, from the angle 0 until retuning back to the same angle at 2π. The y-coordinate begins at 0 and increases up to 1 at the angle $\pi/2$. This is the maximum value of the sine function. Then the y-coordinate decreases from 1, passing through 0 at the angle π and continuing to descend to -1 at the angle of $3\pi/2$. This is the minimum value of the sine function. Finally, the y-coordinate goes from -1 back to 0 as the angle goes from $3\pi/2$ to 2π.

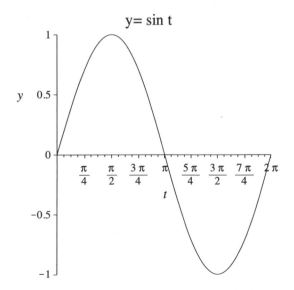

FIGURE 3.5. The graph of $y = \sin t$ over one cycle. The period of the standard sine function is 2π, and the amplitude is 1.

Period and amplitude

The process will then repeat as t moves continuously from 2π to 4π, and the exact same graph will be traced out. This is called *periodicity*, a key property of trigonometric functions. In general, a periodic function is one whose graph repeats itself exactly after some particular amount of time, called the *period*.

Definition 3.3.4. *The period of a function that retraces itself repeatedly is the length of time it takes between successive cycles. The period can be measured from the crest of the wave to the next crest, or from the trough of the wave to the next trough. If $f(t)$ is a periodic function with period P, then $f(t+P) = f(t)$ is satisfied for any time t.*

Thus, the period of $y = \sin t$ is 2π. The period of $x = \cos t$ is also 2π. The period does not have to be measured from the top or the bottom of the wave. It can also be measured between other points on the wave, as long as they represent the same height and have exactly one cycle of the wave completed between them. Since the entire graph of a periodic function retraces itself, the length of time measured between repetitions will be the same regardless of where it is calculated. The equation $f(t+P) = f(t)$ means that the output of the function f is the same at time t as it is one period away, at time $t+P$. This matches our intuitive understanding of the period of a function.

The height of the wave, called the *amplitude*, is defined to be the distance from the top of the wave to its horizontal axis of symmetry. This is the same as half the total vertical distance of the curve (from the highest to the lowest points of the wave). Hence, the amplitude of $y = \sin t$ is 1, which is half of 2, the distance between 1 and -1.

How do we adjust the function $y = \sin t$ to change the amplitude or period of the resulting sine wave? The amplitude can be changed by multiplying the output of the sine function by a scaling factor a. For example, the function $y = 10 \sin t$ has an amplitude of 10 because the maximum still occurs at $t = \pi/2$, where the sine function has a value of 1. Our new function equals 10 at $t = \pi/2$. Similarly, the minimum occurs at $t = 3\pi/2$, just as before, but is now $10 \cdot -1 = -10$. It follows that the amplitude of the new scaled sine function is 10. In general, multiplying a function by a positive constant a will scale the graph vertically by a factor of a.

Changing the period is a bit more subtle. We need to adjust the rate at which t moves around the unit circle. Recall that the period of $y = \sin t$ is 2π. To obtain a longer period, such as 8π, we want to move around the circle *slower*, by a factor of four. A higher period means a slower rate, an inverse relationship. Thus, we replace t by $t/4$; the period of the sine function $y = \sin(t/4)$ is 8π. This can be checked by substituting 8π in for t, giving $\sin(8\pi/4) = \sin(2\pi)$. At time $t = 8\pi$, the angle, which has changed from t to $t/4$, has returned to its starting point for the first time. This inverse relationship is captured in the following formula, which also reminds us how to adjust the amplitude of a sine wave.

> The sine curve $y = a \sin(b\,t)$ has an amplitude of a and a period of $\dfrac{2\pi}{b}$.   (3.4)

In Formula (3.4), we assume that the arbitrary constants a and b are both positive. The formula for the period is derived by solving the equation $b\,t = 2\pi$ for t. The graph of a general sine function $y = a \sin(b\,t)$ is called a *sinusoid*.

Example 3.3.5. *Find the amplitude and period of the function $y = 3\sin(100t)$. Sketch three cycles of the graph.*

Solution: Using Formula (3.4), we have $a = 3$ and $b = 100$, so the amplitude is 3 and the period is $2\pi/100 = \pi/50$. The graph of this sine function is shown in Figure 3.6. □

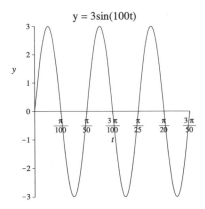

FIGURE 3.6. The graph of three cycles of $y = 3\sin(100t)$. The amplitude is 3 and the period is $\pi/50$.

Phase shift

The last important concept concerning sinusoids is the *phase shift*, which adjusts the starting point of the wave. Consider the effect of replacing t by $t - \pi/2$ in the standard sine function. The angle $t = 0$ shifts to become $-\pi/2$, while the angle $\pi/2$ shifts to 0. In essence, the mapping of t to $t - \pi/2$ rotates the entire unit circle clockwise by $\pi/2$ (90°). The angle $\pi/2$ has become the new starting point at $(1,0)$. It follows that the usual sine wave will now be shifted to the *right* by $\pi/2$. Instead of beginning at $(t = 0, y = 0)$, the wave $y = \sin(t - \pi/2)$ begins at $(t = \pi/2, y = 0)$. The graph of this shifted wave is shown in Figure 3.7. We say that the phase shift of the new wave is $\pi/2$. In general, if t is replaced by $t - c$, then the graph of the new function $f(t - c)$ will be a translation to the right of the old one by c units. Replacing t by $t + c$ will cause the graph to shift left by c units.

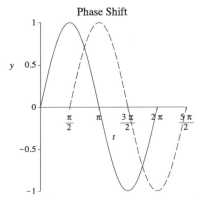

FIGURE 3.7. The graph of the usual sine wave $y = \sin t$ (*solid curve*) alongside its shifted version $y = \sin(t - \pi/2)$ (*dashed curve*).

Frequency and period

Recall that the frequency of a wave is defined to be the number of cycles per second. The pure tone A440 will vibrate at a rate of 440 cycles per second. What is the period of this same wave? If A440 produces 440 cycles in one second, then it only takes 1/440th of a second to do one cycle. Thus, the period of the sine wave corresponding to the pure tone A440 is 1/440. In general, period and frequency are reciprocals of each other, that is,

$$\text{frequency} \;=\; \frac{1}{\text{period}} \quad \text{or} \quad \text{period} \;=\; \frac{1}{\text{frequency}}. \tag{3.5}$$

For a general sine wave $y = a\sin(b\,(t-c))$, a is the amplitude, $2\pi/b$ is the period, and c is the phase shift. The frequency is the reciprocal of the period, or $b/(2\pi)$.

Example 3.3.6. *The vibrations of a tuning fork cause the wave $y = 75\sin(440\pi(t - 0.01))$ to propagate. Find the amplitude, period, phase shift, and frequency of the wave. What note is being sounded by the tuning fork?*

Solution: The amplitude of our wave is $a = 75$, the period is $2\pi/b = 2\pi/(440\pi) = 1/220$, and the phase shift is $c = 0.01$. Since frequency is the inverse of the period, the frequency of the wave is 220 Hz. This is 1/2 of the frequency of A440, which means that it is an octave below A440. This is the A just below middle C on the piano.   □

3.3.3   The harmonic oscillator

To understand why the sine wave, and not some other shaped wave, plays such a prominent role in describing the behavior of a vibrating object, we consider the classic mass-spring system. Take a block of mass m and attach it to a suspended spring. What happens to the position of the block if we stretch or compress the spring? To fully answer this question requires some knowledge of calculus. For readers without a background in calculus, this section may be skimmed over for a general overview. The main point here is that it is possible to construct a mathematical model for the motion of the mass, to use calculus to solve that model, and to interpret the results physically. In the end, the sine function plays a key role in the structure of the solutions to the mass-spring system.

To introduce some variables, we let $y(t)$ measure the distance of the mass from rest position at time t. We will assume that the motion is only in the vertical direction, so y is measured on a vertical axis. If $y = 0$ corresponds to the rest position (known as *equilibrium*), then $y > 0$ means that the spring is compressed and $y < 0$ implies that the spring is stretched (see Figure 3.8). For instance, $y(4) = 2$ means that after 4 seconds, the spring is compressed a distance of two units from the rest position.

To model this situation, we use one of the most famous equations in physics, $F = ma$, formulated by Sir Isaac Newton (1643–1727). Here F is the force on the object, m is its mass, and a represents its acceleration. There are two forces acting on the mass, a spring force and a force due to friction (damping). The force due to gravity has already been accounted for by establishing that $y = 0$ corresponds to the rest position. If the spring is compressed ($y > 0$), then it will exert a downward force on the block, whereas if it is stretched ($y < 0$), it will force the block upward. The formula for the force of the spring

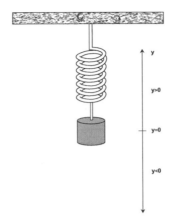

FIGURE 3.8. The mass-spring system, where a block of mass m is attached to a spring. The variable y measures the position of the mass with respect to equilibrium $y = 0$.

comes from *Hooke's Law*, which states that the force is proportional to the distance from equilibrium, but in the opposite direction. This can be modeled as $F_s = -ky$, where k is some positive proportionality constant. The force due to friction is also in the opposite direction (friction tends to slow things down) but is proportional to the velocity v of the object (faster objects encounter more friction). The dampening force is then $F_d = -bv$, where b is another positive constant.

For our model, we will assume that friction is negligible ($b = 0$), a situation more applicable to the physics of musical instruments, and one that captures the importance of the sinusoid. One of the primary goals of calculus is to calculate the slope of a curve, called the *derivative*. If $y(t)$ represents the position at time t, then the velocity $v(t)$ can be found by calculating the slope of $y(t)$. We will use the notation $v = dy/dt$, which stands for the derivative of y with respect to t. Similarly, the slope of the velocity curve is equal to acceleration, which is known as the second derivative of y, denoted $a = d^2y/dt^2$. Substituting into $F = ma$, we have

$$F = ma \implies -ky = m\frac{d^2y}{dt^2} \implies m\frac{d^2y}{dt^2} + ky = 0. \qquad (3.6)$$

The last equation in (3.6) is an example of an ordinary differential equation (ODE).[1] Solving ODEs is beyond the scope of this text, although it is a fascinating and important branch of applied mathematics. The basic goal when solving an ODE is to find a function $y(t)$ that satisfies the differential equation, that is, plugging $y(t)$ and its derivatives into the equation yields a true statement for any value of t. Unless initial conditions are specified on the solution curve, there are typically an *infinite* number of functions that satisfy a given ODE.

For example, consider a mass-spring system with mass $m = 1$, spring constant $k = 4$, and no damping ($b = 0$). The resulting differential equation is

$$\frac{d^2y}{dt^2} + 4y = 0. \qquad (3.7)$$

---

[1] With the damping term included, the differential equation becomes $m\frac{d^2y}{dt^2} + b\frac{dy}{dt} + ky = 0$.

It turns out that $y(t) = \sin(2t)$ is a solution to this differential equation. To check this, we use the chain rule to compute $dy/dt = 2\cos(2t)$ and $d^2y/dt^2 = -4\sin(2t)$. Substituting this last expression and $y(t)$ into Equation (3.7) yields $-4\sin(2t) + 4\sin(2t) = 0$, which is a true statement for any value of t. It follows that $y(t) = \sin(2t)$ is a solution to the ODE, and the resulting motion of the mass is periodic. Since $y(0) = 0$ and $v(0) = 2$ (the value of dy/dt at $t = 0$), the spring starts from the equilibrium position with an initial velocity of 2 in the upward direction (moving toward compression). The mass will oscillate repeatedly about the equilibrium position with a period of π and an amplitude of 1. The function $y(t) = \cos(2t)$ is also a solution to Equation (3.7), but with a different initial position and velocity. In this case, we have $y(0) = 1$ and $v(0) = 0$, so the mass starts compressed by one unit, with no initial velocity.

The complete solution to the ODE in Equation (3.6) is given by

$$y = a\sin(\omega(t - c)), \quad \omega = \sqrt{k/m}.$$

Here the amplitude a and phase shift c are determined by the initial position and velocity of the mass and can assume any value. In other words, an ideal mass-spring system with no friction will oscillate sinusoidally with a frequency of

$$\frac{\omega}{2\pi} = \frac{1}{2\pi} \cdot \sqrt{\frac{k}{m}}. \tag{3.8}$$

Based on Formula (3.8), a heavier mass will oscillate less frequently than a lighter one, and a stronger spring (higher k) encourages more rapid motion. This agrees with our expectations based on the physics of the motion.

The mass-spring system is an example of a *harmonic oscillator*, and the system with no damping force is called a *simple harmonic oscillator*. Many objects, such as tuning forks and simple pendula, exhibit simple harmonic motion described by a sinusoidal curve. Even if there is a little damping present (as there often will be in a real-world system), the motion still mimics a sine curve, except that the amplitude decays exponentially to zero. See Exercise 9 for an example of this type.

Exercises for Section 3.3

1. Convert each of the following angles into radians.

   a. $30°$ **b.** $120°$ **c.** $-135°$ **d.** $270°$ **e.** $900°$

2. Evaluate each of the following without a calculator.

   a. $\sin(50\pi)$ **b.** $\sin\left(-\dfrac{3\pi}{2}\right)$ **c.** $\sin\left(\dfrac{15\pi}{2}\right)$ **d.** $\cos(-20\pi)$

   e. $\cos(8\pi)$ **f.** $\cos(k\pi)$, where k is an odd integer

3. Given a sound wave of the form

   $$y = 17\sin(1000\pi(t - 200)),$$

   where t is measured in seconds, what are the amplitude, period, phase shift, and frequency of the sine wave? Could a dolphin hear this pitch?

4. Given a sound wave of the form

$$y = 2.4 \sin(600{,}000(t - 3.5)),$$

where t is measured in seconds, what are the amplitude, period, phase shift, and frequency of the sine wave? According to Table 3.3, which of the mammals listed could hear this sound?

5. On the same set of axes, sketch the graph of the two sine waves

$$y = 5 \sin(2t) \qquad \text{and} \qquad z = 5 \sin\left(2\left(t - \frac{\pi}{2}\right)\right).$$

Draw two cycles of each wave, with $y(t)$ as a solid curve and $z(t)$ dashed.

6. A mass m is attached to a spring with proportionality constant k. In each case below, assuming that no damping is present ($b = 0$), use Formula (3.8) to give the frequency of the resulting sinusoidal motion.

   a. $m = 1, k = 1$ **b.** $m = 1, k = 4$ **c.** $m = 4, k = 1$ **d.** $m = 3, k = 12\pi^2$

7. **Requires Calculus:** Show that $y_1(t) = \cos(2t)$ and $y_2(t) = 3\sin(2(t - \pi/4))$ are each solutions to the ODE given in Equation (3.7). Describe the motion of the mass for each solution, including the starting behavior (at $t = 0$).

8. **Requires Calculus:** Suppose that a mass-spring system has mass $m = 2$, damping coefficient $b = 0$, and spring constant $k = 18$. Show that $y = \sin(3t)$ is a solution to the corresponding ODE in Equation (3.6). What is the frequency of this solution? Check this against the value obtained by using Formula (3.8).

9. **Requires Calculus:** Suppose that a mass-spring system has mass $m = 1$, damping coefficient $b = 4$, and spring constant $k = 5$. Show that $y = e^{-2t} \sin t$ is a solution to the ODE $m\dfrac{d^2y}{dt^2} + b\dfrac{dy}{dt} + ky = 0$. What is the motion of the mass in this case?

3.4 Understanding Pitch

The way our ear-brain system understands and recognizes pitch is actually quite a complicated subject, with competing theories and debates. Our presentation follows the very readable and comprehensive text *Why You Hear What You Hear* by Harvard professor Eric Heller [2013]. According to Heller, one source of confusion is that pitch is *perceived*; it is a sensation like hot or cold, or spicy or bland. Vast amounts of data about the pressure fluxuations obtained by the ear, and then processed by the auditory cortex, are interpreted by our brain, which makes the best decision it can to determine a pitch.

The frequency of the pitch perceived by our ear is called the *fundamental frequency*. Surprisingly, hearing a pitch such as 200 Hz does not necessarily mean that we are experiencing the result of sine waves with a frequency of 200. Sometimes we perceive pitch for sounds that do not produce periodic sine waves at all, such as a bell or a wood block. There are even instances where several frequencies are present, called *partials*, but the fundamental frequency of the pitch we hear is completely absent.

3.4.1   Residue pitch

This idea of a missing fundamental frequency was first demonstrated in the 1840s by August Seebeck using sirens. It can be clearly understood in the following audio experiment described by Heller.[2] Recall that two pitches whose frequencies are in a $2:1$ ratio are an octave apart.

If we play a pure tone (one sine wave) with a frequency of 200 Hz, then we clearly discern this pitch. Likewise, a pure tone at a frequency of 400 Hz will produce a pitch that is an octave higher. What happens if we play both tones simultaneously? If the amplitudes are the same, then we will hear the pitch of the lower frequency, 200 Hz. The higher frequency is still discernible, but the fundamental pitch is the one at 200 Hz. Note that this is very different from striking two notes on a piano that are an octave apart. On the piano, there is a whole sequence of partials present for any given note, so the sound wave reaching our ears is quite complex. Here we are only producing waves at 200 and 400 Hz. If the amplitude of the 200 Hz tone is decreased significantly, then we will hear the 400 Hz pitch. This seems perfectly logical.

Now consider what happens when we add higher frequencies to the 200 and 400 Hz tones, such as 600, 800, and 1000 Hz. With this combination, we still hear 200 Hz as the primary pitch. However, if we remove the 200 Hz frequency from the group, then remarkably, the pitch heard is still 200 Hz! Even though the fundamental is missing, it is still the pitch perceived by our brain.

Following Heller, we will call the phenomenon of a pitch whose fundamental frequency is absent a *residue pitch*, a term first used by J. F. Schouten in 1940. The idea behind this term is that the partials present in a note create a residue that determines the pitch heard. In the previous example, the residue of the partials 400, 600, 800, and 1000 Hz is 200 Hz, which is the perceived pitch.

Here is another, even more striking example of a residue pitch, as presented by Heller. Consider a tone with frequencies 200, 400, and 600 Hz. The pitch heard in this case is 200 Hz. Next, we add partials at 250, 300, 350, 450, 500, and 550 Hz to the other three. The pitch now perceived is actually 50 Hz, a *full two octaves below the note at 200 Hz*. What is amazing about this example is that not only is the fundamental frequency at 50 Hz missing, but so are its next two partials, $2 \cdot 50 = 100$ Hz and $3 \cdot 50 = 150$ Hz. (In general, partials are integer multiples of the fundamental frequency.) Even with these three key frequencies absent, the pitch heard is still given by 50 Hz.

Graphing the sum of two functions

We can understand the residue pitch effect graphically by plotting the sum of several sine curves. This is a good precursor to the next section on the overtone series. First, we review the concept of adding two functions graphically. If $y = g(t)$ and $y = h(t)$ are each functions of t, then their sum, denoted by $(g + h)(t)$, is found by adding the function values together. For instance, if $g(2) = 1$ and $h(2) = 2$, then $(g + h)(2) = 3$. To draw the graph of $g + h$, we must combine the heights of each graph point by point (see Figure 3.9). Alternatively, starting with the graph of g, raise each point up (or down if $h(t) < 0$) by the height of h. The dashed lines in Figure 3.9 represent the heights of $g(t)$, $h(t)$, and $(g + h)(t)$ at the values $t = 0.5, 1$, and 2.

---

[2]These and other sound experiments can be simulated with Jean-François Charle's MAX patch *Partials*, available at http://www.whyyouhearwhatyouhear.com/subpages/MAX.html.

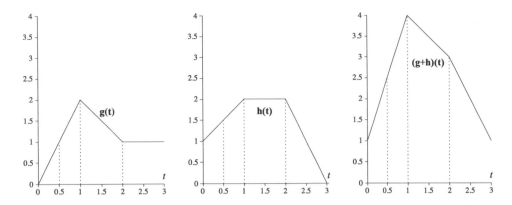

FIGURE 3.9. The sum of two functions $g(t)$ and $h(t)$. The dashed lines indicate the heights of each function for $t = 0.5, 1,$ and 2. For example, since $g(0.5) = 1$ and $h(0.5) = 1.5$, we have $(g + h)(0.5) = 2.5$.

Let us consider the effect of combining sine waves whose frequencies are all multiples of some fundamental frequency f. We begin with the pure wave $y = \sin(2\pi \cdot 100t)$, which has a frequency of 100 Hz and an amplitude of 1 (left graph in Figure 3.10). This wave corresponds to our fundamental. Next, we take the first partial $y = 0.9 \sin(2\pi \cdot 200t)$, with frequency 200 Hz and amplitude 0.9 (center graph in Figure 3.10), and combine it with the fundamental (right graph in Figure 3.10). The result is somewhat counterintuitive. Although the amplitude of the sum has increased to roughly 1.65, there are new mini-humps symmetrically placed between the larger crests. Since the period of the first partial is half that of the fundamental, some of its peaks are reduced in size upon summation of

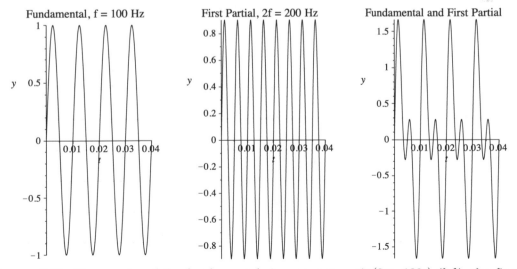

FIGURE 3.10. The graphs of the fundamental sine wave $y = \sin(2\pi \cdot 100t)$ (*left*), the first partial $y = 0.9 \sin(2\pi \cdot 200t)$ (*center*), and the sum of the two waves (*right*). Notice that the period of the sum of the two waves, 0.01, is the same as that of the fundamental.

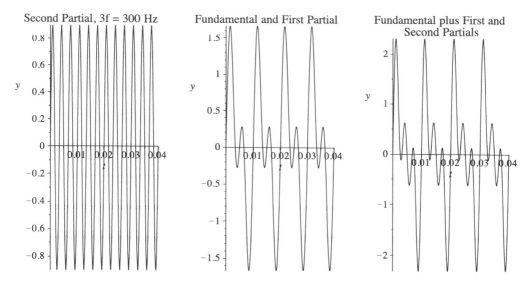

FIGURE 3.11. The graphs of the second partial $y = 0.9\sin(2\pi \cdot 300t)$ (*left*), the sum of the fundamental and first partial (*center*), and the sum of the fundamental and its first two partials (*right*). Notice that the period of the sum of the three waves, 0.01, is the same as that of the fundamental.

the two waves. For instance, when the first partial is at its second maximum peak, the fundamental is negative, which causes the sum to be considerably smaller. This accounts for the mini-peaks in the sum. Notice that the period of the sum of the fundamental and the first partial is $0.01 = 1/100$ seconds, which is the inverse of the fundamental frequency 100 Hz. This explains why we still hear the fundamental frequency (100 Hz) when it is combined with its first partial (200 Hz), even if the amplitudes of the two waves are similar.

Next, suppose that we add the second partial, $y = 0.9\sin(2\pi \cdot 300t)$, with frequency 300 Hz and amplitude 0.9 (left graph in Figure 3.11), to the previous sum (center graph in Figure 3.11). The result (right graph in Figure 3.11) has a larger amplitude (about 2.3) and an additional mini-hump between each of the larger crests. Interestingly, the period remains at $0.01 = 1/100$ seconds, which is the inverse of the fundamental frequency 100 Hz. Even though the summation is not a pure sine wave, its period, and thus its frequency, still agrees with that of the fundamental. Consequently, the resulting note will be heard at a frequency of 100 Hz. The smaller humps in the graph to the right in Figure 3.11 begin to give the sound its timbre.

If we add the next five partials, with frequencies 400, 500, 600, 700, and 800 Hz, respectively, each with an amplitude of 0.7, the underlying pitch is still 100 Hz, but the timbre of the sound is more interesting and developed. To illustrate the residue pitch effect, we first subtract the fundamental tone at 100 Hz (left graph in Figure 3.12). Then we remove both the fundamental and the 200 Hz partial (center graph in Figure 3.12), and finally we remove the fundamental and both the 200 and 300 Hz partials (right graph in Figure 3.12). In each case, even though the lower defining frequencies are missing, the resulting curves are clearly periodic with period 0.01. Consequently, our ear-brain system interprets the pitch as 100 Hz, the *residue* of the higher frequencies.

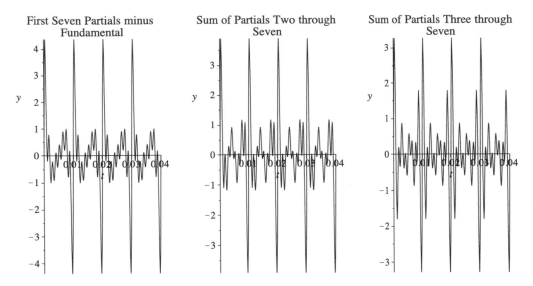

FIGURE 3.12. The graphs of the first seven partials of a 100 Hz wave with the fundamental frequency removed (*left*), the fundamental and the first partial removed (*center*), and the fundamental and the first two partials removed (*right*). In each case, the remaining frequencies produce a residue pitch at 100 Hz, as can be seen by the period of 0.01 in each graph.

The GCD and the residue pitch

Notice that in each of the previous examples, the perceived pitch was the greatest common divisor (GCD) of the set of frequencies being sounded. In the last example, 100 is the GCD of 400, 500, 600, 700, and 800, and 100 Hz was the frequency of the pitch heard. Likewise, 50 is the greatest common factor of 200, 250, 300, 350, . . . , 600 in Heller's example, and this was the frequency of the perceived pitch. In the opening example, the GCD of 200 and 400, which is 200, established the frequency of the discerned pitch. These results certainly depend on the amplitude of the partials; in each example we have assumed that the strength of each partial is similar.

The fact that the GCD gives the frequency of the residue pitch can be explained in part with a straightforward calculation. Recall that the 2π periodicity of the sine function can be expressed by the formula $\sin(t + 2\pi) = \sin(t)$, which holds for any time t. In other words, the value of the sine function is unchanged if the angle is increased by 2π. Moreover, if we replace t by $t + 2\pi$, then $\sin(t + 4\pi) = \sin(t + 2\pi + 2\pi) = \sin(t + 2\pi) = \sin(t)$. This makes sense intuitively; if we travel around the unit circle twice and return to the same spot, the y-coordinate stays the same. By extension, we have the formula

$$\sin(t + k \cdot 2\pi) = \sin(t), \quad \text{for any angle } t, \text{ and for any integer } k. \tag{3.9}$$

Theorem 3.4.1. *Suppose that partials with frequencies $f_1, f_2, \ldots, f_n$ are sounded together as pure sine waves with no phase shift, where each $f_i \in \mathbb{N}$. Suppose further that the positive integer $d > 1$ is the greatest common divisor of all n frequencies, that is, d divides evenly into each frequency and is the largest such integer to do so. Then the sum of the sine waves repeats itself every $1/d$ time units.*

Proof: We denote each partial as $y_1 = a_1 \sin(2\pi f_1 t), y_2 = a_2 \sin(2\pi f_2 t)$, etc., and let the combined signal be $f(t) = y_1 + y_2 + \cdots + y_n$. We want to show that $f(t + 1/d) = f(t)$ for any angle t. By assumption, d divides evenly into each frequency, so there exists integers k_i such that $f_1 = d \cdot k_1$, $f_2 = d \cdot k_2$, etc. We then have

$$
\begin{aligned}
f\left(t + \tfrac{1}{d}\right) &= a_1 \sin\left(2\pi f_1\left(t + \tfrac{1}{d}\right)\right) + a_2 \sin\left(2\pi f_2\left(t + \tfrac{1}{d}\right)\right) + \cdots + a_n \sin\left(2\pi f_n\left(t + \tfrac{1}{d}\right)\right) \\
&= a_1 \sin\left(2\pi\left(f_1 t + \tfrac{f_1}{d}\right)\right) + a_2 \sin\left(2\pi\left(f_2 t + \tfrac{f_2}{d}\right)\right) + \cdots + a_n \sin\left(2\pi\left(f_n t + \tfrac{f_n}{d}\right)\right) \\
&= a_1 \sin(2\pi f_1 t + k_1 2\pi) + a_2 \sin(2\pi f_2 t + k_2 2\pi) + \cdots + a_n \sin(2\pi f_n t + k_n 2\pi) \\
&= a_1 \sin(2\pi f_1 t) + a_2 \sin(2\pi f_2 t) + \cdots + a_n \sin(2\pi f_n t) \\
&= f(t),
\end{aligned}
$$

as desired. The penultimate step in the proof follows from Equation (3.9).   □

Note that we have not shown that $1/d$ is the actual period of the sum of the sine waves, but rather that the summation curve repeats itself after $1/d$ units of time. To conclude that $1/d$ is the actual period of the summation, we would need to show that there is no *smaller* value $P < 1/d$ such that $f(t + P) = f(t)$.[3]

Nevertheless, Theorem 3.4.1 helps explain why the greatest common divisor is a likely candidate for the frequency of the residue pitch. If the combination of sine waves repeats after $1/d$ time units, then it will appear to act as a single wave with frequency d. We are *not* claiming that this implies that the pitch we hear is d Hz. The amplitudes of the corresponding partials play an important role that could alter the perceived pitch, or d could be too low for our ears to hear (less than 20 Hz). Heller describes an example where the frequencies are 120, 220, 320, 420, 520, and 620 Hz, but the perceived pitch was 104.6 Hz, not the GCD of the group, 20 Hz. We also mention that the frequencies in Theorem 3.4.1 do not necessarily have to be integers. The proof works perfectly well if the frequencies are rational numbers (the ratio of two integers).

Autocorrelation

The adjustment by our brain to identify the missing fundamental when the residue pitch effect is present is an example of *autocorrelation*. Using Fourier analysis, a signal can be decomposed into a sum of sine waves with varying frequencies and amplitudes. The plot of these key frequencies, with heights determined by their amplitudes, is called a *power spectrum*. From the power spectrum, a mathematical calculation can be made to generate an autocorrelation function. In essence, a peak in the autocorrelation curve at time P implies that the original signal tends to repeat itself every P seconds, that is, P appears to be the period of the wave. If the signal is perfectly periodic (e.g., partials of 200, 400, and 600 Hz), then the autocorrelation function predicts the same period as the natural period of the signal (in this case, $1/200$). However, when the signal is aperiodic, the autocorrelation function can still produce peaks that determine the perceived pitch. This case is more qualitative, that is, the peaks are estimated and identified with descriptive phrases such as tallest, most isolated, and first to appear. Moreover, the peaks will usually decline in height over time, making the pitch perceived less definitive. In other words, different listeners may identify different pitches, and preference may depend on duration

---

[3]Thanks to Alex Barnett for pointing this technicality out to the author.

as well as context. For instance, if a listener recently heard a particular pitch, then they may be predisposed to selecting it again as a perceived pitch.

Based on joint work with J. Zink, Heller presents a formula for pitch based on locating the first maximum of the autocorrelation function. This formula, which was first applied to molecular spectroscopy in 1982, depends on the amplitudes and the frequencies found in the power spectrum of a signal. If the n frequencies $f_1, f_2, \ldots, f_n$ are played together, with corresponding amplitudes $a_1, a_2, \ldots, a_n$, then the pitch heard will have an approximate frequency $\bar{f}$ given by

$$\bar{f} = \frac{a_1^2 f_1^2 + a_2^2 f_2^2 + \cdots + a_n^2 f_n^2}{a_1^2 k_1 f_1 + a_2^2 k_2 f_2 + \cdots + a_n^2 k_n f_n}. \tag{3.10}$$

Here each k_i in the denominator is an integer defined as the closest integer to the ratio $f_i / \bar{f}$. Equation (3.10) is somewhat confusing because it is circular (self-referential); the output of the formula $\bar{f}$ is itself part of the formula, as it enters through the integers k_i. Nevertheless, several successful applications of Formula (3.10) are presented in Heller's text that give a very strong argument in favor of autocorrelation.

Let us demonstrate how to apply Equation (3.10) to one of these examples. Suppose that the three frequencies $f_1 = 820$, $f_2 = 1020$, and $f_3 = 1220$ are played together with equal amplitudes $a_1 = a_2 = a_3 = 1$. The GCD of these three frequencies is 20 Hz, but this is too low a candidate for the perceived pitch. We would like to find three integers k_1, k_2, and k_3 such that $k_1 \bar{f} \approx f_1$, $k_2 \bar{f} \approx f_2$, and $k_3 \bar{f} \approx f_3$ are all very good approximations. The problem with finding these integers is that we have no *a priori* idea of the value of $\bar{f}$. However, if we solve each of these approximations for the unknown frequency $\bar{f}$, we find

$$\bar{f} \approx \frac{f_1}{k_1} \approx \frac{f_2}{k_2} \approx \frac{f_3}{k_3}, \quad \text{which implies that} \quad \frac{f_2}{f_1} \approx \frac{k_2}{k_1} \quad \text{and} \quad \frac{f_3}{f_1} \approx \frac{k_3}{k_1}.$$

In other words, to find good candidates for the integers k_i, divide each of the higher frequencies by the lowest one, and approximate the results using fractions with small-valued integers in the numerator and denominator. For our example, we have

$$\frac{1020}{820} \approx 1.244 \approx \frac{5}{4} \quad \text{and} \quad \frac{1220}{820} \approx 1.488 \approx \frac{3}{2} = \frac{6}{4}.$$

Thus, we can choose $k_1 = 4, k_2 = 5$, and $k_3 = 6$. Substituting these values into Equation (3.10) gives a frequency of $\bar{f} = 203.9$. In *The Science of Musical Sound*, author John Pierce discusses this particular example and reports that the frequency perceived is 204 Hz, remarkably close to the value predicted using the autocorrelation formula given in Equation (3.10).[4]

Another example comes from research using volunteer listeners which was conducted by Rausch and Plomp and presented in a chapter in the book *The Psychology of Music* [Rausch and Plomp, 1982]. Partials at frequencies of 850, 1050, 1250, 1450, and 1650 Hz were played together with equal amplitudes. While the GCD of these numbers is 50 Hz, the pitch perceived by listeners was 208.3 Hz. Heller and Zink's autocorrelation formula predicts a frequency of 207.9 Hz, an imperceptible difference (see Exercise 2 at the end of this section).

Of particular merit is the fact that Equation (3.10) also works well in the case of instruments that do not produce equally spaced partials (the aperiodic case), such as bells,

---

[4]Pierce, *Science of Musical Sound*, p. 95.

chimes, or drums. Citing the Hosanna Bell in Freiburg, Germany, and a kettle drum as examples, the autocorrelation formula does an excellent job at predicting the pitches heard. The theory of autocorrelation can also be used to explain the famous auditory phenomenon of *Shepard tones*, where a rising chromatic scale somehow ascends back to the starting pitch after 12 steps. Akin to an impossible Escher-like staircase, Shephard tones can be explained via adjustments in amplitude and phase shift in the first two peaks of the autocorrelation function as the scale is ascended.

3.4.2   A vibrating string

We have primarily been discussing the effects of combining (and removing) frequencies of the form $f, 2f, 3f, 4f, \ldots$, that is, a fundamental frequency f, and its partials $2f, 3f, 4f, \ldots$. Inherent in our development has been the assumption that there actually exist instruments that produce such sequences of frequencies. One well-studied example is a vibrating string.

Suppose we have a string of length L that is tied down at both ends (e.g., the string on a monochord, or a violin). If the string is set in motion by plucking or bowing, how can we describe the ensuing vibrations? The mathematical theory behind a vibrating string involves the subject of partial differential equations (PDEs), another important branch of applied mathematics, but one that is well beyond the scope of this text. Nonetheless, we provide a summary of the main results here, since they explain the existence of this special sequence of frequencies. A complete understanding of this material requires a background in the calculus of functions of several variables.

The tricky part of modeling a vibrating string is that there are now *two* independent variables, the time t and the position along the string x. Before we were only concerned with an individual sine wave and its height at time t. We are still interested in the height, but it will vary based on both position and time. Let the function $u(x, t)$ denote the position of the string above or below equilibrium at position x ($0 \leq x \leq L$) and time t. For example, $u(L/2, 1) = 5$ means that after 1 second, the point halfway along the string is five units above equilibrium. The expressions $u(0, t) = 0$ and $u(L, t) = 0$, called boundary conditions, mean that the string is fixed at the ends for all time.

The differential equation that governs the motion of the string is called the *wave equation*, a famous PDE given by

$$u_{tt} = c^2 u_{xx}.$$

Here c is a constant defined as $c = \sqrt{T/\rho}$, where T represents the tension of the string and ρ its density. The terms u_{tt} and u_{xx} are partial derivatives representing the acceleration and concavity of the string, respectively.

Applying Fourier analysis, the solution of the wave equation turns out to be not just one sine wave, but an infinite sum of sine waves. Similar to the infinite geometric series we studied back in Section 1.1.2, the solution to the wave equation consists of an infinite sum of sine functions whose amplitudes and frequencies vary. As we add more and more sine functions, the amplitudes must quickly head to zero so that the resulting infinite series actually settles down (converges) to a specific function. If we fix a particular point on the string (fix x), then the motion of that point is determined by the infinite sum

$$a_1 \sin(2\pi \cdot f(t - \phi_1)) + a_2 \sin(2\pi \cdot 2f(t - \phi_2)) + a_3 \sin(2\pi \cdot 3f(t - \phi_3)) + \cdots,$$

where $f = c/(2L)$ is the fundamental frequency and $2f, 3f, 4f, 5f, \ldots$ are its partials. In other words, the sound wave generated from one particular point on a vibrating string is precisely the sum of sine waves with frequencies $f, 2f, 3f, 4f, \ldots$. This special list of frequencies, called the *overtone series*, is critical to our perception of pitch, the timbre of a note, and the theory of harmony.

Before discussing the overtone series, we make some important observations about the fundamental frequency

$$f = \frac{c}{2L} = \frac{\sqrt{T}}{2L\sqrt{\rho}},$$

a formula first discovered by Mersenne. First, note that increasing the tension T, which can be accomplished by tightening the string, will increase the fundamental frequency (and thereby all of its partials). This makes the pitch higher. Likewise, decreasing the tension of a string will lower the pitch. This observation is second nature to string players and guitarists, but here we have given a mathematical explanation. Second, note that cutting the length of the string in half, replacing L by $L/2$, changes the fundamental frequency from f to $2f$ and doubles all the frequencies of the partials. This means that the fundamental pitch of a string with half its original length is an octave higher than the original pitch. The length L of a string and the fundamental frequency f are inversely proportional. It is worth checking these two facts on a stringed instrument such as a monochord (see the suggested Lab Project in Section 3.5).

3.4.3 The overtone series

Why do different instruments playing the same pitch sound different to our ears? What qualities of sound allow us to differentiate between a violin and an oboe, even if they are both playing an A440? Why do some combinations of notes sound more pleasing to our ears than others? The answers to these questions largely concern the overtone series.

The overtone series is simply the sequence of multiples of the fundamental frequency. For example, if the fundamental frequency is $f = 100$ Hz, then the overtone series is given by the sequence

$$100, 200, 300, 400, 500, 600, 700, 800, 900, 1000, \ldots.$$

Note that mathematically speaking, this is clearly not a series (a sum), but rather a *sequence*, an infinite list of numbers. The fundamental frequency is 100 Hz, and the *overtones* (also called the *partials* or *harmonics*) are 200 Hz, 300 Hz, 400 Hz, etc. When a musical note such as 100 Hz is produced by a vibrating string, the overtones are also created, although the amplitudes may not be as large as that of the fundamental. These higher-frequency partials are critical because they give the note its timbre—its characteristic sound. As can be seen from Figures 3.10, 3.11, and 3.12, the overtones do not alter the period of the wave; they give it additional complexity. Varying the strengths of the partials creates different sounds.

Our amazing ear-brain system does two things with the overtone series. It can identify the fundamental pitch so we know what note is being sounded, but it also collects the information from the partials and interprets the result as timbre. Each individual partial is not heard distinctly, but rather the collection is grouped in a way that helps us distinguish an oboe from a violin, even if they are both sounding the same fundamental frequency.

The overtone series not only influences the sound we hear but also plays a key role in understanding some of our preferences for certain intervals over others. Recall that two notes whose fundamental frequencies are in a 2:1 ratio are an octave apart. Consider the overtone series of both $f = 100$ Hz and $2f = 200$ Hz:

$$f \quad : \quad 100, \mathbf{200}, 300, \mathbf{400}, 500, \mathbf{600}, 700, \mathbf{800}, 900, \mathbf{1000}, \ldots$$
$$2f \quad : \quad \mathbf{200}, \mathbf{400}, \mathbf{600}, \mathbf{800}, \mathbf{1000}, \mathbf{1200}, \mathbf{1400}, \mathbf{1600}, \mathbf{1800}, \mathbf{2000}, \ldots.$$

Here we have highlighted the frequencies that the two different overtone series have in common. Note that all of the partials (including the fundamental) for $2f$ are also included in the overtone series for f. This will be true regardless of the actual value of f. In other words, the note an octave above f contains every other overtone in the series for f. Since f and $2f$ have so many partials in common, playing them together leads to a very pleasing sound. In fact, two notes an octave apart often sound the "same" to our ears, even though two different pitches are being played. The agreement of so many overtones helps create a resonance that is very satisfying to our brain. This special consonance makes the octave the key musical interval.

Building on this idea of an overlap between the overtone series of different notes, consider playing two notes with frequencies in the ratio 3:2. The overtone series for $f = 100$ Hz and $\frac{3}{2}f = 150$ Hz are

$$f \quad : \quad 100, 200, \mathbf{300}, 400, 500, \mathbf{600}, 700, \ 800, \ \mathbf{900}, \ 1000, \ 1100, \ \mathbf{1200}, \ldots$$
$$\tfrac{3}{2}f \quad : \quad 150, \mathbf{300}, 450, \mathbf{600}, 750, \mathbf{900}, 1050, \mathbf{1200}, 1350, \mathbf{1500}, \ldots.$$

In this case, we see that every other overtone of $\frac{3}{2}f$ agrees with every third partial of the original frequency f. This consonance is also quite pleasing to our ear. It results in the musical interval of a perfect fifth. Here the two notes do not sound quite the same, but their overtone series are in strong enough agreement to make a very consonant interval. As a general rule, combinations of pitches whose overtone series overlap tend to sound more consonant than those that do not, a theory developed by the famous German scientist Hermann von Helmholtz (1821–1894). Helmholtz's theory of consonance and dissonance helps explain the strengths of different tuning systems such as *just intonation*, to be discussed in Section 4.2.

We can now identify the pitches associated with the first two partials in the overtone series for f. The first overtone, $2f$, corresponds to the note an octave above the fundamental, while the second partial, $3f$, corresponds to the note an octave plus a fifth above the fundamental. This last fact follows since the ratio between $3f$ and $2f$ is simply 3:2. For example, if f corresponds to middle C, then the first two overtones correspond to the note an octave above middle C, notated C′, and then a fifth above that, notated G′. The Pythagoreans recognized the musical importance of these two simple ratios, creating a musical scale very close to our modern major scale based entirely on the ratios 2:1 and 3:2. The so-called *Pythagorean scale* will be discussed at great length in the next chapter. It is also the focus of the lab project at the end of this chapter. We will determine the next several notes associated with the partials in the overtone series after we learn about this important scale (see Section 4.1.3).

We close this section by describing a few acoustical examples where the overtone series is evident. One fun experiment involving partials can be tried on a piano. Pick one note to be the fundamental, such as middle C, and play it so you have the note in your head. Then, slowly push down the keys for an octave above your note (in this case C′) and a perfect fifth above that (G′). By gently pushing down the keys for these two higher notes, we are removing the dampers on the strings, allowing them to vibrate freely without actually sounding their pitches. Keeping the keys for the higher notes depressed, play the fundamental note (middle C), and release it. After the fundamental note stops sounding, the higher pitches corresponding to the octave and octave plus a fifth above the fundamental should be audible. Even though we are not actually playing these notes, the first and second overtones of the fundamental are powerful enough to force these higher notes to sound!

A similar phenomenon due to overtones can occur when voices come together in perfect harmony. This is particularly noticeable with a good choir in a space with strong reverberation. When the chorus holds a major chord for a long enough period of time, and the singers are well in tune with each other, a listener can often hear the overtones "pop" above the choir. If the acoustical setting is cooperative, a performer may even feel like an overtone is being sung even though no member of the chorus is actually singing it. Choral conductors often listen for the overtones in the concert hall to help determine whether their choir is singing in tune.

3.4.4   The starting transient

Given the importance of the overtones in creating the timbre of a sound, why does a synthesized sound of an instrument, where the amplitudes and frequencies of the higher partials can be reproduced precisely, still sound "electronic" to our ears? One reason has to do with the irregularities present in a signal coming from a real instrument. Although the initial sound wave may appear regular and repetitive, over time minute pressure changes and a decrease in strength can dramatically affect the shape of the signal. Our brain recognizes this irregularity, particularly after repeated hearings of a given instrument, and therefore identifies a fake copy when these imperfections are absent.

Another explanation for why we can tell the difference between a real instrument and its synthesized copy has to do with how the sound is initiated. A vibrating system needs to first be forced and then amplified to produce a sound. The vibrating string is plucked or bowed, and then the waves interact with the body of the instrument to propagate the sound. Likewise, the lips of a brass player vibrate into the mouth piece and the horn amplifies the note. This interaction of competing processes can be modeled by a *forced harmonic oscillator*, where the natural oscillator (e.g., the mass connected to a spring) is driven by an external force. If we assume that the damping force due to friction is negligible, the ODE modeling this system is

$$m\frac{d^2y}{dt^2} + ky = F(t),$$

which is the simple harmonic oscillator from Equation (3.6) with the right-hand side changed from 0 (no external force) to $F(t)$ (the external force). Typically, the solution to this type of ODE can be broken into two parts, the *starting transient* and the *steady state*. As the terms suggest, the transient piece dies off quickly, while the steady state remains.

For example, in the solution

$$y(t) = 2e^{-10t}\sin(200\,t) + \sin(100\pi\,t),$$

the first term in the sum is a sine curve whose amplitude is determined by a decaying exponential function (left graph in Figure 3.13). Consequently, it will head to zero very

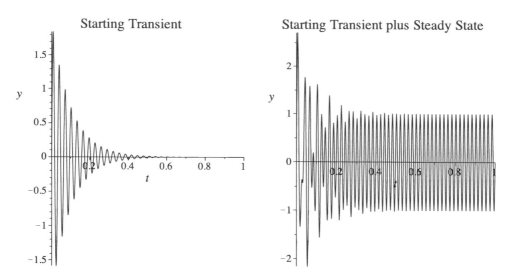

FIGURE 3.13. On the left is the graph of the starting transient $y(t) = 2e^{-10t}\sin(200\,t)$, which quickly approaches zero. The right graph shows the sum of the starting transient and the steady state $y = \sin(100\pi\,t)$. After about half a second, the steady state is the only function remaining.

quickly. The second term in the solution is the steady state, which produces a pure tone at a frequency of 50 Hz. The graph on the right in Figure 3.13 shows the combination of the starting transient and the steady state. After roughly half a second, the starting transient has essentially vanished, leaving the steady state to produce a pure tone.

Even though it dies off quickly, the starting transient is what helps our brains recognize that an actual instrument is being played. Based on how they create sound, different instruments will have different starting transients, and our magnificent brains can store this information and recall it instantly. Given the complexities in the physics of how a note is produced, it is too challenging a task (at least thus far) for an electronic synthesizer to convincingly reconstruct the origins of a note.

A simple way to demonstrate the importance of the starting transient is to record an instrument playing a note and then play it backward. If the overtones and corresponding timbre are all that are required to hear and enjoy the instrument, then the order in which the sound is heard should not matter. However, it certainly does. In their book *The Musician's Guide to Acoustics*, authors Campbell and Greated describe the result of playing a piano sound backward as transforming the instrument into a "leaky old harmonium" [Campbell and Greated, 1994]. The starting transient is not just familiar; it also provides our brains with a context that allows us to enjoy the particular sound of that instrument.

3.4.5 Resonance and beats

One of the most important properties in the study of acoustics is *resonance*. It can be found in practically every musical instrument as a way of producing and reinforcing sound. The resonance in the throat of a horn, called proximity resonance, helps amplify the sound before it passes through the wider bell. Resonance can also occur when the natural frequency of an oscillator is driven by an external force with the same frequency. When the internal and external frequencies match or are very close, a resonance is created that can strongly amplify the natural vibrations of the object.

Resonance can be useful, as in the case of a musical instrument or voice, or very dangerous. There are some notorious physical examples where resonance led to serious problems and, in some cases, outright failure. One recent example involves the construction of London's Millennium Bridge, first opened in June of 2000. The pedestrian bridge spans the river Thames and was the first solely pedestrian bridge to be built over London's iconic river since 1894. Naturally, there was much anticipation, and Londoners flocked to the bridge in great numbers on opening day. As people walked across the bridge, a noticeable sway in the horizontal direction occurred that effected their walking patterns, as well as the stability of the bridge. Quickly nicknamed the "Wobbly Bridge," it was shut down a few days after it opened, remaining closed for two years as stabilizers were installed to make it safe for the public.

The engineers who designed the bridge were not concerned with a detrimental resonance effect due to walking because pedestrians typically walk at random rates. Unless a marching band or military outfit were present to encourage a synchronous pattern, it was thought that resonance was unlikely to be a factor.[5] What transpired on the Millennium Bridge is that the initial force of the pedestrians influenced the bridge to sway slightly, and in response to this motion, pedestrians had to widen their steps and find a common pace that made walking feasible. This common speed, although small, created an external force whose frequency (around 1.7 Hz) was in resonance with the natural frequency of the bridge. The two frequencies reinforced each other, making the bridge sway quite noticeably.

Here is a simple example where an oscillator is being driven by an external force whose frequency perfectly matches the natural frequency of the oscillator:

$$\frac{d^2y}{dt^2} + 4y = 4\cos(2t). \tag{3.11}$$

The general solution to this ODE is

$$y(t) = c_1 \cos(2t) + c_2 \sin(2t) + t \sin(2t),$$

where c_1 and c_2 are arbitrary constants. For those who know some calculus, this can be verified by substituting $y(t)$ into Equation (3.11) and checking that both sides are equivalent. To understand how this relates to the phenomenon of resonance, we note that the first part of the solution, $c_1 \cos(2t) + c_2 \sin(2t)$, is a sum of trigonometric functions with frequency $2/(2\pi) = 1/\pi$, the natural frequency of the oscillator. On the other hand, the second part, $t \sin(2t)$, corresponds to a particular solution that arises from the forcing term on the right-hand side of Equation (3.11). Resonance occurs here because the natural frequency agrees with the frequency of the forcing term $4\cos(2t)$. At first the resonance is barely felt, but over time, the external force is stimulating the oscillator to grow in amplitude, which can cause it to come apart. This is captured mathematically by

---

[5]Actually, marching bands have been known to cause problems in sports stadiums. Apparently, one major university in the United States had to ban the playing of the song *Hey, Jude* during its football games because of a resonance between the stomping rate of the fans and the natural frequency of the stadium (see problem #22 in Blanchard, Devaney, and Hall, *Differential Equations*, 4th ed., p. 426).

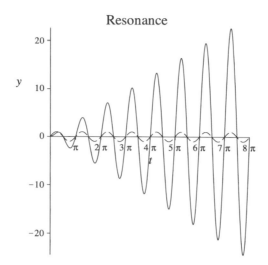

FIGURE 3.14. The graph of $y = \sin(2t)$ (*dashed*) versus $y = t\sin(2t)$ (*solid*), demonstrating the phenomenon of resonance. Although similar at first, the two curves separate over time since the amplitude of $t\sin(2t)$ grows linearly.

the $t\sin(2t)$ term in the solution. As t grows, the amplitude grows as well, creating an oscillation that will grow without bound (see Figure 3.14).

One experiment demonstrating resonance involves two tuning forks of the same frequency, each attached to small wooden boxes to amplify their vibrations. Place the two forks next to each other so that the open ends of their respective boxes are facing each other. Strike one of the tuning forks. After a second, stop the tuning fork that was struck. The other fork should be vibrating strongly, even though it was never struck itself!

Beats

The previous experiment with two equal-frequency tuning forks suggests a related one. What happens when the two tuning forks are not quite equal in frequency, but are very close? For example, what if a tuning fork of frequency A440 is struck simultaneously with one that is of frequency A441? To answer this question, we need to understand what happens when we add two sine functions with very similar frequencies. In this particular case, how do we simplify the sum $\sin(880\pi t) + \sin(882\pi t)$?

To accomplish this, we need to use some trig addition formulas. The following argument is a nice application demonstrating the importance of these formulas. Recall that the sine of the sum (or difference) of two angles α and β, pronounced "alpha" and "beta," respectively, is given by

$$\sin(\alpha + \beta) = \sin\alpha\cos\beta + \cos\alpha\sin\beta, \tag{3.12}$$
$$\sin(\alpha - \beta) = \sin\alpha\cos\beta - \cos\alpha\sin\beta. \tag{3.13}$$

If we add Equations (3.12) and (3.13) together, we obtain

$$\sin(\alpha + \beta) + \sin(\alpha - \beta) = 2\sin\alpha\cos\beta. \tag{3.14}$$

Next we perform a substitution to put this equation into a more useful form. Let $u = \alpha + \beta$ and $v = \alpha - \beta$. Some quick algebra then yields

$$u + v = 2\alpha \quad \text{or} \quad \alpha = \frac{1}{2}(u + v), \quad \text{and}$$

$$u - v = 2\beta \quad \text{or} \quad \beta = \frac{1}{2}(u - v).$$

Substituting these values for α and β back into Equation (3.14) yields

$$\sin u + \sin v = 2\sin\left(\frac{u+v}{2}\right)\cos\left(\frac{u-v}{2}\right). \tag{3.15}$$

Letting $u = 880\pi t$ and $v = 882\pi t$ in Equation (3.15), we find that

$$\sin(880\pi t) + \sin(882\pi t) = 2\sin\left(\frac{880\pi t + 882\pi t}{2}\right)\cos\left(\frac{880\pi t - 882\pi t}{2}\right)$$

$$= 2\sin(881\pi t)\cos(-\pi t)$$

$$= 2\cos(\pi t)\sin(881\pi t),$$

where the final step uses the fact that cosine is an even function ($\cos(-t) = \cos(t)$ for any angle t). This last formula shows that the sum of the two sine waves $\sin(880\pi t)$ and $\sin(882\pi t)$ results in a sine wave with frequency 440.5 and a *varying* amplitude described by the function $2\cos(\pi t)$. The fact that the amplitude is changing is crucial; it results in an interesting acoustical phenomenon known as *beating*, or simply just *beats*.

If we consider the first few cycles of the graphs of the two original sine waves, as well as their sum, it does not appear to be very interesting (see Figure 3.15). The sum simply looks like A440 with twice the amplitude. However, graphing the sum of the waves over

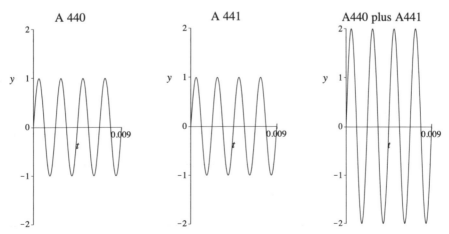

FIGURE 3.15. The graphs of A440, A441, and their sum over the first four cycles. Notice that the graph of the sum appears to simply be A440 with twice the amplitude.

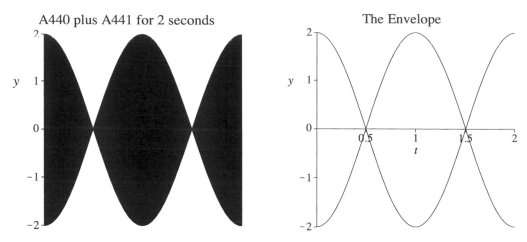

FIGURE 3.16. The graph of $y = \sin(880\pi t) + \sin(882\pi, t) = 2\cos(\pi t)\sin(881\pi t)$ (left) over 2 seconds. The graph appears to be solid because the frequency of the wave is 440.5 Hz, implying that it has 881 cycles in 2 seconds. Notice that in 1 second, the resulting wave changes in amplitude from a maximum of 2 to a minimum of 0 (at $t = 0.5$) and then back to a maximum of 2. The graph of the wave's envelope, $y = \pm 2\cos(\pi t)$, is shown to the right.

a much longer period, say, 2 seconds, reveals the beats (see Figure 3.16). The outline of the curve $y = 2\cos(\pi t)\sin(881\pi t)$, called the *envelope*, is given by the two cosine curves $y = 2\cos(\pi t)$ and $y = -2\cos(\pi t)$. The $\sin(881\pi t)$ term represents the pitch we hear (440.5 Hz), while the envelope creates the beating phenomenon. Our ear-brain system follows the oscillating amplitude of the combined sine waves, resulting in a wave that quickly alternates between loud (double the amplitude) and soft (no amplitude). This quick change in amplitude causes the "beats," a pulsating sound that comes and goes at a rate of one beat per second.

Equation (3.15) provides a general rule for beats. Suppose that two pitches with close frequencies of f and g Hz are sounded simultaneously, each with an amplitude of 1. Let us assume that $f > g$. Applying Formula (3.15), we find that

$$\sin(2\pi f t) + \sin(2\pi g t) = 2\cos\left(2\pi \frac{f-g}{2}t\right)\sin\left(2\pi \frac{f+g}{2}t\right).$$

Thus, the combined waves produce a sine wave with frequency $(f + g)/2$, with a varying amplitude between 0 and 2. Although the frequency of the cosine term is $(f - g)/2$, the beats actually appear with a frequency that is twice this value, namely, $f - g$. This doubling occurs because the overall shape of the resulting wave will have twice as many peaks per period as the cosine function (see Figure 3.16 for an example). Here the envelope of the resulting wave involves both $y = 2\cos(2\pi(f - g)t/2)$ and its negative, $y = -2\cos(2\pi(f - g)t/2)$, so there are twice as many humps. We summarize our fundings in the following rule.

Theorem 3.4.2 (The Rule of Beats). *Two pitches with close frequencies f and g Hz, with $f > g$, combine together to produce a pitch with the average frequency $(f + g)/2$ Hz, but they will beat at a frequency of $f - g$ Hz.*

Observe that the number of beats per second increases as the two notes move farther apart. This is one of the techniques used by piano tuners when adjusting the length and tension in the strings of a given note. The tuner aims to minimize the beating phenomenon by making the strings vibrate with the same frequency. This is also the governing principle behind Helmholtz's theory of dissonance. If two pitches have similar partials in their overtone series which do not match up precisely, then the beating between these partials causes dissonance in the resulting sound. According to Helmholtz, a rate of 35 beats per second between two pitches leads to the maximum dissonance. This value and the effect of the dissonance vary depending on the location of the two notes. It is more pronounced for notes in the lower octaves than the higher ones.

Example 3.4.3. *Suppose that two tuning forks are sounded simultaneously, one with a frequency of 600 Hz, the other with a frequency of 605 Hz. Determine the frequency of the resulting wave and the number of beats per second.*

Solution: We have $\alpha = 605$ and $\beta = 600$. Applying Theorem 3.4.2, the resulting wave will have a frequency of 602.5 Hz, with five beats per second.   □

Exercises for Section 3.4

1. In each of the following examples, a group of partials is sounded together, each at approximately the same amplitude. Determine the frequency of the residue pitch that is likely heard in each case.

   a. $150, 300, 450, 600$ Hz   **b.** $150, 225, 300, 375, 450, 525, 600$ Hz

   c. $100, 200, 400, 800, 1600$ Hz   **d.** $180, 270, 360, 540, 720, 810$ Hz

2. Suppose that the frequencies 850, 1050, 1250, 1450, and 1650 Hz are sounded together, each with amplitude equal to 1. By approximating the fractions 1050/850, 1250/850, 1450/850, and 1650/850 with small-valued integer ratios, deduce from Equation (3.10) that the perceived pitch is approximately 207.9 Hz. Give the integer values of k_1, k_2, k_3, k_4, and k_5 used to obtain this result.

3. Suppose that the frequencies $f, 2f, 3f$, and $4f$ are played simultaneously with amplitudes each equal to 1. Show that the autocorrelation formula given in Equation (3.10) yields $\bar{f} = f$, the residue pitch.

4. The formula for the fundamental frequency of a vibrating string is $f = \sqrt{T}/(2L\sqrt{\rho})$.

   a. If the length L of the string is doubled, what is the effect on the fundamental frequency f? By what interval does the corresponding note change?

   b. Suppose that the tension T is doubled and the density ρ is cut in half. What is the effect on the fundamental frequency, and by what interval does the corresponding note change?

   c. By what factor should we shorten the length of the string in order to raise the pitch up by two octaves?

   d. By what factor should we shorten the length of the string in order to raise the pitch up by an octave and a perfect fifth?

5. Compare the overtone series for 300 Hz and 400 Hz.

    a. List the first 12 frequencies (including the fundamental) in the overtone series for 300 Hz.

    b. List the first 12 frequencies (including the fundamental) in the overtone series for 400 Hz.

    c. Circle the frequencies that the overtone series for 300 Hz and 400 Hz have in common.

    d. Form a concise statement concerning the overlap of two overtone series with frequencies $3f$ and $4f$.

6. Suppose that a wave has the form $y(t) = -0.25e^{-0.05t}\cos(2000\pi t) + \sin(1500\pi t)$. Which part of the function is the starting transient, and which part corresponds to the steady-state solution? What is the frequency of the resulting pitch heard?

7. Residents living on the upper floors of an old, tall building have been reporting feeling nauseous due to a swaying motion of the building. Upon investigation, it is discovered that the swaying always occurs during a popular aerobic dance class, held in the top-floor gymnasium. Students in the class receive a vigorous workout while stepping in time to driving techno music.[6] Using some of the concepts discussed in this section, explain why the building is swaying during the dance class. What is an easy solution to the problem which avoids having to cancel the dance class altogether?

8. A piano tuner comparing two of three strings on the same note of the piano hears four beats per second. If one of the two notes is 330 Hz, the E above middle C, what are the two possibilities for the frequency of the other string?

9. Use Equation (3.15) to simplify the sum $\sin(310\pi t) + \sin(318\pi t)$. What is the frequency of the resulting wave, and how many beats per second would you expect to hear if the two waves were sounded together?

10. Recall that

$$\sin(\alpha + \beta) = \sin\alpha\cos\beta + \cos\alpha\sin\beta, \qquad (3.16)$$
$$\cos(\alpha + \beta) = \cos\alpha\cos\beta - \sin\alpha\sin\beta. \qquad (3.17)$$

    a. Set $\alpha = \beta = \theta$ in Equation (3.16) and derive the double-angle formula for the sine function, $\sin(2\theta) = 2\sin\theta\cos\theta$.

    b. Set $\alpha = \beta = \theta$ in Equation (3.17) and derive a double-angle formula for the cosine function, $\cos(2\theta) = 2\cos^2\theta - 1$. Note: $\cos^2\theta = (\cos\theta)^2$.
       Hint: Somewhere in your calculation you will need to use the famous identity between $\cos^2\theta$ and $\sin^2\theta$.

    c. Derive a formula for $\cos(3\theta)$ in terms of only $\cos\theta$ and higher powers of $\cos\theta$.

11. Following the derivation in the text for the sum $\sin u + \sin v$, derive a similar formula for the sum of two cosines:

$$\cos u + \cos v = 2\cos\left(\frac{u+v}{2}\right)\cos\left(\frac{u-v}{2}\right).$$

---

[6]Apparently this exact scenario took place in a 40-story building in New York City (see Heller, *Why You Hear What You Hear*, p. 196).

3.5   The Monochord Lab: Length versus Pitch

Note for instructors: Since the point of this project is to derive (and hear) the Pythagorean scale, this lab should be taught before proceeding to the material in the next chapter. The lab requires the use of a monochord (a one-stringed instrument) containing a movable fret that easily allows different string lengths to be plucked. Monochords vary in prices depending on quality. The company Artec Educational sells a kit to construct a monochord for under $15. It is also important to have a ruler running parallel to the string to determine string lengths. Students can work in small groups, with one monochord per group. The project has two parts; the second part should not be started (or viewed) until the first part is completed correctly.

The goal of this lab project is to explore the connection between the length of a string and the note sounded when the string is plucked. In order to investigate this relationship, we will use a simple, one-string instrument called a *monochord*. The monochord was commonly used in European classrooms during the Middle Ages and Renaissance to teach interval recognition and to study the connection between ratios of whole numbers and pitch. With some simple calculations and observations, we will re-create a famous and ancient musical scale known as the *Pythagorean scale*.

Be careful when handling the monochord, as overtightening the string may cause it to break. Keep your face away from the string so that you do not get hit in the eye by a breaking string. Take care when restringing the instrument. Also, a monochord should be plucked gently, not strummed aggressively. Vigorous playing may generate a louder sound, but the string life is reduced significantly.

Part 1: Simple Ratios

1. The tension of the string may be adjusted by tightening the peg at one end of the monochord. What happens to the pitch as the string is tightened?

2. Place the sliding fret at exactly half the length of the string. Hold one finger over the fret and pluck the string gently. What is the relationship between the pitch of this note and the note produced without the fret? In other words, what musical interval lies between the two notes? It may help to check the notes against a piano or another musical instrument, although it is necessary to first "tune" the monochord (on the open, unfretted string) to match some particular note of your instrument.

3. Place the sliding fret at exactly 1/4 of the string's length. Hold the string down over the fret and pluck the smaller portion of the string. What is the musical relationship between this note and the one sounded by plucking half the string? What is the relationship between this note and the one sounded on the open string?

4. Place the sliding fret at exactly 1/3 of the string's length. Hold the string down over the fret and pluck the two pieces to either side of the fret. What is the musical interval between these two pitches? Draw an important conclusion:

   Cutting the length of the string in half _____ (raises or lowers) the pitch by _____ (what musical interval?).

5. Next, investigate what happens if the ratio of the string lengths is 2 : 3. In other words, what effect does changing the string length to 2/3 its original value have on the pitch?

6. Finally, investigate what happens if the ratio of the string lengths is 3 : 4. In other words, what effect does changing the string length to 3/4 its original value have on the pitch? Given that $\frac{3}{4} = \frac{1}{2} \cdot \frac{3}{2}$, how could this result have been predicted using the answers to the previous two questions?

Before proceeding to the next part of the lab, check with your instructor to make sure that you have answered the questions above correctly.

Part 2: Ratios for the Pythagorean Scale

Given the ground work accomplished above, we are now ready to construct the entire Pythagorean scale. Although it sounds like the usual major scale, it differs in some important ways from the scale found on a modern piano. These differences are explored in great detail in Chapter 4.

Thus far, we have uncovered three important facts concerning the relationship between the ratios of string lengths and the corresponding musical interval.

1) String lengths in a 1 : 2 ratio are an octave apart. Specifically, cutting the length of the string in half raises the pitch by an octave. Conversely, doubling the length of the string lowers the pitch an octave.

2) String lengths in a 2 : 3 ratio are a perfect fifth apart. Specifically, cutting the length of the string by a factor of 2/3 raises the pitch by a perfect fifth.

3) String lengths in a 3 : 4 ratio are a perfect fourth apart. Specifically, cutting the length of the string by a factor of 3/4 raises the pitch a perfect fourth.

Notice the musical and mathematical simplicity of these facts. Simple ratios lead to the "perfect" intervals of an octave, fifth, and fourth. It is no accident that these are the primary intervals of early music (e.g., Gregorian chant used octaves, while Medieval polyphony used fourths and fifths). They also underscore the critical tonic-dominant harmonic relationship and the V–I and I–IV–V chord progressions commonly found in music.

It is also important to recognize that fact 3 can be derived from the first two facts. Multiplying the length of the string by 1/2 raises the pitch an octave. Then, multiplying the length of the new string by a factor of 3/2 will *lower* the pitch by a perfect fifth (longer strings have lower pitches). Going up an octave and then down by a perfect fifth is equivalent to going up by a perfect fourth (e.g., middle C up an octave to the next highest C and then down a perfect fifth gives F, and the interval between middle C and this F is a perfect fourth). Since

$$\frac{1}{2} \cdot \frac{3}{2} = \frac{3}{4}, \tag{3.18}$$

the net effect of shortening the string length by a factor of 3/4 is to raise the pitch by a perfect fourth. Notice that we multiply the two ratios in Equation (3.18), as opposed to adding them. We are always interested in the ratios of string lengths, not the difference.

The goal for Part 2 of this lab is to discover the remaining ratios in the Pythagorean scale. The scale is essentially the same as the major scale, except the string lengths are slightly different from the modern ones. We will derive this scale using only facts 1 and 2 above and the circle of fifths (see Section 2.4). For example, to find the second note of the major scale, we go up two perfect fifths from the tonic and then down an octave. The net effect is raising the pitch by a whole step. If the starting note is middle C, then up a perfect fifth gives G, and up another perfect fifth gives D' (the first two "hours" of the circle of fifths). We then go down an octave to find the note D just above the starting C. Using the ratios given in facts 1 and 2, what fraction of the full string do we take to obtain the second note of the scale? Use a calculator to find the actual length of the string (round to two decimal places) and play it with the open string to hear the first two notes of the Pythagorean scale (Do and Re).

The next note to find would be the sixth of the scale since that is a perfect fifth above the second. From there we can find the third and finally the seventh note of the scale, the leading tone. Make sure that each note is obtained using only the ratios from facts 1 and 2. Complete Table 3.4 and then try playing the full scale, perhaps as a class. Do you hear any differences between your scale and the major scale on a modern instrument?

Scale Degree	Interval	Ratio $\cdot L$	Actual Length
1	Unison	$(1/1) \cdot L$	
2	Major Second		
3	Major Third		
4	Perfect Fourth	$(3/4) \cdot L$	
5	Perfect Fifth	$(2/3) \cdot L$	
6	Major Sixth		
7	Major Seventh		
8 = 1	Octave	$(1/2) \cdot L$	

TABLE 3.4. The lengths and ratios of the Pythagorean scale. L represents the length of the full string. Give lengths to two decimal places.

Some Concluding Questions

1. Find the prime factorization of the numerator and denominator in each ratio in the third column of Table 3.4. For example, $3/4 = 3/2^2$. What do you notice about each factorization?

2. According to Table 3.4, what is the ratio of string lengths that are a whole step apart? Be sure to check *all* five whole steps in the scale (they should be identical). This ratio, call it W, is the factor used to shorten the string length in order to raise the pitch by a whole step. Express W as a ratio of two integers.

3. According to Table 3.4, what is the ratio of string lengths that are a half step apart? Be sure to check *both* half steps in the scale (they should be identical). This ratio, call it H, is the factor used to shorten the string length in order to raise the pitch by a half step. Express H as a ratio of two integers.

4. Recall that two half steps are equivalent to one whole step. Hence, it should be the case that $H \cdot H = H^2 = W$. Is this equation valid? Compute the quantity H^2/W, giving your answer as a ratio of two integers and in decimal form (to five decimal places). Also give the prime factorization of the numerator and denominator. This value is important in the theory of tuning and is known as the *Pythagorean comma*. We will discuss this comma, as well as the problems it creates, in Section 4.1.

References for Chapter 3

Blanchard, P., Devaney, R. L. and Hall, G. R.: 2012, *Differential Equations*, 4th edition, Brooks/Cole, Cengage Learning.

Buehler-McWilliams, K. and Murray Jr., R. E.: 2013, The Monochord in the Medieval and Modern Classrooms, *Journal of Music History Pedagogy* 3(2), 151–172.

Campbell, M. and Greated, C.: 1994, *The Musician's Guide to Acoustics*, Oxford University Press.

Hartmann, B.: n.d., Consonance and Dissonance—Roughness Theory. Ohio State University, School of Music, http://www.music-cog.ohio-state.edu/Music829B/roughness.html.

Heller, E. J.: 2013, *Why You Hear What You Hear: An Experiential Approach to Sound, Music and Psychoacoustics*, Princeton University Press.

Helmholtz, H. L. F.: 1954, *On the Sensations of Tones as a Physiological Basis for the Theory of Music*, Dover Publications, Inc., New York. Reprint; original German version (1877) translated in 1885 by A. J. Ellis.

Johnston, I.: 2009, *Measured Tones: The Interplay of Physics and Music*, 3rd edition, CRC Press.

Pierce, J. R.: 1992, *The Science of Musical Sound* (revised edition), W. H. Freeman and Company.

Rausch, R. A. and Plomp, R.: 1982, The Perception of Musical Tones, *in* D. Deutsch (ed.), *The Psychology of Music*, Academic Press.

Sautoy, M., Thomas, R. and Woolley, T.: n.d., The Millennium Bridge, London, Maths in the City (University of Oxford). http://www.mathsinthecity.com.

Taylor, C.: 2003, The Science of Musical Sound, *in* J. Fauvel, R. Flood and R. Wilson (eds), *Music and Mathematics: From Pythagoras to Fractals*, Oxford University Press, pp. 47–59.

Chapter 4

Tuning and Temperament

Music is true. An octave is a mathematical reality. So is a 5th. So is a major 7th chord. And I have the feeling that these have emotional meanings to us, not only because we're taught that a major 7th is warm and fuzzy and a diminished is sort of threatening and dark, but also because they actually do have these meanings. It's almost like it's a language that's not a matter of our choosing. It's a truth. The laws of physics apply to music, and music follows that. So it really lifts us out of this subjective, opinionated human position and drops us into the cosmic picture just like that.

— James Taylor[1]

Why does the 12-note Western musical scale work particularly well? Why does a major triad sound "happy," while a minor chord seems "sad"? Is there any reason why singers think they are in tune, but their accompanying instrumentalists often feel otherwise? Why are musicians in small ensembles such as string quartets taught to flat the third of a major chord? How do we place the frets correctly on a guitar?

These and related questions are best answered through an investigation into the construction of the different scales and tuning systems that have been used in the evolution of Western classical music. We shall discover that the subject is actually quite deep. At its core is a fundamental mathematical truth; there exist numbers that cannot be expressed as the ratio of two integers. Some of the problems encountered by different tuning systems have ultimately run afoul of this mathematical truism. Striving for a rational system sometimes requires the use of *irrational* numbers. In many ways James Taylor is correct. There are physical and mathematical reasons that can help explain why a major chord is nice and a diminished chord is creepy. To understand why, we start with one of the earliest musical scales ever created, the Pythagorean scale.

4.1 The Pythagorean Scale

As legend goes, the great Greek mathematician and philosopher Pythagoras was walking by a blacksmith's shop when he noticed that certain sounds of the hammers banging against the anvils were more consonant than others. Upon investigation, he discovered

---

[1]Hutchinson, Interview with James Taylor.

that the more consonant sounds came from hammers whose weights were in simple proportions like 2 : 1, 3 : 2, and 4 : 3. After further musical experiments with strings and lutes, he and his followers (the Pythagoreans) became convinced that basic musical harmony could be expressed through simple ratios of whole numbers. Their general belief was that ratios consisting of smaller numbers lead to more pleasing sounds. Thus, the natural building blocks of mathematics (small whole numbers) were aligned with the basic interval relationships used to create harmonious music. This helps explain the devotion of the Pythagoreans to rational numbers, numbers that can be expressed as the ratio of two integers.

4.1.1   Consonance and integer ratios

We have already seen the importance of the ratios 2 : 1 and 3 : 2 in the overtone series. Recall that two notes whose frequencies are in a 2 : 1 ratio are an octave apart, while notes in a 3 : 2 ratio are a perfect fifth away. Since the numbers in these fractions are low, the respective overtone series have a large amount of overlap, leading to a consonant sound.

As we discuss the different tuning systems throughout this chapter, it is important to remember that it is the *ratio* between the frequencies that determines the musical interval, not the distance between them.

Key Fact: The musical interval between two notes is determined by the **ratio** between the frequencies of each note.

For example, the musical interval between the notes corresponding to 100 and 200 Hz is an octave, as is the interval between notes at 1500 and 3000 Hz, since each pair is in a 2 : 1 ratio. Even though the frequencies in the second pair are much farther apart, the musical intervals are the same. To raise the pitch up an octave, we multiply the frequency by 2; to lower the pitch an octave, we must divide the frequency by 2. Similarly, to raise the pitch by a perfect fifth, we multiply the frequency by $\frac{3}{2}$, while to lower it by a perfect fifth, we divide the frequency by $\frac{3}{2}$ (or multiply by $\frac{2}{3}$). Two notes that are the same frequency are in a 1 : 1 ratio, which corresponds to a unison interval. Larger ratios, assuming they are greater than 1, correspond to wider musical intervals.

The Pythagoreans' clever idea was to create an entire musical scale solely based on the ratios 2 : 1 and 3 : 2. By going up and down by octaves and perfect fifths, they generated new notes within the octave. If a pitch was located outside the desired octave, then it was multiplied or divided by powers of 2 until it settled back within the correct range of notes.

For example, suppose we wanted to find the ratio for creating a perfect fourth (P4). If we first go up by an octave (multiply by 2) and then go *down* by a perfect fifth (multiply by $\frac{2}{3}$), we will have arrived at a note that is a perfect fourth above the original. In terms of musical notes, one such example starts on middle C, goes up an octave to C′ (we will use C′, C″, and so forth to denote the Cs in higher octaves), and then moves down a perfect fifth to F. The resulting ratio for a perfect fourth is thus

$$2 \cdot \frac{2}{3} \; = \; \frac{4}{3}.$$

Two notes whose frequencies are in the ratio 4 : 3 will produce the musical interval of a perfect fourth.

Another way to check that $4:3$ is the correct ratio for a perfect fourth is to use the fact that a perfect fourth and a perfect fifth combine to make an octave. Thus, multiplying the respective ratios for a P4 and a P5, we have

$$\frac{4}{3} \cdot \frac{3}{2} = \frac{2}{1},$$

which is the correct ratio to raise the pitch by an octave.

The next note we will find is simply a whole step above the tonic, or an interval of a major second (M2). This can be found by going up two perfect fifths from the tonic (multiply by $\frac{3}{2}$ twice) and then going down by an octave (divide by 2). A musical example is to start on middle C, go up a perfect fifth to G, go up another perfect fifth to D′, and finally move down an octave to D, ending up a whole step above the original C. The resulting ratio for a whole step or major second in the Pythagorean scale is thus

$$\frac{3}{2} \cdot \frac{3}{2} \cdot \frac{1}{2} = \frac{9}{8}.$$

At this point, we know the ratios (sometimes called *multipliers*) for creating the musical intervals of a major second, a perfect fourth, a perfect fifth, and an octave. The remaining three intervals can be found in a similar fashion. For example, the ratio for a major sixth is

$$\frac{9}{8} \cdot \frac{3}{2} = \frac{27}{16},$$

because a major sixth is a perfect fifth plus a major second (D up a perfect fifth to A gives a note that is a major sixth away from C). Remember, we use *multiplication* (not addition) here because we are interested in the ratios between frequencies (not their difference). The derivations for a major third and major seventh are left as exercises.

The full set of ratios used for the Pythagorean scale are given in Table 4.1. We make five important observations based on the table:

1. The numerator and denominator of each ratio are always a power of 2 or 3.

2. All perfect fifths are in the ratio $3:2$.

3. All five whole steps in the major scale are in the ratio $W = 9:8$.

4. Both half steps in the major scale are in the ratio $H = 256:243$.

5. Two half steps do *not* equal one whole step, that is, $H^2 \neq W$!

The first two facts follow by construction, since the Pythagorean scale was created using only the ratios $2:1$ and $3:2$. Observations 3 and 4 are also easily verified. The five whole steps can be checked directly. If $W = 9/8$, then

$$\frac{1}{1} \cdot W = \frac{9}{8}, \quad \frac{9}{8} \cdot W = \frac{81}{64}, \quad \frac{4}{3} \cdot W = \frac{3}{2}, \quad \frac{3}{2} \cdot W = \frac{27}{16}, \quad \text{and} \quad \frac{27}{16} \cdot W = \frac{243}{128}.$$

The ratio H for each half step is similarly confirmed. Finally, we check that

$$H^2 = \frac{256}{243} \cdot \frac{256}{243} = \frac{2^{16}}{3^{10}} \neq \frac{3^2}{2^3} = \frac{9}{8} = W.$$

Scale Degree	Interval	Ratio
1	Uni.	1
2	M2	9/8
3	M3	81/64
4	P4	4/3
5	P5	3/2
6	M6	27/16
7	M7	243/128
8 = 1	Oct.	2/1

TABLE 4.1. The ratios or multipliers used to raise a note (increase the frequency) by a given musical interval in the Pythagorean scale. For example, to find the note that is a perfect fourth above 300 Hz, we multiply 300 by 4/3, giving a frequency of 400 Hz.

The fact that two half steps in Pythagorean tuning does not equal one whole step is quite problematic. How should we determine a minor second? If we are to find the frequency for C$\sharp$, we could begin at C and go up a half step (multiply by H), or we could start at D and go down by a half step (multiply by W and then divide by H). Unfortunately, these two methods will yield different results since $H \neq W/H$, or rather $H^2 \neq W$. This difference is known as the *Pythagorean comma*, an inconsistency that is one of the major drawbacks of using the Pythagorean scale.

In general, a *comma* is a gap or a discrepancy. There are different kinds of commas depending on the tuning system. The specific value of the Pythagorean comma is defined as the gap between two half steps and one whole step. To measure this gap, we divide the ratio for a whole step (W) by the ratio for two half steps (H^2) and compare the result to 1. (Equivalent ratios would give a quotient of 1.) This yields

$$\frac{W}{H^2} = \frac{9/8}{(256/243)^2} = \frac{9 \cdot 243 \cdot 243}{8 \cdot 256 \cdot 256} = \frac{3^{12}}{2^{19}} = \frac{531{,}441}{524{,}288} \approx 1.013643265.$$

We see that raising the pitch by two half steps is just a little shy of raising the pitch by one whole step.

Definition 4.1.1. *The Pythagorean comma is defined to be the ratio* $3^{12} : 2^{19}$. *It equals the gap between two half steps and one whole step in the Pythagorean scale.*

4.1.2 The spiral of fifths

The Pythagorean comma arises in another important musical situation, namely, the circle of fifths. This demonstrates a rather disconcerting issue with the Pythagorean scale.

Recall that the circle of fifths is obtained by raising the pitch repeatedly by perfect fifths until we "return" to the starting note. Musically speaking, raising the pitch by 12 perfect fifths is equivalent to going up by seven octaves. From a mathematical perspective, if we use the basic Pythagorean ratios, $2:1$ and $3:2$, then it should be the case that

$$\left(\frac{3}{2}\right)^{12} \overset{?}{=} 2^7, \quad \text{or} \quad 3^{12} \overset{?}{=} 2^{19} \tag{4.1}$$

after cross-multiplication. However, although it is close, the last equation in (4.1) is clearly false; an integer power of 3 is always odd, yet an integer power of 2 is always even.

Note the presence of the Pythagorean comma here. If we use Pythagorean tuning to construct our usual scale, we have an inconsistency: seven octaves is not equivalent to 12 perfect fifths. We are forced to introduce a new note that is a Pythagorean comma away from the note we want. Starting on C (think near the bottom of the piano keyboard) and successively rising by perfect fifths, we obtain C, G, D, ..., A♯, E♯, and finally B♯. As musicians, we should identify B♯ with its enharmonic equivalent C, but as mathematicians using Pythagorean tuning, we cannot. If f represents the frequency of the starting note C, then the new note B♯ (way at the other end of the piano) will have the frequency $(\frac{3}{2})^{12}f$, while its enharmonic equivalent will be at $2^7 f$. The note B♯ will be slightly higher in pitch than our C. By taking the ratio of these two notes, $(3^{12}f/2^{12})/(2^7 f) = 3^{12}/2^{19}$, we see that the gap between them is precisely the Pythagorean comma.

Since B♯ $\neq$ C in the Pythagorean scale, the circle of fifths does not close up; it becomes a growing and never-ending spiral. The problem persists as we continue to increase by perfect fifths, multiplying each frequency by $3/2$, as F$^{\times}$ $\neq$ G (recall that F$^{\times}$ is F with two sharps), C$^{\times}$ $\neq$ D, G$^{\times}$ $\neq$ A, etc. After transposing up by another 12 perfect fifths, we will have created 12 new notes, each a Pythagorean comma away from their enharmonic equivalents. When we reach A$^{\times}$♯ (A with three sharps), we will now be two Pythagorean commas away from C, and the spiral continues (see Figure 4.1).

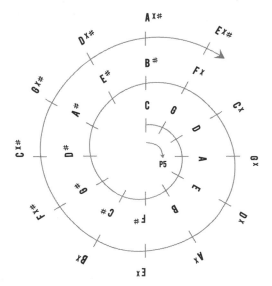

FIGURE 4.1. For the Pythagorean scale, the circle of fifths does not close up; it spirals around, creating a new gap equal to the Pythagorean comma each time it passes by C.

This might appear to be just a fun but impractical mathematical exercise. After all, aren't the pitches we are discussing octaves above the highest note on the piano, in frequency ranges only dolphins can hear? In some sense that is true, but remember that all of these new pitches can be dropped back down into the middle of the piano by dividing by two repeatedly (going down an octave at a time). Our former 12-note chromatic scale just got extremely crowded!

Here is another example that illustrates the problem. Suppose we started on a C with frequency f and wanted to find the frequency for a note precisely a tritone above C. There are two equally plausible methods to find this note by moving in different directions around the circle of fifths. We can go up by six perfect fifths to F$\sharp$ and then drop down by three octaves, giving us a note with a frequency of

$$\left(\frac{3}{2}\right)^6 \cdot \frac{1}{2^3}\, f \;=\; \left(\frac{3^6}{2^9}\right) f. \tag{4.2}$$

On the other hand, we can go down by six perfect fifths to G$\flat$ and then go up by four octaves to return to a tritone above the original C. In this case, we obtain a frequency of

$$\left(\frac{2}{3}\right)^6 \cdot 2^4\, f \;=\; \left(\frac{2^{10}}{3^6}\right) f. \tag{4.3}$$

Once again, we have the troubling issue that F$\sharp \neq$ G$\flat$, with F$\sharp$ being a Pythagorean comma above G$\flat$. There is nothing special about F$\sharp$ and G$\flat$ here. The same argument will hold true for other enharmonically equivalent notes.

The basic problem inherent in Pythagorean tuning arises from the fact that it is mathematically impossible to guarantee both 2:1 octaves and 3:2 perfect fifths simultaneously throughout the piano. While the Pythagorean scale will work fine for chords and intervals featuring octaves, fifths, and fourths, it runs into some serious trouble as soon as our intervals become smaller and our melodies more chromatic. Changing keys by half steps quickly leads to incongruities. As music developed to contain more and more sophisticated harmonic structure, the Pythagorean scale needed to be discarded in favor of a more flexible system. This new tuning system, known as *equal temperament*, was a harmonic compromise that used a simple mathematical idea to rectify issues such as the spiral of fifths and the fact that two half steps did not equal a whole step. We discuss the equal temperament tuning system in Section 4.3.

4.1.3   The overtone series revisited

Before describing other tuning systems, we return to discuss the overtone series, applying the knowledge gleaned from the ratios of the Pythagorean scale. Recall that the overtone series consists of successive multiples of the fundamental frequency. Suppose we begin with the pitch $f = 110$ Hz. This is the A two octaves below the standard concert A440 Hz. Let us find the pitches associated with the first 12 notes in the overtone series

$$110,\ 220,\ 330,\ 440,\ 550,\ 660,\ 770,\ 880,\ 990,\ 1100,\ 1210,\ 1320,\ \ldots.$$

Since the ratio 2:1 corresponds to an octave, we know that $2f = 220$ is A$'$, an octave below A440. Similarly, we have that $8f = 880$ is A$'''$, an octave above A440. Next, using the fact that the ratio 3:2 corresponds to a perfect fifth, we know that $3f = 330$ is the pitch E$'$, since

$330/220 = 3/2$. This note is an octave and a fifth above A110, and a perfect fourth below A440. As confirmation, we mention that the fraction $440/330$ simplifies to $4/3$, the ratio in Pythagorean tuning for a perfect fourth. Since $660/330 = 1320/660 = 2/1$, we also have that $6f = 660$ Hz corresponds to E″, a perfect fifth above A440, and $12f = 1320$ Hz corresponds to E‴, an octave above E″.

Next, consider the overtone $5f = 550$. We claim that this note corresponds to the C♯ a major third above A440. This follows because $550/440 = 5/4 = 80/64$ is very close to $81/64$, the ratio used for a major third in the Pythagorean scale. Given that 550 Hz is C♯″, it follows that $10f = 1100$ Hz corresponds to the note C♯‴, which is a major 10th above A440. The overtone $9f = 990$ is in the ratio $9:8$ with $8f = 880$, so this note must be a whole step above A‴, which gives the note B‴.

The two remaining overtones on our list, $7f = 770$ and $11f = 1210$, do not fit very well with the Pythagorean scale, nor, for that matter, do they have a natural location in our current Western chromatic scale. The logical choices for 770 Hz are a major second or a minor third above E660. The ratio for a minor third in Pythagorean tuning is $32/27$ (see Exercise 6). Since

$$\frac{9}{8} \cdot 660 = 742.5 \quad \text{and} \quad \frac{32}{27} \cdot 660 = 782.222\overline{2},$$

we will take 770 Hz to be a minor third above E660. This yields the note G″, although 770 Hz is substantially flatter than this pitch. A similar analysis shows that $11f = 1210$ Hz is closer to the note a whole step above $10f = 1100$ Hz than a half step above, so we identify $11f$ with D♯‴. Alternatively, if we take $11f$ to be D‴, then the interval between $11f$ and $12f$ would be a major second, *larger* than the interval between $10f$ and $11f$, a minor second. This is contrary to what we would expect, since the fraction $12/11$ is *smaller* than $11/10$. Thus, choosing D♯‴ to represent $11f = 1210$ Hz is the more logical choice. As with $7f$, this overtone is also substantially flat when compared with its approximate note. The results of our calculations and estimations are best captured on the musical staff (see Figure 4.2).

There is nothing unique to the overtone series we just created. The same process will work regardless of which frequency f we start on because the *ratios* between successive overtones remain the same. The crucial aspect of the overtone series is the musical intervals between successive notes. In order from the bottom, the first 11 intervals of the overtone series are

Oct., P5, P4, M3, m3, m3*, M2, M2, M2, M2*, m2,

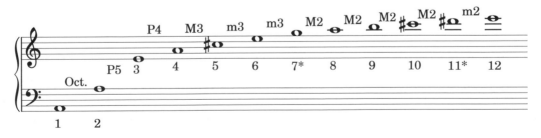

FIGURE 4.2. The first 12 notes corresponding to the overtone series of A110 Hz, along with the intervals between consecutive notes. The notes indicated for $7f$ and $11f$ are approximations. Both of these overtones are flat with respect to the corresponding note.

where the $*$ indicates the fact that the 7th and 11th overtones are approximations. The successive ratios between the first 12 frequencies in the overtone series are

$$\frac{2}{\mathbf{1}}, \frac{3}{\mathbf{2}}, \frac{4}{\mathbf{3}}, \frac{5}{\mathbf{4}}, \frac{6}{\mathbf{5}}, \frac{7}{6}, \frac{8}{7}, \frac{9}{\mathbf{8}}, \frac{10}{9}, \frac{11}{10}, \frac{12}{11}. \tag{4.4}$$

Here we have used boldface to highlight those ratios that are in perfect agreement with the Pythagorean scale (see Table 4.1).

Notice that the first six notes of the overtone series, as well as notes 8, 10, and 12, are the pitches of a major triad (1, 3, and 5 of the major scale), with root equal to the fundamental frequency. For the example in Figure 4.2, these are the notes A, C♯, and E, where the root A appears four times, the fifth E three times, and the third C♯ twice. This helps explain why a major chord sounds so pleasing to our ear. The large amount of overlap in the overtones of the notes in a major chord produces a particularly satisfying consonance.

According to the overtone series, the ratio for a major triad in root position is simply $4\!:\!5\!:\!6$, since these are the ratios that correspond to $4f, 5f$, and $6f$ (A440, C♯550, and E660 in our example). In other words, a major third has the ratio $5\!:\!4$, a minor third corresponds to $6\!:\!5$, and a perfect fifth is $6\!:\!4 = 3\!:\!2$. But in Pythagorean tuning, this ratio involves much larger integers, namely, $64:81:96$. Why not create a scale and corresponding tuning system that aligns itself better with the overtone series and such simple ratios as $4\!:\!5\!:\!6$? This is precisely the intention of the style of temperament known as *just intonation*, discussed in the next section.

Exercises for Section 4.1

1. Verify that the ratio for a major third in the Pythagorean scale is $81\!:\!64$. This can be accomplished in several different ways. Show that you obtain the same ratio whether you raise the fundamental by two whole steps (e.g., from C to D and then D to E) or you go up four perfect fifths and then down two octaves (e.g., from C up a P5 to G, up a P5 to D′, up a P5 to A′, up a P5 to E″, down an octave to E′, and finally down another octave to E).

2. Verify that the ratio for a major seventh in the Pythagorean scale is $243\!:\!128$. There are many different combinations of musical intervals to obtain a major seventh. Choose two different methods and show that you obtain the same ratio.

3. Check that the two half steps in the Pythagorean scale are each determined by the ratio $H = 256\!:\!243$.

4. Show that the frequencies obtained for F♯ and G♭ in Equations (4.2) and (4.3), respectively, can also be found by using H. Specifically, check that F♯ is equivalent to going up from C by a perfect fifth and then going down by one half step, while G♭ can be found by going up from C by a perfect fourth and then up again by a half step.

5. Find the frequencies for an E major Pythagorean scale (one octave) assuming that the starting note is 330 Hz (the E just above middle C). Round your answers to one decimal place.

6. For the Pythagorean scale, show that raising the pitch by a minor third corresponds to multiplying the frequency by 32/27. Due to the Pythagorean comma, this value is not unique. Therefore, we are stating the ratio with the smallest possible numerator and denominator. For example, going up by three half steps yields a ratio with very large (too large) numbers in the numerator and denominator.

7. Find the best possible ratio (smallest values in the numerator and denominator) in the Pythagorean scale used to raise the pitch by a minor sixth.

8. Beginning on the F at the bottom of the piano staff in the bass clef, write the first 12 notes corresponding to the frequencies in the overtone series. Give the musical interval between consecutive notes.

9. Beginning on the note two octaves below middle C, use the piano staff to write the first 12 notes corresponding to the frequencies in the overtone series. Start in the bass clef. Give the musical interval between consecutive notes.

4.2 Just Intonation

To create a tuning system that builds on the resonances of the overtone system, we choose frequency ratios that use smaller numbers in the numerator and denominator such as $5:4$ as opposed to $81:64$. At the end of the previous section, we learned that a major triad will align itself nicely with the overtone series if it is in the ratio $4:5:6$. Thus, we keep the primary ratios used for the perfect intervals and octave in the Pythagorean scale but change the intervals of a major third, major sixth, and major seventh to agree with the important major triad ratio $4:5:6$.

The ratio for our new major third changes to $5:4$, called a *just major third*. The reason for "just" is that this new major third is more consonant than its Pythagorean counterpart $81:64$, as it is based on the overtone series and the way we hear. Intervals whose ratios are well aligned with the overtone series are called *just* or *true* intervals.

A *just major sixth* corresponds to the ratio $5:3$. This follows from Figure 4.2, as the interval between the third and fifth notes in the overtone series (from E' to $C\sharp''$) is a major sixth. Alternatively, we know that a *just minor third* has the ratio $6:5$, since that is the interval for the top half of a major triad. Since a minor third plus a major sixth yields an octave, the product of the two ratios for these intervals should be 2. Solving

$$\frac{6}{5} \cdot r = \frac{2}{1}$$

for r quickly yields $r = 5/3$. Note that the ratio for a major sixth in Pythagorean tuning is $27/16 = 1.6875$, which is close to our new ratio of $5/3 = 1.666\overline{6}$.

The final interval to be determined is the major seventh. The simplest way to determine this ratio is to use the fact that a perfect fifth combined with a major third yields a major seventh. (Recall that the odd definition of a musical interval leads to anomalous equations such as $5 + 3 = 7$.) Consequently, the ratio for a major seventh is

$$\frac{3}{2} \cdot \frac{5}{4} = \frac{15}{8}.$$

Scale Degree	Interval	Ratio
1	Uni.	1
2	M2	9/8
3	M3	**5/4**
4	P4	4/3
5	P5	3/2
6	M6	**5/3**
7	M7	**15/8**
8 = 1	Oct.	2/1

TABLE 4.2. The ratios or multipliers used to raise a note (increase the frequency) by a given musical interval in just intonation. The ratios in bold are changes from the Pythagorean scale. For example, to find the note that is a major third above 200 Hz using just intonation, we multiply 200 by 5/4, giving a frequency of 250 Hz.

The ratios for each interval of the major scale using this new system are shown in Table 4.2. In general, the numerators and denominators of the ratios are much smaller in size than those of the Pythagorean scale. They are also no longer just powers of 2 and 3. In fact, the numerators of the three new ratios $5:4$, $5:3$, and $15:8$ all contain 5 as a prime factor.

We will refer to this new system of tuning as *just intonation*, an umbrella term for scales that use just intervals such as the $5:4$ just major third and the $5:3$ just major sixth. There are different ways to create scales using just intonation, but most of this difference lies with the remaining notes of the chromatic scale.

The idea of adjusting the Pythagorean major third to the just ratio of $5:4$ dates as far back as medieval times, when it was proposed by two British theorists, Theinred of Dover and Walter Odington. In 1482, Bartolomé Ramos de Pareja's theoretical monochord was proposed using just intonation, thereby acquainting musical scholars of the Renaissance with this new system. As music was becoming more harmonically complicated, the use of perfect fourths and fifths expanded to include other major intervals such as the major third and sixth. The desire for a more consonant scale erupted across Europe, with notable scientific luminaries such as Kepler, Mersenne, and Euler weighing in with their own versions (sometimes more than one) of a justly tuned scale.[2]

One of the advantages of just intonation is that its lower-numbered ratios for the major intervals are designed to agree with the ratios inherent in the overtone series, giving a richer, more resonant sound. This is particularly noticeable when examining the three

---

[2]See Section 5.10 of Benson, *Music*, for a nice overview of the many different scales created using just intonation.

major chords of the major scale (the I, IV, and V chords). By construction, the major chord on the tonic (the I chord) has ratios

$$\frac{1}{1}, \frac{5}{4}, \frac{3}{2} \quad \text{or} \quad \frac{4}{4}, \frac{5}{4}, \frac{6}{4}.$$

If we consider the major chord built on the fourth scale degree (the IV chord, such as F A C in the key of C major), then the three ratios are

$$\frac{4}{3}, \frac{5}{3}, \frac{2}{1} \quad \text{or} \quad \frac{4}{3}, \frac{5}{3}, \frac{6}{3}.$$

Once again, we see the important $4:5:6$ ratio between the three notes. One can check that the dominant major chord (the V chord, such as G B D in the key of C major) is also in the ratio $4:5:6$ (see Exercise 2). Music that relies predominantly on these three major chords will sound particularly pleasing in just intonation.

4.2.1   Problems with just intonation: The syntonic comma

There are several problems that arise when using just intonation. Some of these are the same problems we encountered with the Pythagorean scale, namely, that the circle of fifths still doesn't close. This difficulty persists with just intonation because the $3:2$ ratio for the perfect fifth and $2:1$ for the octave remain unchanged. Going up by 12 perfect fifths is still not equivalent to seven octaves.

Furthermore, it is still the case that two half steps are not equivalent to one whole step. In fact, this problem is exacerbated, as just intonation actually has *two* different ratios for a whole step within its major scale. For example, the ratio between the first and second scale degrees is still $9:8$, the same as in the Pythagorean system. However, the ratio between the second and third scale degrees is $5/4 \div 9/8 = 10/9$. The discrepancy here is quite immediate; within the first three notes of the major scale in just intonation, we encounter two different whole steps. One can check that of the remaining three whole steps in the major scale, two are in the ratio $9:8$, while the other is $10:9$ (see Exercise 4). The gap between the two different whole steps is $9/8 \div 10/9 = 81/80$, a ratio known as the *syntonic comma*.

Definition 4.2.1. *The syntonic comma is defined to be the ratio $81:80$. It measures the gap (as a ratio) between the two different whole steps in the just intonation tuning system. It is also the gap between the three contrasting ratios (major third, sixth, and seventh) between the Pythagorean scale and just intonation.*

The last item in the definition for the syntonic comma is readily checked by comparing the different ratios in Tables 4.1 and 4.2. For example, if we consider the major third, then the gap between the two systems is $81/64 \div 5/4 = 81/80$, the syntonic comma. The same gap arises when considering the different values for a major sixth, major seventh, and minor third (see Exercise 6).

If we examine the two half steps in the major scale using just intonation, we see from Table 4.2 that they are each equivalent to $H = 16/15$. While it is good that there is only one value for raising the pitch a half step, we still have the problem that two half steps do not equal a whole step (either one), since

$$H^2 = \frac{256}{225} = 1.13\bar{7} > \frac{9}{8} = 1.125 > \frac{10}{9} = 1.11\bar{1}.$$

The fact that we have *three* different ways to raise the pitch a whole step in just intonation is obviously a major drawback with this tuning system.

Another problem, which arises from changing the sixth scale degree (but not the second), is that the perfect fifth between the second and sixth scale degrees (from Re to La) is no longer in the "perfect" or just ratio $3:2$. The perfect fifth between these two notes is $5/3 \div 9/8 = 40/27$, a syntonic comma below the true ratio $3:2$. This presents more problems for just intonation, particularly if we change keys by shifting a whole step above the tonic. In the new key, the major chord on the tonic will no longer have the pleasing $4:5:6$ relationship.

Finally, there is the issue of *melodic drift*. Without too much effort, it is possible to construct short lines of music that, when played using the interval ratios of just intonation, lead to the same note (the tonic) having two different frequencies. Here is a simple example. Consider the sequence of notes C F D G C, all lying within a perfect fifth of the tonic C (see Figure 4.3). In terms of scale degrees, this is simply 1, 4, 2, 5, 1, or I–IV–ii–V–I as a chord progression. This particular progression is quite common. For example, it is the main progression in the chorus of Sister Hazel's *All For You*.

FIGURE 4.3. A simple melody that drifts downward under just intonation.

Let us compare the frequency of the opening C, call it f, to the closing C. We go up by a perfect fourth (multiply by $4/3$), down by a minor third (divide by $6/5$), up by a perfect fourth again, and then finally down by a perfect fifth (divide by $3/2$). The result is

$$f \cdot \frac{4}{3} \cdot \frac{5}{6} \cdot \frac{4}{3} \cdot \frac{2}{3} = \frac{80}{81}f,$$

so the closing C is a syntonic comma lower than the opening C. Our simple melody has drifted downward and out of tune in a very short amount of time. Dunne and McConnell describe a similar drift in the chorus of Madonna's *Borderline*, where they predict, assuming that the band and singer are in tune using just intonation, the entire song to drift upward by more than a half step [Dunne and McConnell, 1999].

In summary, although the I, IV, and V major chords in just intonation are perfectly aligned with the overtone series, the problems created by achieving this harmony typically outweigh the benefits, particularly with music that has key changes or more dissonant intervals. The problems with just intonation include the same issues as the Pythagorean scale, namely, that the circle of fifths doesn't close up and that two half steps do not equal a whole step. Additional concerns include the creation of two different whole steps, the altered perfect fifth between the second and sixth scale degrees, and melodic drift. Although it certainly has some harmonic strengths, just intonation seems to cause more problems than it solves.

4.2.2   Major versus minor

At this point we have acquired enough information to consider the difference between the major and minor tonalities. To understand why minor chords sometimes sound "sad," we offer a hypothesis that combines the intervals of just intonation with the pitch residue effect discussed in Section 3.4.1.

Recall that a major triad consists of a major third between the bottom two notes and a minor third between the top two notes. In just intonation, this gives the simple ratio of $4:5:6$. In a minor triad the intervals are flipped, with a minor third on the bottom and a major third on top. In both cases we assume that the chords are in root position (e.g., scale degrees 1, 3, and 5). Alternatively, a minor triad simply flattens the third of a major chord by one half step. Why does such a small adjustment have such a big impact on how we perceive the combination of pitches?

Consider the ratios of frequencies in a minor triad in just intonation. Since the bottom interval is the 6:5 minor third, we have

$$\frac{1}{1}, \frac{6}{5}, \frac{3}{2} \quad \text{or} \quad \frac{10}{10}, \frac{12}{10}, \frac{15}{10}.$$

This means that a minor triad has the ratio $10:12:15$, numbers that are substantially higher than the corresponding values in a major triad. It follows that the overtones of the notes in a minor triad overlap less often than their counterparts in a major triad (see Exercise 9 for an example). According to the Helmholtz theory of consonance and dissonance (discussed previously in Sections 3.4.3 and 3.4.5), the minor sound will be less consonant than its major counterpart.

However, less alignment of overtones does not necessarily explain the associated sadness of the minor chord, nor the joy in a major chord. One thing to observe is that there is a definitive residue pitch when the frequencies $4f, 5f$, and $6f$ are played together, namely, f. Recall the experiment described by Heller (see Section 3.4.1), where the partials 200, 250, 300, 350 Hz, and so forth, produced a clear residue pitch of 50 Hz, a full two octaves below the first partial of 200 Hz. It stands to reason that a major chord exhibits the same type of phenomenon, as the residue pitch two octaves below the tonic of the chord helps reinforce the connection between the notes of the triad. In other words, in addition to the overlap between the different overtone series of each note in a major triad, there is also the understanding by our brain that we are hearing the 4th, 5th, and 6th partials of a note precisely two octaves below the tonic. These two effects together help explain why the just major chord is so pleasing to our ears.

Now consider the minor chord and its $10:12:15$ ratio. First off, the residue pitch is much too low to be of an impact, let alone be discerned. For frequencies of 500, 600, and 750 Hz (all above A440), the residue pitch is 50 Hz, way down at the bottom of the piano. Not only is this unlikely to be heard, but it does not even correspond to the same note as the root of the chord. This pitch is actually three octaves and a major third below the root.

The pitch that may contribute to our perception of a minor chord is the first overtone common to all three notes. Notice that

$$6 \cdot 10 = 5 \cdot 12 = 4 \cdot 15 = 60,$$

which means the 6th partial of the root, the 5th partial of the third, and the 4th partial of the fifth of the chord all agree. For instance, if the minor chord corresponds to the frequencies 300, 360, and 450 Hz, then the first common overtone is 1800 Hz. This note is two octaves above the highest note in the triad, far closer to the chord than the residue pitch. For example, in the C minor chord C E♭ G, it is the note G″. Benson [2007] reports that we tend to hear this partial an octave lower than it should sound. Thus, this higher common partial has the greatest effect on the feel of the chord. Instead of providing stability as a lower note whose overtones match the notes in the chord (as is the case with the residue pitch in a major triad), this high partial matches the fifth of the chord, not the tonic, thereby contributing to its instability and solemn nature.

Exercises for Section 4.2

1. What famous scientist tried to solve the musical difficulties of just intonation by using symmetry and linking the seven notes of the major scale to the seven colors of the spectrum?

2. Recall that in the just intonation tuning system, the notes of the major chords starting on the first and fourth scale degrees are in the simple ratio $4:5:6$. Show that this same ratio defines the notes for the major chord beginning on the fifth scale degree (the dominant) as well.

3. In contrast to the previous exercise, show that the notes for the three major chords I, IV, and V are in the ratio $64:81:96$ in the Pythagorean scale. Although this looks "worse" than the ratio for just intonation, note that amending the 81 to 80 in the ratio $64:81:96$ reduces it to the simple ratio $4:5:6$.

4. Show that the whole steps between the fourth and fifth scale degrees, and the sixth and seventh scale degrees, using just intonation, are each equal to $9/8$, while the ratio for the whole step between the fifth and sixth scale degrees is $10/9$.

5. Find the frequencies for an E major scale (one octave) in just intonation, assuming that the starting note is 330 Hz (the E just above middle C). Round your answers to one decimal place.

6. For the musical intervals of a major sixth, major seventh, and minor third, check that the gap between the Pythagorean scale and just intonation is equal to the syntonic comma.

7. Find the ratio, with the smallest possible values in the numerator and denominator, used to raise the pitch a minor sixth in just intonation.

8. Recall that a minor chord in just intonation has the ratio $10:12:15$. Suppose that the root of a minor triad has a frequency of 250 Hz. What are the frequencies of the other two notes? What is the frequency of the first partial common to the overtone series of all three notes?

9. In this problem we will compare the overtone series for the notes in a major and minor chord (in just intonation) with the same root. Let the root of the chord have a frequency of 200 Hz, and assume that both chords are in root position (scale degrees 1, 3, and 5).

   a. What are the three frequencies of the major triad with a root of 200 Hz?

   b. What are the three frequencies of the minor triad with a root of 200 Hz?

    c. List the first 12 frequencies (including the fundamental) in the overtone series for each note in the major triad.

    d. List the first 12 frequencies (including the fundamental) in the overtone series for each note in the minor triad.

    e. Focusing on the middle note of each chord (the third), circle all the overtones for that note which also appear as overtones for either the root or the fifth. Do this for both the major and minor chords.

    f. Which chord has a greater number of overtones from its third which match an overtone from either the root or the fifth?

4.3 Equal Temperament

To address the deficiencies in Pythagorean tuning and just intonation, we focus on choosing a half step H that will satisfy several properties at once. Ideally, we want two half steps to equal a whole step, seven half steps to equal a perfect fifth $3:2$, and 12 half steps to make an octave $2:1$. In order to easily change keys and create more chromatic harmonies and melodies, we should also ask for a uniform half step throughout the chromatic scale. Mathematically speaking, we want a "magic" ratio H satisfying the following three properties:

$$H^2 = W \qquad \text{(two half steps equal one whole step)}$$

$$H^7 = \frac{3}{2} \qquad \text{(seven half steps equal a perfect fifth)}$$

$$H^{12} = 2 \qquad \text{(12 half steps in an octave).}$$

Unfortunately, there is no value of H that satisfies all three properties!

4.3.1 A conundrum and a compromise

The Flemish mathematician and music theorist Simon Stevin (1548–1620) ignored the just $3:2$ perfect fifth and focused solely on the last equation, $H^{12} = 2$. Apparently, Stevin did not subscribe to the belief that a $3:2$ ratio was the most desirable for a perfect fifth. Taking the 12th root of both sides, we have

$$H^{12} = 2 \quad \Longrightarrow \quad H = \sqrt[12]{2} = 2^{1/12}.$$

Stevin's idea was to let the half step equal the 12th root of 2, the number that when multiplied by itself 12 times equals 2. He then defined *all* of the other musical intervals using this same half step. In other words, a whole step (major second) was defined to have the "ratio" equal to two half steps,

$$W = H^2 = \left(2^{1/12}\right)^2 = 2^{1/6} = \sqrt[6]{2},$$

while a perfect fifth, with seven half steps, became

$$P5 = H^7 = \left(2^{1/12}\right)^7 = 2^{7/12} = \sqrt[12]{128} \approx 1.4983.$$

Although the perfect fifth is no longer a $3:2$ ratio, it is not very far off. In fact, the gap between the just perfect fifth $3/2$ and the number $2^{7/12}$ is

$$\frac{3}{2} \div 2^{7/12} \approx 1.00113.$$

This is substantially closer to 1, by a factor of 10, than either the Pythagorean comma (roughly 1.0136) or the syntonic comma (1.0125). The closeness of this new perfect fifth to the just $3:2$ ratio is one of the reasons Stevin's system was eventually adopted. Since the octave has been equally divided into 12 parts, the resulting tuning system is known as *equal temperament*. The word "tempered" is used whenever adjustments are made to address some of the deficiencies of the Pythagorean or just scales.

It is worth noting that the first scholar to advocate equal temperament was likely the Chinese theorist Zhu Zaiyu (1536–1611). Zhu discussed equal temperament in his treatise *Lüxue Xinshuo* (New Theory of Musical Pitches), which included calculations of different pitches using equal temperament. His treatise was completed by 1584, while Stevin's work *Vande Spiegheling der Singconst* (On the Theory of Music; unpublished) was written between 1605 and 1608. In practice, equal temperament took a few centuries to become firmly established as the tuning method of choice. For instance, none of the British organs were tuned using equal temperament for the Great Exhibition of 1851, held in Hyde Park, London.

The multipliers used to raise the pitch for different musical intervals in equal temperament are shown in Table 4.3. Here we have intentionally left the exponents unreduced to stress the simplicity of this system. The numerator in each exponent corresponds to the number of half steps in each interval. If an interval consists of n half steps, then the multiplier for raising the frequency is simply $H^n = 2^{n/12}$. This rule holds for both major and minor intervals. For example, since a perfect fourth contains five half steps, the multiplier is $H^5 = 2^{5/12}$. Similarly, the multiplier for a minor sixth, which consists of eight half steps, is $H^8 = 2^{8/12} = \sqrt[3]{4}$.

It is important to note that equal temperament has solved two of the major problems that plagued the previous tuning systems. First, two half steps now equal one whole step (by construction). Second, the circle of fifths closes up, as 12 perfect fifths are now equivalent to seven octaves:

$$\text{twelve P5s} \;=\; \left(2^{7/12}\right)^{12} \;=\; 2^7 \;=\; \text{seven octaves}.$$

Furthermore, since we have a uniform half step throughout the entire octave, there will be no discrepancies when we change keys; all intervals across all octaves will be equivalent. For example, all minor thirds will consist of three half steps, $H^3 = 2^{3/12} = \sqrt[4]{2}$, while all major thirds will have four half steps, $H^4 = 2^{4/12} = \sqrt[3]{2}$. The consistency of equal temperament makes it a very practical system, particularly for music with multiple key changes, chromatic melodies, or dissonant harmony. It is desirable for more modern and atonal music, such as 12-tone, where each note of the chromatic scale is just as relevant as any other.

On the other hand, equal temperament has discarded the particularly pleasing ratios of integers. The only ratio that has been maintained is that of the octave, $2:1$. Gone are the "nice" simple ratios of $3:2$ and $4:3$ for the perfect fifth and fourth, respectively, not to mention the pleasing $4:5:6$ ratio of the major chords in just intonation. This

Scale Degree	Interval	Multiplier
1	Uni.	1
2	M2	$2^{2/12}$
3	M3	$2^{4/12}$
4	P4	$2^{5/12}$
5	P5	$2^{7/12}$
6	M6	$2^{9/12}$
7	M7	$2^{11/12}$
8 = 1	Oct.	2

TABLE 4.3. The multipliers used to raise a note (increase the frequency) by a given musical interval in equal temperament. For example, to find the note that is a major third above 200 Hz using equal temperament, we multiply 200 by $2^{4/12}$, giving a frequency that is approximately 251.98 Hz. The numerators of the exponents in the third column are easily found by calculating the number of half steps in each interval, a rule that holds for both major and minor intervals.

is the compromise of equal temperament. Any connection to the overtone series and the manner in which our ear-brain system appreciates harmony has been abandoned in favor of the practical and universal half step. While it certainly resolves some of the key issues of Pythagorean tuning and just intonation, equal temperament fails to preserve the harmonic and physical strengths of those systems.

4.3.2 Rational and irrational numbers

To fully appreciate the compromise of equal temperament, it is crucial to understand the difference between rational and irrational numbers. Both Pythagorean tuning and just intonation use only *rational* numbers to construct their musical scales. In contrast, equal temperament is based on the *irrational* number $2^{1/12}$, and the only interval with integers in the numerator and denominator is the octave, $2:1$. The irony here is that to create a practical working system, one that had logical rules such as two half steps equaling a whole step and a closed circle of fifths, it was necessary to utilize *irrational* numbers in order to solve the inherent problems created by the *rational*, more harmonious intervals of the previous tuning systems. From a mathematical perspective, this compromise is required because there is only one solution to $H^{12} = 2$, and it is irrational.

Definition 4.3.1. *A number is* <u>rational</u> *if it can be expressed as the ratio of two integers. A number is* <u>irrational</u> *if it is not rational.*

For example, the numbers

$$2 = \frac{2}{1}, \quad 0, \quad -\frac{4}{3}, \quad \frac{22}{7}, \quad \frac{3^{12}}{2^{19}}, \quad \frac{81}{80}$$

are all rational, while

$$\pi, \quad \sqrt{2}, \quad \sqrt{3}, \quad \sqrt[3]{2}, \quad \sin(1), \quad \log_2(3), \quad 2^{1/12}$$

are all examples of irrational numbers. It is typical to express a rational number as p/q, where p and q are assumed to be integers. Rational numbers are the only numbers with a finite or repeating decimal expansion. For example,

$$2 = 2.0, \quad \frac{7}{4} = 1.75, \quad \frac{1}{3} = 0.3\overline{3}3.$$

The decimal expansion of an irrational number never ends and never repeats. Perhaps the most famous example is the ratio between the circumference of a circle and its diameter, namely,

$$\pi = 3.141592653589793238462643383328\ldots,$$

which lists the first 30 digits of π.

For a long time in human history it was thought that the most important numbers and ratios, particularly those found in geometry and the natural world, were rational. In the Bible, π is inferred to be precisely 3. The infamous Indiana Pi Bill of 1897 contains a statement equivalent to $\pi = 3.2$. The bill actually passed through the Indiana General Assembly, but thankfully was later discarded by the Indiana Senate. In 1761, Johann Heinrich Lambert used properties of the tangent function to prove conclusively that π is irrational.[3]

The Pythagoreans were firm believers in the supremacy of the rational numbers. They held strongly to the belief that all important quantities arising in nature (e.g., music) were the result of whole-numbered ratios. Thus, the discovery of the extreme consonance between notes in a frequency ratio of $2\!:\!1$ or $3\!:\!2$ was celebrated mightily by the Pythagoreans. That theory was correctly challenged one day by a Pythagorean philosopher named Hippasus. As legend goes, Hippasus discovered that a simple right triangle with two congruent legs equal to 1 has a hypotenuse that must be irrational. His proof may have been too much for the Pythagoreans to bear, as they supposedly dumped Hippasus overboard to drown after he announced his discovery. While the story of Hippasus is enticing as a tragic mathematical drama, there is little ancient evidence to corroborate it.

A simple application of the Pythagorean theorem shows that the length of the hypotenuse of the right triangle with two congruent legs equal to 1 is $\sqrt{2}$. This length is indeed irrational; it cannot be expressed as the ratio of two integers. Here we have a clearly constructible length (it's the diagonal of the unit square) that is irrational. The proof we give here uses an important technique known as *proof by contradiction*. It is one of the most useful methods in a mathematician's bag of tricks. We will first assume that $\sqrt{2}$ is rational, and then we deduce a clear contradiction relying on established principles.

---

[3]More fun facts about π can be found in Blatner, *Joy of π*.

If our principles are correct, then the assumption of the initial statement must have been false, and hence $\sqrt{2}$ is irrational.

The proof we give below is not the traditional one taught to most undergraduates. It relies on the Fundamental Theorem of Arithmetic and has the advantage of being easily generalizable to other types of irrational numbers such as $\sqrt{3}$ and $\sqrt[12]{2}$. We begin by recalling that a *prime number* is a positive integer larger than 1 whose only factors are 1 and itself. The numbers $2, 3, 17$, and 101 are prime numbers, while $9 = 3 \cdot 3$, $44 = 4 \cdot 11$, and $144 = 16 \cdot 9$ are not. Prime numbers are the basic building blocks of the integers. We can always factor a natural number into a product of primes. Moreover, there is only one way (assuming that the primes are written from smallest to largest) to write this factorization.

Theorem 4.3.2 (The Fundamental Theorem of Arithmetic). *Every natural number* $1, 2, 3, \ldots$ *has a unique factorization in terms of primes.*

By unique, we mean that the types of primes and the number of each prime occurring in the prime factorization of a natural number are special to that number. No other number can have a factorization with the exact same set of primes along with the same number of those primes. For example, the prime factorization of 600 is

$$600 = 2^3 \cdot 3 \cdot 5^2.$$

No other number can have a prime factorization with just three 2s, one 3, and two 5s. This is unique to the number 600.

Next, notice what happens to the prime factorization of a number when it is squared. Compare the factorizations of 12 and $12^2 = 144$:

$$12 = 2^2 \cdot 3 \quad \text{versus} \quad 144 = 2^4 \cdot 3^2.$$

While 12 has two 2s and one 3 in its prime factorization, 144 has four 2s and two 3s. This is precisely *twice* the amount of the primes in the factorization of 12. This will always occur when we square a natural number. The number of primes for a perfect square is always double the number of those particular primes found in the original factorization. This fact can easily be verified using the laws of exponents. Suppose that

$$n = p_1^{k_1} \cdot p_2^{k_2} \cdots p_j^{k_j}$$

is the prime factorization for n, where p_i represents a prime factor and k_i indicates how many times that factor occurs. Then we have

$$n^2 = \left(p_1^{k_1} \cdot p_2^{k_2} \cdots p_j^{k_j} \right)^2 = p_1^{2k_1} \cdot p_2^{2k_2} \cdots p_j^{2k_j}.$$

This shows that n^2 and n have the same primes in their respective prime factorizations, but that the number of those primes (shown by the exponent) is always even.

Theorem 4.3.3. *The $\sqrt{2}$ is irrational.*

Proof: First, by contradiction, suppose that $\sqrt{2}$ was rational. Then we can write it as the ratio of two positive integers, $\sqrt{2} = p/q$, where p and q are natural numbers. If we square both sides of this equation, we find that $2 = p^2/q^2$ or

$$2q^2 = p^2. \tag{4.5}$$

Consider the prime factorization of the number $2q^2$. It clearly has at least one 2 in it. We don't know how many other 2s are in the factorization of q or q^2, but we do know that there will be an *even* number of 2s in the prime factorization of q^2. Since there's one more 2 outside, we know that $2q^2$ will have an *odd* number of 2s in its prime factorization. But this leads to a problem with Equation (4.5) because the prime factorization of p^2 must have an even number of 2s (or there could be no 2s). Either way, the fact that we have an odd number of 2s on the left-hand side of Equation (4.5) but an even number on the right-hand side contradicts the Fundamental Theorem of Arithmetic, since the number of 2s is supposed to be unique. This contradiction means that our original assumption about $\sqrt{2}$ is false, and thus $\sqrt{2}$ is irrational.   □

One of the best ways to understand a proof is to reproduce it in a slightly different setting. In the exercises you are asked to prove that $\sqrt{3}$ and $\sqrt[12]{2}$ are also irrational. In fact, every single "ratio" used to raise a pitch in equal temperament (minor third, perfect fourth, perfect fifth, etc.) is irrational, except for the octave ratio 2:1.

4.3.3   Cents

Given the importance of equal temperament and its construction based on $2^{1/12}$, it is natural to utilize a different unit of measurement to describe the system, one that exploits its simple structure. This unit is known as a *cent*. Just as there are 100 cents in a U.S. dollar, there are 100 cents in an equally tempered half step, and 1200 cents in an octave.

Definition 4.3.4. *A <u>cent</u> is a musical unit of measurement, based on a logarithmic scale, to compare two different pitches with frequencies f_1 and f_2. If we assume that $f_1 < f_2$ and we denote r as the ratio between the frequencies, $r = f_2/f_1$, then the number of cents between the two pitches is given by the following formula:*

$$\# \text{ of cents } = 1200 \log_2(r) = 1200 \log_2\left(\frac{f_2}{f_1}\right). \tag{4.6}$$

Cents were introduced by the mathematician Alexander Ellis (1804–90) around 1880. They are now a commonly used unit of measurement when comparing different tuning systems, or discussing nontraditional tunings. A typical listener can distinguish pitches that are between 4 and 8 cents apart. This can vary greatly depending on the frequency of the pitches, the duration of the notes, loudness, and the musical training of the listener.

Cents are particularly well suited for equal temperament because they involve the logarithm to the base 2, and frequencies in equal temperament are found by using powers of 2. Recall that 2^x and $\log_2(x)$ are inverse functions of each other. This means that

$$\log_2(2^x) = x \quad \text{and} \quad 2^{\log_2(x)} = x.$$

To compute the number of cents in one half step in equal temperament, we use $r = 2^{1/12}$ in Formula (4.6) to find

$$1200 \log_2(2^{1/12}) = 1200 \cdot \frac{1}{12} = 100 \text{ cents}.$$

The number of cents in an equally tempered perfect fifth is 700. This follows from

$$1200 \log_2(2^{7/12}) = 1200 \cdot \frac{7}{12} = 700 \text{ cents.}$$

The important feature to understand when using the cents system is that it is a *logarithmic* scale. This means that raising the pitch in terms of cents uses *addition* as opposed to multiplication. Previously, we have been comparing the ratios of frequencies to determine the musical interval. With cents, multiplication by these ratios is transformed into addition. This follows by the usual properties of logarithms, namely, that $\log(a \cdot b) = \log(a) + \log(b)$.

For example, there are 200 cents in an equally tempered whole step because two half steps equal one whole step ($200 = 100 + 100$). Mathematically speaking, the following steps demonstrate how multiplication of ratios is transformed into addition of cents:

$$
\begin{aligned}
1200 \log_2(2^{2/12}) &= 1200 \left(\log_2(2^{1/12} \cdot 2^{1/12}) \right) \\
&= 1200 \left(\log_2(2^{1/12}) + \log_2(2^{1/12}) \right) \\
&= 1200 \left(\frac{1}{12} + \frac{1}{12} \right) \\
&= 100 + 100 \\
&= 200 \text{ cents.}
\end{aligned}
$$

Since all half steps are equal in equal temperament, it is straightforward to compute the number of cents in any musical interval. Once we know the number of half steps in the given interval, we simply multiply that number by 100. Thus, there are 300 cents in a minor third because a minor third consists of three half steps, while there are $12 \cdot 100 = 1200$ cents in an octave because an octave equals 12 half steps. One of the advantages of equal temperament is that it is easily expressible in terms of cents.

Computing the number of cents for ratios that are not expressed in base 2 requires computations involving the logarithm to the base 2, which is not readily available on basic scientific calculators. However, we can use the formula $\log_2(r) = \ln(r)/\ln(2)$, since the natural logarithm $\ln(x)$ is typically found on most calculators. This formula is straightforward to derive using the standard properties of logarithms. We have

$$
\begin{aligned}
\log_2(r) = x &\implies 2^x = r \\
&\implies \ln(2^x) = \ln(r) \\
&\implies x \cdot \ln(2) = \ln(r) \\
&\implies x = \frac{\ln r}{\ln 2}.
\end{aligned}
$$

Thus, we have two equivalent expressions, one using $\log_2(x)$ and the other using $\ln(x)$, for converting ratios or multipliers into cents:

$$
\boxed{\text{\# of cents} = 1200 \log_2(r) = 1200 \cdot \frac{\ln r}{\ln 2}.} \tag{4.7}
$$

Example 4.3.5. *Compute the number of cents in a major third using Pythagorean tuning and just intonation.*

Solution: In the Pythagorean scale, the ratio used to raise a pitch by a major third is $r = 81/64$. To convert this ratio into cents, we apply Equation (4.7) to find that the number of cents in a Pythagorean major third is

$$1200 \cdot \frac{\ln(81/64)}{\ln 2} \approx 407.8 \text{ cents.}$$

In just intonation, the ratio for a major third is 5/4, so we have

$$1200 \cdot \frac{\ln(5/4)}{\ln 2} \approx 386.3 \text{ cents}$$

in a just major third. Note the discrepancy between these values and the equally tempered major third of 400 cents. $\square$

To gauge how a cent compares with the commas of the two previous tuning systems, we remark that the Pythagorean comma is approximately 23.5 cents while the syntonic comma is roughly 21.5 cents.

Exercises for Section 4.3

1. Using equal temperament, what factor do you multiply a frequency by to raise the note up a tritone? What factor raises the note by a minor seventh? What factor raises the note by a major 10th? Be sure to simplify your answers.

2. Recall that the *geometric mean* of two numbers a and b is defined as $\sqrt{ab}$. In Section 4.1.2, we demonstrated that using Pythagorean tuning, the note F♯, above a C with frequency f, has a frequency of $(3^6/2^9)f$, while its enharmonic equivalent G♭ has a different frequency of $(2^{10}/3^6)f$. Show that the geometric mean of these two frequencies is equivalent to the frequency obtained for F♯ = G♭ (they are now equal) using equal temperament.

3. How many cents are in an equally tempered tritone? How many cents are in an equally tempered major 10th? What musical interval consists of 800 cents?

4. How many cents are in a just perfect fifth? In terms of cents, how close are the equally tempered and just perfect fifth?

5. Amend the proof of Theorem 4.3.3 to show that $\sqrt{3}$ is irrational. *Hint:* Focus on the number of 3s in the prime factorization, not 2s.

6. Amend the proof of Theorem 4.3.3 to show that $2^{1/12}$ is irrational. *Hint:* $q^{12} = (q^6)^2$.

7. Prove that the sum of any two rational numbers is rational. Mathematically, this means that the set of rational numbers is *closed* under addition. *Hint:* How do we add fractions? Show that the result of adding p_1/q_1 and p_2/q_2 is always a rational number.

8. Prove that the sum of a rational and an irrational number is always irrational. *Hint:* Use proof by contradiction along with the result of the previous problem.

9. Are the irrational numbers closed under addition? In other words, is it true that the sum of two irrationals is *always* irrational? Explain with a proof or provide a counterexample.

4.4 Comparing the Three Systems

We have discussed three different tuning systems: Pythagorean, just intonation, and equal temperament. To recap, the Pythagorean scale was constructed so that all octaves and perfect fifths have the frequency ratios of $2:1$ and $3:2$, respectively. Just intonation kept some of the ratios of the Pythagorean system but altered others to better align with the overtone series. Both systems use only rational numbers. Equal temperament abandons rational frequency ratios by requiring all half steps to be $2^{1/12}$ apart.

A comparison of the ratios or multipliers used for each musical interval in the major scale is given in Table 4.4. The decimal values shown for the first two tuning systems, which use only rational numbers, are exact. On the other hand, those indicated for equal temperament, which uses only irrational multipliers (except for unison and the octave), are approximations rounded to four decimal places. We can use the values in Table 4.4 to determine the frequencies of different notes.

Scale Degree	Interval	Pythagorean	Just Intonation	Equal Temp.
1	Uni.	1	1	1
2	M2	$\frac{9}{8} = 1.125$	$\frac{9}{8} = 1.125$	$2^{2/12} \approx 1.1225$
3	M3	$\frac{81}{64} = 1.265625$	$\frac{5}{4} = 1.25$	$2^{4/12} \approx 1.2599$
4	P4	$\frac{4}{3} = 1.3\bar{3}$	$\frac{4}{3} = 1.3\bar{3}$	$2^{5/12} \approx 1.3348$
5	P5	$\frac{3}{2} = 1.5$	$\frac{3}{2} = 1.5$	$2^{7/12} \approx 1.4983$
6	M6	$\frac{27}{16} = 1.6875$	$\frac{5}{3} = 1.6\bar{6}$	$2^{9/12} \approx 1.6818$
7	M7	$\frac{243}{128} = 1.8984375$	$\frac{15}{8} = 1.875$	$2^{11/12} \approx 1.8877$
8 = 1	Oct.	2	2	2

TABLE 4.4. The ratios or multipliers used to raise a note (increase the frequency) by a major interval in the three different tuning systems: Pythagorean scale, just intonation, and equal temperament.

Example 4.4.1. *Find the frequency of the note C♯ above A440 Hz in each of the three different tuning systems.*

Solution: Since C♯ is a major third above A440, we use the multipliers for scale degree 3 listed in Table 4.4. For the Pythagorean scale, we multiply 440 by 1.265625 to find that C♯ is 556.875 Hz. Using just intonation, we should tune C♯ to $440 \cdot 1.25 = 550$ Hz. Finally, in equal temperament, C♯ is given by $440 \cdot 2^{4/12} \approx 554.365$ Hz.   □

Example 4.4.2. *Find the frequency of the note F just below A440 Hz in each of the three different tuning systems.*

Solution: In this case, since the note F is a major third *below* A440, we must *divide* 440 by the multipliers for scale degree 3 listed in Table 4.4. For the Pythagorean scale, we calculate that the F just below A440 is $440 \div 1.1.265625 \approx 347.654$ Hz. Using just intonation, we tune the F just below A440 to $440 \div 1.25 = 352$ Hz. Finally, in equal temperament, the F just below A440 is $440 \div 2^{4/12} \approx 349.228$ Hz.

Note that if we had tried using the A an octave below A440, namely, A220 Hz, then we would need to raise the pitch by a *minor* sixth above A220 to find the F we seek. This means that we need ratios from the minor scale. As discussed in the previous sections, these ratios are not uniquely defined in Pythagorean tuning or just intonation since they depend on the type of just scale being used. On the other hand, using equal temperament is straightforward and consistent; dropping the pitch by an octave and then raising it by a minor sixth is equivalent to lowering the pitch by a major third. Mathematically speaking, this follows since $\frac{1}{2} \cdot 2^{8/12} = 1/2^{4/12}$. For the other two tuning systems, obtaining a consistent answer to this particular problem requires using the correct ratio for the minor sixth (see Exercise 7).   □

A comparison of the three tuning systems using cents as the unit of measurement is given in Table 4.5. Recall that good listeners can distinguish pitches that are 5 cents apart. The table demonstrates some of the strengths and weaknesses of equal temperament as a tuning system. On the one hand, the values for a major second, perfect fourth, perfect fifth, and octave are equal to or sufficiently close to those of just intonation. This means that an equally tempered instrument will do a good job aligning with the overtone series for these particular intervals. On the other hand, the major third and sixth are substantially sharp from their counterparts in just intonation by roughly 14 and 16 cents, respectively. This is a noticeable difference, one that many musicians and listeners can discern.

The fact that the equally tempered major third is substantially sharp when compared to a just third presents complications when playing major chords, particularly for string or wind instruments. When playing in small ensembles, such as string quartets or brass ensembles, musicians are taught to "flat the third" whenever they have the third in a major chord. This is particularly important if the notes are being held for some length, for example, at the end of a piece. The longer a major chord is held, the more time our ear/brain system hopes to find that nice $4:5:6$ resonance of the overtone series. Singers, who can easily adjust pitch, will tend to gravitate toward just intonation, although this can present problems when accompanied by equally tempered instruments (as nearly all modern musical instruments are tuned).

Scale Degree	Interval	Pythagorean	Just Intonation	Equal Temp.
1	Uni.	0	0	0
2	M2	203.9	203.9	200
3	M3	407.8	386.3	400
4	P4	498.0	498.0	500
5	P5	702.0	702.0	700
6	M6	905.9	884.4	900
7	M7	1109.8	1088.3	1100
8 = 1	Oct.	1200	1200	1200

TABLE 4.5. A comparison of the three tuning systems using cents, rounded to one decimal place. Note that equal temperament does a good job approximating a perfect fifth but is noticeably sharp of a just major third.

Ross Duffin advocates for a more informed approach to tuning our musical instruments and voices in his book *How Equal Temperament Ruined Harmony (and Why You Should Care)* [Duffin, 2007]. In the "Prelude" of his book, Duffin recounts a telling anecdote about the famous orchestral conductor Christoph von Dohnányi. While rehearsing the opening of Beethoven's 9th Symphony with the Cleveland Orchestra, Dohnányi was routinely frustrated by the orchestra's inability to move from the key of D minor to a B♭ major chord. Here were some of the best musicians in the world, unable to reconcile the tuning issues inherent with equal temperament. Likely, the persistent tonic on D that is featured for the first two minutes of the symphony could not be flattened sufficiently (and collectively) when it became the third of the B♭ major chord. Duffin's book was inspired in part by a perceived lack of understanding of the tuning issues we have been discussing in this chapter.

An excellent source for hearing some of the key differences between tuning systems is the Wolfram Demonstrations Project, which features some free, online, interactive demonstration tools.[4] Two programs worth exploring are *Pythagorean, Meantone, and Equal Temperament Musical Scales* and *Exploring Musical Tuning and Temperament*. The first program allows the user to compare a just perfect fifth (click Pythagorean for the scale) with an equally tempered fifth, where the tonic and dominant are sounded together along with their respective overtones. A clear beating phenomenon is noticeable when equal temperament is used. (It is even more pronounced for the quarter-comma meantone scale, to be discussed in Section 4.6.2.) The second program allows the listener to compare chords and scales in different tuning systems. Contrasting a major chord built on A440 in Pythagorean tuning, just intonation, and equal temperament is quite revealing. Be sure to set the duration to last at least 5 seconds so the beats are discernible.

---

[4]See http://demonstrations.wolfram.com/

Exercises for Section 4.4

1. In a brief essay, compare and contrast the three tuning systems: Pythagorean tuning, just intonation, and equal temperament. What are the strengths and weaknesses of each system? Which intervals are the same and which are different?

2. Suppose you are writing a piece of music for the classical guitar and you plan to use only one key (say, C major), only the notes in the major scale, and an abundance of major I, IV, and V chords. Which of the three tuning systems would you use to tune your guitar? Explain why.

3. Make a chart showing the frequencies (rounded to one decimal place) of all the notes in the A major scale starting on A220 Hz. Do this for each of the three different tuning systems: Pythagorean tuning, just intonation, and equal temperament.

4. Assuming that A440 Hz is the note A above middle C, find the following frequencies in Hz (round to the nearest 10th) using the given tuning system.

    a. Middle C using just intonation

    b. Middle C using equal temperament

    c. G above middle C using Pythagorean tuning

    d. G above middle C using equal temperament

    e. A tritone above A440 using equal temperament

5. Find the frequency for the note a tritone above A440 using Pythagorean tuning. There are several valid answers to this question. Compare with your answer to Exercise 4e.

6. Assuming that A440 Hz is the note A above middle C, find the frequency of the lowest and highest notes on the piano in equal temperament.

7. Consider Example 4.4.2 in the text for Pythagorean tuning or just intonation. Show that for each tuning system, if the ratio with the least possible numerator and denominator is used for a minor sixth, then the value of the frequency for F below A440 can be obtained in two different ways: (1) going down a major third, or (2) dropping an octave to A220 and then going up by a minor sixth.

4.5 Strähle's Guitar

Suppose we were to construct a guitar from scratch. Where should we place the frets on the fretboard in order to tune the guitar in equal temperament? A fret is a thin metal strip lying beneath the guitar strings indicating where a player should locate their finger in order to shorten the length of a particular string, thereby changing its pitch. On a given string, the musical interval between each fret should be equal to one half step; however, the distance between successive frets will become smaller as you move toward the center of the guitar because the ratio (not difference) of successive string lengths should remain constant (see Exercise 1 at the end of this section). Most modern classical guitars have 19 frets, but for simplicity, let us assume that our guitar has 12.

In 1743, the Swedish craftsman Daniel Strähle published an article in *Proceedings of the Swedish Academy* describing a simple and accurate method to position the frets on a guitar [Strähle, 1743]. Using similar triangles, Strähle's method is easily generalized to any stringed instrument with a fingerboard (e.g., banjo, lute, ukulele, viola da gamba). The geometer and economist Jacob Faggot studied Strähle's method in detail and concluded that it contained errors of up to 1.7% when measured against equal temperament. He appended his findings to Strähle's original paper. Since this was a large enough error to be noticed by a musician—it is slightly higher than 29 cents when converted—the method was discarded, particularly because Faggot was a founding member of the Swedish Academy and later ranked fourth in the Academy. Unfortunately, Faggot had made a tragic error, calculating one of the angles incorrectly. This was discovered much later, by the mathematician J. Murray Barbour [1957]. It turns out that Strähle's method is remarkably accurate, an accuracy that can be explained using the mathematical tool of *continued fractions*.[5]

4.5.1   An ingenious construction

To explain Strähle's method, we will focus on one particular string running down the center of the fretboard of one particular guitar. To make the numbers work out nicely, we will assume that the length of the guitar string is $2\sqrt{151} \approx 24.5764$ units. Although this seems rather peculiar, keep in mind that by using similar triangles, it is possible to apply the same method for any size string (on any fretted instrument), as long as the ratios between the relevant sides are preserved.

Take an isosceles triangle with congruent sides of length 24 and a base of length 12. Place the guitar so that the neck sits at one vertex of the triangle between a leg and the base, and position the midpoint of the string 7 units up the opposite side of the triangle (see Figure 4.4). Divide the base of the triangle into 12 equally spaced subintervals and then draw 12 lines from the top of the triangle to each equally spaced point on the base (one of these lines is simply a leg of the triangle). The intersection of these 12 lines with the fretboard gives the location of the frets. That's it!

Strähle's elementary construction provides the location of the frets with a minimal amount of measurement and no dependence on irrational numbers. Moreover, the method easily extends to any size string, as long as the proportion between the base of the isosceles triangle and its congruent legs is $1:2$, and the midpoint of the string lies $7/24$ of the way up one leg of the triangle. If these proportions are maintained, then Strähle's method will provide an excellent approximation to equal temperament for any size string, making the technique of great practical value.

To understand why Strähle's method is such a good approximation, we need to calculate the location of Strähle's frets precisely and then compare them with equal temperament. To do this, we will place coordinates on the fretboard of the guitar. Since the base of the triangle has been subdivided equally, let us consider the base the x-axis, with point B in Figure 4.4 corresponding to $x = 0$ and point C corresponding to $x = 1$. Since the base has been equally subdivided, each of the remaining tick marks on the x-axis will be of the form $x = n/12$ for some value of $n \in \{1, 2, \ldots, 11\}$.

Next, treat the fretboard as the y-axis and rescale the guitar string to have a length of 1. Specifically, let the point B correspond to $y = 1$, M (the midpoint of the string) to $y = 1/2$, and G (where the string is fastened to the body of the guitar) to $y = 0$. These values make it easier to compare with the tuning multipliers used in equal temperament. For example,

---

[5]Much of the material for this section is based on Stewart, Faggot's Fretful Fiasco.

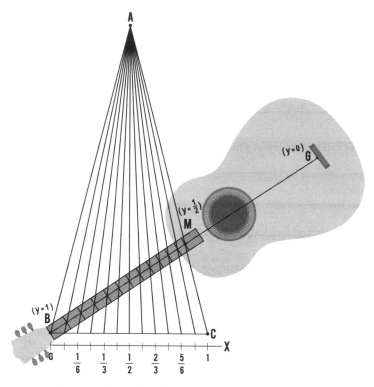

FIGURE 4.4. Strähle's construction to place the frets on a guitar. The outer triangle is isosceles with congruent legs AB and AC of length 24 and base BC of length 12. M is the midpoint of the guitar string, and CM has a length of 7.

placing your finger at M gives a string with half the length of the original. When plucked, the new note will be an octave higher than the fundamental of the full string. Thus, plucking at $y = 1/2$ corresponds to the musical interval of an octave. Additionally, $y = 1$ (the point B) corresponds to the original string and thus represents a unison interval.

Note that it is $y = 1/2$ that corresponds to an octave, rather than our usual 2/1 ratio. This occurs because we are comparing ratios of string *lengths* rather than frequencies, and length is inversely related to frequency. (Recall that shortening the length of a vibrating string raises the pitch, that is, it raises the frequency.) We will need to compare the y-values of Strähle's method with the *reciprocal* of the equally tempered values in Table 4.3.

Ideally, we would like to find a function $y = f(x)$ that maps the values along the equally subdivided x-axis to the newly formed y-axis corresponding to Strähle's construction. In this way the value of y will be the ratio between the lengths of the original string and the plucked string when placing a finger down on the fret at point y. Moreover, since we have subdivided the base of the triangle into 12 equal parts, with $(x = 0, y = 1)$ corresponding to the fundamental note and $(x = 1, y = 1/2)$ corresponding to an octave higher, the value of y for each x-value of the form $x = n/12$ gives Strähle's frequency ratio for moving up n half steps in pitch.

Strähle's construction is an example of a projection, where one line (the x-axis) is being projected onto another (the y-axis). It turns out that this type of projection can always be

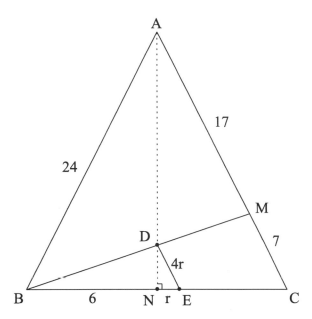

FIGURE 4.5. The location of the 6th fret can be found by using two pairs of similar ($\sim$) triangles: DNE $\sim$ ANC (with common vertex N), and DBE $\sim$ MBC (with common vertex B).

represented by a *linear fractional transformation*, which is a function of the form

$$y = \frac{a + bx}{c + dx}, \quad \text{for some constants } a, b, c, \text{ and } d. \tag{4.8}$$

We want to calculate the unknown constants a, b, c, d that correspond to our specific projection. We already know two points that satisfy this function, $(x = 0, y = 1)$ and $(x = 1, y = 1/2)$. We can find a third point by using similar triangles.

Let D be the point corresponding to the 6th fret on our special guitar (the middle fret), and let E be the point on the base so that segment DE is parallel to the leg AC. Denote the center of the base of the isosceles triangle as N. Then, the small right triangle DNE is similar to the right triangle ANC, with common vertex N. If r denotes the length of segment NE, then using similar triangles, the hypotenuse of triangle DNE has length $4r$ (see Figure 4.5).

Next, consider the similar triangles DBE and MBC with common vertex B. By aligning similar sides, we have

$$\frac{4r}{6 + r} = \frac{7}{12} \quad \Longrightarrow \quad r = \frac{42}{41},$$

as well as

$$\frac{BD}{BM} = \frac{4r}{7} = \frac{24}{41}.$$

In our setup, the length of BM is $1/2$, which means that BD has length $12/41$. This implies that GD has a length of $1 - 12/41 = 29/41$. Thus, we have shown that the 6th fret is located at $y = 29/41$. We have found an additional point on our function, $(x = 1/2, y = 29/41)$.

We now calculate the unknown constants a, b, c, and d in Equation (4.8). Substituting the point $(x = 0, y = 1)$ into Equation (4.8) yields $1 = a/c$ or $c = a$. Plugging in the point $(x = 1, y = 1/2)$ yields

$$\frac{1}{2} = \frac{a+b}{c+d} = \frac{a+b}{a+d} \implies d = a + 2b, \tag{4.9}$$

while plugging in the point $(x = 1/2, y = 29/41)$ gives

$$\frac{29}{41} = \frac{a+b/2}{c+d/2} = \frac{a+b/2}{a+d/2} \implies d = \frac{24}{29}a + \frac{41}{29}b. \tag{4.10}$$

Equating the two expressions for d in Equations (4.9) and (4.10), we obtain $b = (-5/17)a$, which, in turn, implies that $d = (7/17)a$. At this point, each of the unknowns b, c, and d has been determined as some constant times a. Changing the value of a is equivalent to multiplying the numerator and denominator of Equation (4.8) by an arbitrary constant. In other words, a is arbitrary. To make our linear fractional transformation look nice, we choose $a = 17$. It follows that $c = 17, b = -5$, and $d = 7$. We have calculated that Strähle's placement of the frets on the guitar corresponds to the linear fractional transformation

$$\boxed{y = \frac{17 - 5x}{17 + 7x}.} \tag{4.11}$$

As an example, consider the interval of a perfect fifth, consisting of seven half steps. To determine the frequency ratio in Strähle's construction, we plug $x = 7/12$ into Equation (4.11). This yields $169/253 \approx 0.66798$, which is quite close to the just perfect fifth of $2/3$.

Table 4.6 compares the ratios for Strähle's method with those of equal temperament. The fractions in the third column are found by substituting $x = n/12$ into Equation (4.11) for $n \in \{1, 2, \ldots, 11\}$. Since we are focusing on the ratios of the lengths of a plucked string rather than frequencies, the values are all less than or equal to 1. The last column of the table shows the relative error of Strähle's method when measured against equal temperament. For example, to compute the relative error in a perfect fifth, we subtract $2^{-7/12}$ (the multiplier for an equally tempered P5) from $169/253$ and then divide by $2^{-7/12}$. This gives an error of

$$\frac{169/253 - 2^{-7/12}}{2^{-7/12}} \approx 0.0008454 \approx 0.085\%.$$

Although Strähle's method consists of numerators and denominators that are not particularly memorable, nor small, the maximum absolute error when compared with equal temperament is about 0.152%, a much better result than the 1.7% incorrectly computed by Faggot. As discussed by Barbour [1957], note the sine-like pattern to the error in the last column of Table 4.6. The error begins at 0, dips down to a minimum of -0.111%, returns near to zero halfway along (tritone), rises to a maximum of 0.152%, and then ends at 0.

Table 4.7 shows the frequency ratios in terms of cents, with just intonation included for comparison. Recall that the number of cents in a musical interval determined by the ratio r is given by $1200 \log_2(r)$ or, equivalently, $1200 \ln r / \ln 2$. The reciprocals of the values for Strähle's method were used since we are comparing ratios r satisfying $r > 1$. Remarkably, the discrepancy between Strähle's method and equal temperament is never more than 3 cents, virtually an imperceptible difference even to a trained musician.

Note	Interval	Strähle's Method	Equal Temp.	% Error
C	Unison	1	1	0
D♭	Minor second	$199/211 \approx 0.9431$	$2^{-1/12} \approx 0.9439$	-0.079
D	Major second	$97/109 \approx 0.8899$	$2^{-1/6} \approx 0.8909$	-0.111
E♭	Minor third	$21/25 = 0.84$	$2^{-1/4} \approx 0.8409$	-0.107
E	Major third	$23/29 \approx 0.7931$	$2^{-1/3} \approx 0.7937$	-0.075
F	Perfect fourth	$179/239 \approx 0.7490$	$2^{-5/12} \approx 0.7492$	-0.027
G♭	Tritone	$29/41 \approx 0.7073$	$2^{-1/2} \approx 0.7071$	0.030
G	Perfect fifth	$169/253 \approx 0.6680$	$2^{-7/12} \approx 0.6674$	0.085
A♭	Minor sixth	$41/65 \approx 0.6308$	$2^{-2/3} \approx 0.6300$	0.128
A	Major sixth	$53/89 \approx 0.5955$	$2^{-3/4} \approx 0.5946$	0.152
B♭	Minor seventh	$77/137 \approx 0.5620$	$2^{-5/6} \approx 0.5612$	0.145
B	Major seventh	$149/281 \approx 0.5302$	$2^{-11/12} \approx 0.5297$	0.098
C	Octave	$1/2$	$1/2$	0

TABLE 4.6. Comparing the ratios of Strähle's method with those used for equal temperament. Since we are dealing with string lengths rather than frequencies, the ratios and multipliers are all less than 1. Note that the largest absolute error is approximately 0.152%, far less troublesome than the 1.7% error incorrectly calculated by Faggot.

4.5.2 Continued fractions

Tables 4.6 and 4.7 demonstrate the effectiveness of Strähle's construction for placing the frets on a string instrument. In this section we dig a little deeper, exploring the mathematical topic of *continued fractions* to understand why Strähle's method is so accurate.[6]

The famous mathematical constant π is irrational, with a nonrepeating, nonterminating decimal expansion. To 40 decimal places, π is

$$\pi = 3.1415926535897932384626433832795028841972\ldots.$$

Since π is irrational, it cannot be written as the ratio of two integers. However, we could try to *approximate* π using rational numbers. How do we find good approximations to irrational numbers, particularly with small numerators and denominators?

We begin by writing π as

$$\pi = 3 + 0.1415926535897\ldots.$$

One approximation for π is thus 3, but that is not very impressive. We could continue by approximating $0.14\ldots$ as $14/100 = 7/50$ in order to obtain the first two decimal places

---

[6]The material for this section is based on the wonderful exposition in Chapters 39 and 40 of Silverman, *Friendly Introduction to Number Theory*. Proofs of the theorems in this section can be found in that text.

Note	Interval	Strähle's Method	Equal Temp.	Just Int.
C	Unison	0	0	0
D♭	Minor second	101.4	100	111.7
D	Major second	201.9	200	203.9
E♭	Minor third	301.8	300	315.6
E	Major third	401.3	400	386.3
F	Perfect fourth	500.5	500	498.0
G♭	Tritone	599.5	600	590.2
G	Perfect fifth	698.5	700	702.0
A♭	Minor sixth	797.8	800	813.7
A	Major sixth	897.4	900	884.4
B♭	Minor seventh	997.5	1000	996.1
B	Major seventh	1098.3	1100	1088.3
C	Octave	1200	1200	1200

TABLE 4.7. Comparing the frequency ratios (multipliers) for Strähle's method, equal temperament, and just intonation in terms of cents. Note that Strähle's method is never more than 3 cents away from equal temperament.

of π. However, there is a more clever approach. Instead of approximating $0.14\ldots$ as $7/50$, consider inverting it twice, writing

$$0.1415926535897\ldots = \cfrac{1}{\cfrac{1}{0.1415926535897\ldots}} = \frac{1}{7.062513305931\ldots}.$$

Ignoring the decimals in $7.06\ldots$ gives the approximation $3 + 1/7 = 22/7$, a famous estimate of π. As with the fraction $157/50 = 3 + 7/50$, this gives π to two decimal places, but notice that $22/7$ has a denominator about 7 times *smaller* than $157/50$. This is more in line with our goal of finding low-numbered rational approximations.

We can continue this process. Inverting the new "remainder" decimal $0.062513305931\ldots$ twice gives

$$0.062513305931\ldots = \cfrac{1}{\cfrac{1}{0.062513305931\ldots}} = \frac{1}{15.99659441\ldots}.$$

This last number in the denominator is very close to 16. If we approximate at this point, we find

$$\pi \approx 3 + \cfrac{1}{7 + \cfrac{1}{16}} = 3 + \frac{16}{113} = \frac{355}{113} = 3.14159292\ldots.$$

Remarkably, the approximation $355/113$ agrees with π to six decimal places. If we tried to obtain this approximation by truncating π to six decimal places, we would have a reduced fraction with denominator 125,000. That's roughly 1100 times larger than 113.

If we continue this process of inverting the remainder decimal, after two more iterations we obtain the following multilayered fraction:

$$\pi \approx 3 + \cfrac{1}{7 + \cfrac{1}{15 + \cfrac{1}{1 + \frac{1}{293}}}} = \frac{104{,}348}{33{,}215} = 3.14159265392142\ldots, \tag{4.12}$$

which approximates π accurately to nine decimal places.

These crazy, multistory fractions are called *continued fractions* and are an important topic in the subject of number theory, although they find their way into all sorts of applications in other fields.

Definition 4.5.1. *Given a real number α, the continued fraction expansion of α is*

$$\alpha = a_0 + \cfrac{1}{a_1 + \cfrac{1}{a_2 + \cfrac{1}{a_3 + \cfrac{1}{\ddots}}}} = [a_0; a_1, a_2, a_3, \ldots],$$

where each a_i (except possibly a_0) is a positive integer.

The notation $\alpha = [a_0; a_1, a_2, a_3, \ldots]$ is much easier to use than writing the multistory fraction. The first integer a_0 is just the number to the left of the decimal point (e.g., $a_0 = 3$ for $\alpha = \pi$). Notice the semicolon after the a_0. The succeeding list of integers are the denominators of each fraction since we know that each numerator is always 1. For example, from Equation (4.12), we have that

$$\pi = [3; 7, 15, 1, 292, \ldots].$$

The reason for 292, as opposed to 293, is that the final approximation we found for π was $292.63459\ldots$. Although we rounded this to 293 to obtain the approximation $104{,}348/33{,}215$, the actual integer *before* rounding is 292.

Observe that if the number we are approximating is irrational, then this process of inverting the remainder decimal twice will continue forever; otherwise, the approximation would be exact and we would obtain a rational number.

Theorem 4.5.2. *The continued fraction expansion of a real number is finite if and only if that number is rational. In other words, the continued fraction expansion of an irrational number is infinite.*

To see how this process works on a rational number, consider the following example.

Example 4.5.3. *Compute the continued fraction expansion for the rational number $\alpha = 37/13$.*

Solution: We start by dividing 37 by 13 to obtain 2 with a remainder of 11. Thus,

$$\frac{37}{13} = 2 + \frac{11}{13} = 2 + \frac{1}{13/11} \, .$$

Notice that the final fraction in the denominator is bigger than 1. This will always be the case since we are inverting a number less than 1 (the remainder). Continuing, we divide 13 by 11 to obtain 1 with a remainder of 2. Thus, we have

$$\frac{37}{13} = 2 + \frac{11}{13} = 2 + \frac{1}{13/11} = 2 + \frac{1}{1 + \dfrac{2}{11}} = 2 + \frac{1}{1 + \dfrac{1}{11/2}} \, .$$

Next, we divide 11 by 2 to obtain 5 with a remainder of 1. This gives

$$\frac{37}{13} = 2 + \frac{11}{13} = 2 + \frac{1}{13/11} = 2 + \frac{1}{1 + \dfrac{2}{11}} = 2 + \frac{1}{1 + \dfrac{1}{11/2}} = 2 + \frac{1}{1 + \dfrac{1}{5 + \dfrac{1}{2}}} \, .$$

The process terminates at this point because we are left with the integer 2 in the denominator rather than a fraction. There is nothing left to divide when the remainder is 1. The last fraction of $11/2$ is simply $5 + 1/2$, which is already in the required form because of the 1 in the numerator. Thus, we have shown that the continued fraction expansion for the rational number $37/13$ is

$$\frac{37}{13} = [2; 1, 5, 2] \, .$$

$\square$

Note that before we reached the final step in the example, each of the previous steps did not have a 1 for a remainder. This is a useful fact.

> **Key Fact:** When writing the finite continued fraction expansion of a rational number, the process terminates precisely when a remainder of 1 occurs.

Periodic expansions

Sometimes, instead of terminating, the continued fraction expansion will repeat a certain pattern forever. For example,

$$\sqrt{8} = [2; 1, 4, 1, 4, 1, 4, 1, 4, \ldots]$$

has a repeating pattern of period 2 since it ends with the sequence $1, 4, 1, 4, \ldots$. We will refer to this kind of sequence as a *periodic sequence of period 2*. When a continued fraction expansion has a periodic sequence, we write

$$\sqrt{8} = [2; \overline{1, 4}].$$

Another example is

$$\sqrt{7} = [2; 1, 1, 1, 4, 1, 1, 1, 4, 1, 1, 1, 4, \ldots] = [2; \overline{1, 1, 1, 4}].$$

Example 4.5.4. *Find the irrational number that corresponds to the continued fraction expansion*

$$\alpha = [1; 2, 2, 2, \ldots] = [1; \overline{2}].$$

Solution: One way to approach the problem is to approximate α by terminating the expansion at different locations. These approximations are called *convergents*. In general, the

$$n\text{th convergent to } \alpha = \frac{p_n}{q_n} = [a_0; a_1, a_2, a_3, \ldots, a_n].$$

As n increases, we move farther out in the expansion of α and thereby obtain a better approximation.

For $\alpha = [1; 2, 2, 2, \ldots]$, the first five convergents are given below:

$$\frac{p_0}{q_0} = 1 \quad \text{(this is just } a_0\text{)},$$

$$\frac{p_1}{q_1} = 1 + \frac{1}{2} = \frac{3}{2} = 1.5,$$

$$\frac{p_2}{q_2} = 1 + \cfrac{1}{2 + \cfrac{1}{2}} = \frac{7}{5} = 1.4,$$

$$\frac{p_3}{q_3} = 1 + \cfrac{1}{2 + \cfrac{1}{2 + \cfrac{1}{2}}} = \frac{17}{12} = 1.41\bar{6}, \tag{4.13}$$

$$\frac{p_4}{q_4} = 1 + \cfrac{1}{2 + \cfrac{1}{2 + \cfrac{1}{2 + \cfrac{1}{2}}}} = \frac{41}{29} \approx 1.4138. \tag{4.14}$$

Savvy readers may recognize the last approximation of 1.414, which points toward the value $\alpha = \sqrt{2}$. Notice that not only are the convergents getting closer to $\sqrt{2}$, but they also oscillate about it (below, above, below, above, etc.). This will always be the case, no matter the value of α.

There is a clever trick to see that $\alpha = \sqrt{2}$. Since the continued fraction pattern is periodic, the expression for α actually *reappears* inside its own expansion. We have

$$\alpha = 1 + \cfrac{1}{2 + \cfrac{1}{2 + \cfrac{1}{2 + \cfrac{1}{\ddots}}}} = 1 + \cfrac{1}{1 + 1 + \cfrac{1}{2 + \cfrac{1}{2 + \cfrac{1}{\ddots}}}} = 1 + \cfrac{1}{1 + \alpha}.$$

This last expression may seem strange, but if we know that the infinite continued fraction converges to a real number (this requires a proof), then the continued fraction expansion in the denominator of the penultimate expression is equal to the original continued fraction expansion for α. The main concept being invoked here is known as *self-similarity*, a key ingredient in the fascinating field of *chaos theory*.

Thus, we have that

$$\alpha = 1 + \frac{1}{1+\alpha} \quad \implies \quad \alpha - 1 = \frac{1}{1+\alpha}$$
$$\implies \quad \alpha^2 - 1 = 1 \quad \text{(by cross-multiplying)}$$
$$\implies \quad \alpha^2 = 2$$
$$\implies \quad \alpha = \sqrt{2}.$$

Note that we chose the positive square root because it is clear from the continued fraction expansion that $\alpha > 0$. $\qquad\qquad\square$

This example is illustrative of an important fact about continued fractions.

Theorem 4.5.5. *The continued fraction expansion of an irrational number is periodic if and only if α is of the form $\alpha = r + s\sqrt{n}$, where r, s are rational numbers and n is a positive integer not equal to a perfect square. In this case, periodic means that the continued fraction expansion ends with a periodic, repeating sequence of positive integers.*

Convergents and their accuracy

Convergents are rational numbers obtained by truncating the continued fraction expansion at some location. Specifically, if $\alpha = [a_0; a_1, a_2, a_3, \dots]$, we define the *nth convergent* to α to be the rational number

$$\frac{p_n}{q_n} = [a_0; a_1, a_2, a_3, \dots, a_n] = a_0 + \cfrac{1}{a_1 + \cfrac{1}{a_2 + \cfrac{1}{\ddots + \cfrac{1}{a_n}}}}. \qquad (4.15)$$

As demonstrated in Example 4.5.4, the convergents are excellent approximations to the number α, and these approximations improve as n gets larger.

Instead of having to compute the value of the convergents from scratch each time, there are some convenient recursive formulas that give the next convergent in terms of the previous ones. The formulas are

$$\begin{aligned} p_0 &= a_0 \\ p_1 &= a_1 a_0 + 1 \\ p_n &= a_n p_{n-1} + p_{n-2} \quad \text{if } n \geq 2, \end{aligned}$$

and

$$\begin{aligned} q_0 &= 1 \\ q_1 &= a_1 \\ q_n &= a_n q_{n-1} + q_{n-2} \quad \text{if } n \geq 2. \end{aligned}$$

For example, if $\alpha = \sqrt{2} = [1; 2, 2, 2, \ldots]$, then $p_0 = 1, p_1 = 2 \cdot 1 + 1 = 3$, and

$$
\begin{aligned}
p_2 &= a_2\, p_1 + p_0 &= 2 \cdot 3 + 1 &= 7 \\
p_3 &= a_3\, p_2 + p_1 &= 2 \cdot 7 + 3 &= 17 \\
p_4 &= a_4\, p_3 + p_2 &= 2 \cdot 17 + 7 &= 41.
\end{aligned}
$$

A similar set of calculations can be used to find the denominators q_n. Try checking them with $\sqrt{2}$ and comparing with the convergents we obtained in Example 4.5.4.

Before closing our discussion of continued fractions, we make one final point concerning the accuracy of the convergents. It turns out that the rational approximations to an irrational number α obtained by using a continued fraction expansion are the *best* possible for a given denominator. Moreover, each convergent is closer to α than the preceding one, and the convergents oscillate about α, with the even convergents ($n = 0, 2, 4, \ldots$) lying below α and the odd convergents ($n = 1, 3, 5, \ldots$) lying above. Specifically, we have the following theorem:

Theorem 4.5.6. *If $\{p_n/q_n\}$ is the sequence of convergents to an irrational number α, then*

$$
\frac{p_n}{q_n} - \frac{p_{n-1}}{q_{n-1}} = \frac{(-1)^{n+1}}{q_{n-1}\, q_n}.
$$

Moreover, any particular convergent p_n/q_n is closer to α than any other fraction whose denominator is less than q_n.

For example, when approximating $\alpha = \sqrt{2}$, choosing the fourth convergent $41/29$ gives the best possible rational approximation to $\sqrt{2}$ with denominator less than 29. This is precisely why the convergents of a continued fraction expansion are considered to be the *best* rational approximations.

4.5.3 On the accuracy of Strähle's method

Recall that Strähle's method of placing the guitar frets corresponds to evaluating the function

$$
y = \frac{17 - 5x}{17 + 7x}
$$

at the points $x = n/12$, where $n \in \{0, 1, 2, \ldots, 11, 12\}$. As shown in Tables 4.6 and 4.7, this is an exceptionally good approximation to equal temperament, which corresponds to the function $y = 2^{-x}$ evaluated at the same x-values as above. The graphs of these two functions over $0 \leq x \leq 1$ are virtually indistinguishable (see Figure 4.6). It is clear that Strähle's construction is a very accurate approximation. Why is it so good?

For starters, consider the interval of a tritone (six half steps), for which Strähle's method gives a ratio of $29/41$ and equal temperament produces $1/2^{6/12} = 1/\sqrt{2}$. Inverting each fraction, we should find that

$$
\frac{41}{29} \approx 1.4138 \quad \text{approximates} \quad \sqrt{2} \approx 1.4142,
$$

which is clearly true.

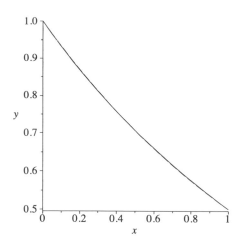

FIGURE 4.6. The graphs of $y = (17 - 5x)/(17 + 7x)$ (Strähle's method, *solid curve*) and $y = 2^{-x}$ (equal temperament, *dashed curve*) over $0 \le x \le 1$ are practically identical.

We have seen the number $41/29$ before. It is the fourth convergent p_4/q_4 in the continued fraction expansion of $\sqrt{2}$ (see Equation (4.14)). Recall that the convergents obtained from the continued fraction expansion of an irrational number are the *best* approximating rational numbers with small denominators. The fact that Strähle's method just happens to use a convergent from the continued fraction expansion for $\sqrt{2}$ is one of the main reasons the method is so effective.

For further explanation regarding the success of Strähle's method, let us examine the rational approximation $y = (17 - 5x)/(17 + 7x)$ to the irrational function $y = 2^{-x} = 1/2^x$. (The terms "rational" and "irrational" being used here refer to the values obtained when the functions are evaluated at $x = n/12$, where $n \in \{1, 2, \ldots, 11\}$.) Consider the reciprocals of each function:

$$\frac{17 + 7x}{17 - 5x} \approx 2^x. \tag{4.16}$$

Suppose we wanted to find the best linear fractional transformation for approximating $y = 2^x$. Using the three points $(x = 0, y = 1)$, $(x = 1/2, y = \sqrt{2})$, and $(x = 1, y = 2)$, we obtain the function

$$y = \frac{(2 - \sqrt{2})x + \sqrt{2}}{(1 - \sqrt{2})x + \sqrt{2}} \tag{4.17}$$

(see Exercise 7). Note that the coefficients used in this linear fractional transformation are irrational numbers. Using continued fractions, if we replace $\sqrt{2}$ in Equation (4.17) by its third convergent $17/12$ (see Equation (4.13)), we obtain

$$y = \frac{(2 - 17/12)x + 17/12}{(1 - 17/12)x + 17/12} = \frac{(7/12)x + 17/12}{(-5/12)x + 17/12} = \frac{17 + 7x}{17 - 5x},$$

which is precisely the left-hand side of the approximation in (4.16). In other words, the theory of continued fractions shows that one of the best linear fractional transformations for approximating the irrational values of equal temperament is the rational formula that corresponds to Strähle's method.

Exercises for Section 4.5

1. Let L be the length of a guitar string. Suppose that frets are to be placed beginning at the end of the neck of the guitar so that the musical interval between each fret is a half step. Let d_1 equal the distance between the end of the neck and the first fret, d_2 equal the distance between the first and second frets, d_3 equal the distance between the second and third frets, and so forth. Denote by H the ratio or multiplier used to *lower* the pitch by a half step. In other words, $H = 15/16 < 1$ for just intonation, while $H = 2^{-1/12} < 1$ in equal temperament.

   a) Show that $d_1 = L(1 - H)$.

   b) Show that $d_2 = L(1 - H) \cdot H$ and conclude that $d_2 < d_1$.

   c) Deduce that $d_j = L(1 - H) \cdot H^{j-1}$ and conclude that the frets are getting successively closer as you move toward the center of the guitar.

   d) Assuming that $H = 2^{-1/12}$, find the sum $d_1 + d_2 + d_3 + \cdots + d_{12}$. Why does this value make sense musically speaking? *Hint:* What type of series is being summed?

2. Recall the Law of Cosines from trigonometry:

$$c^2 = a^2 + b^2 - 2ab \cos \gamma,$$

   where a, b, c are the lengths of the three sides of a triangle and γ is the angle opposite the side of length c. This law can be interpreted as an extension of the Pythagorean theorem, since $\gamma = \pi/2 = 90°$ yields $a^2 + b^2 = c^2$. Use the Law of Cosines to verify that the length of the guitar string in Figure 4.4 is precisely $2\sqrt{151}$.

3. *Faggot's Error:* The mistake that Faggot made in checking Strähle's construction was his computation of $\angle MBC$ in Figure 4.4, for which he incorrectly obtained $40°14'$. This in turn led to the incorrect location of each fret because Faggot used the Law of Sines repeatedly with an incorrect angle each time. Show that the correct value of $\angle MBC$ is $33°28'$.[7]

4. Compute the continued fraction expansion for the rational number $\alpha = 53/14$. Give your answer in the form $[a_0; a_1, a_2, \ldots]$.

5. Compute the continued fraction expansion for the irrational number $\alpha = \sqrt{6}$. Give your answer in the form $[a_0; a_1, a_2, \ldots]$. *Hint:* Your answer should be periodic.

6. Consider the irrational number α with continued fraction expansion $[1; 1, 1, 1, \ldots]$.

   a) Compute the first seven convergents of α, that is, compute p_n/q_n for $n = 0, 1, \ldots, 6$. What do you notice about the numbers in the numerator and denominator?

   b) Find the exact value of α using the method of self-similarity demonstrated in Example 4.5.4. What special name is given to the number α?

7. Use the three points $(x = 0, y = 1)$, $(x = 1/2, y = \sqrt{2})$, and $(x = 1, y = 2)$ to derive the linear fractional transformation $y = (a + bx)/(c + dx)$ for approximating 2^x, given in Formula (4.17).

---

[7] Ironically, Barbour (and later Stewart) also give this angle incorrectly in their analyses, finding it to be $33°32'$ (a very small error, but an error nonetheless). Apparently, this is a particularly tricky angle to compute.

4.6 Alternative Tuning Systems

In the Western musical scale, the octave is divided into 12 equal parts, that is, 12 equal half steps. Why does this work so particularly well? What happens for scales constructed with a different number of equal subdivisions?

4.6.1 The significance of $\log_2(3/2)$

Part of the reason that an equally spaced 12-note octave succeeds is that the next most prized musical interval after the octave is the perfect fifth. Recall that the frequencies for two notes an octave apart have a ratio of 2:1, while frequencies for notes in a perfect fifth have a 3:2 ratio. The Pythagorean comma arises because 12 perfect fifths is not equivalent to 7 octaves. Mathematically speaking,

$$\left(\frac{3}{2}\right)^{12} \neq 2^7 \quad \text{or} \quad \frac{3}{2} \neq 2^{7/12}, \tag{4.18}$$

where the second inequality is obtained from the first by taking the 12th root of each side. Musically speaking, the right-hand inequality in (4.18) reminds us that the just 3:2 perfect fifth is being approximated by the irrational $2^{7/12}$ in equal temperament.

Using the logarithm to the base 2, the right-hand inequality in (4.18) can be rewritten as

$$\log_2\left(\frac{3}{2}\right) \approx 0.5849625007\ldots \neq \frac{7}{12} = 0.5833\overline{3}. \tag{4.19}$$

The fact that $7/12$ is a good approximation to $\log_2(3/2)$ is why equal temperament is so effective. The gap is only about 2 cents. In other words, dividing the 2:1 octave into 12 equal parts allows seven half steps to be very close to the 3:2 perfect fifth. Consequently, the other important musical intervals (P4, M2, M3, M6, etc.), which were first tuned by relying solely on the 3 : 2 perfect fifth and the 2 : 1 octave of Pythagorean tuning, are also well approximated by the irrational values of equal temperament. There are certainly issues (particularly the sharp major third and sixth), but the compromise of giving up the true perfect fifth does not corrupt the rest of the scale. Thus, the uniform half step of equal temperament creates a coherent and workable musical system.

What happens if we subdivide the octave into a different number of equal parts? How well would such a system work? Let n be a positive integer other than 12. If we maintain our focus on the perfect fifth, and we assume a subdivision of the octave into n different but equally spaced pitches, then we want to find an integer m so that the mth tone in our scale is a good approximation to a perfect fifth. One step of our new musical scale would require a multiplier of $2^{1/n}$, so m steps requires $(2^{1/n})^m = 2^{m/n}$ as the frequency multiplier. We seek values of the integers m and n for which our new perfect fifth $2^{m/n}$ is a reasonable approximation to the true perfect fifth of $3/2$. Thus, the generalization of Equation (4.19) is

$$\log_2\left(\frac{3}{2}\right) \approx \frac{m}{n}, \tag{4.20}$$

for some positive integers m and n, with $m < n$ required.

It should be noted that Equation (4.20) has no exact solution. This follows because $\log_2(3/2)$ is irrational (see Exercise 1). Finding good rational approximations to the important irrational number $\log_2(3/2)$ leads us back to the topic of continued fractions.

Recall that the convergents in the continued fraction expansion of an irrational number are the best rational approximations to that number. It should therefore come as little surprise to learn that $7/12$ is the third convergent in the continued fraction expansion for $\log_2(3/2)$. The sequence of convergents for $\log_2(3/2)$ is

$$1, \frac{1}{2}, \frac{3}{5}, \frac{7}{12}, \frac{24}{41}, \frac{31}{53}, \frac{179}{306}, \frac{389}{665}, \dots \tag{4.21}$$

Not only does this show that equal temperament is good at approximating the $3:2$ perfect fifth, but it also explains why other similar subdivisions of the octave (e.g., $n = 10, 11, 13, 14$) will not be as effective. The rational number $7/12$ isn't just a good approximation to $\log_2(3/2)$; it is one of the best.

For example, if we try dividing the octave into 10 equal parts, then the best value for the perfect fifth would be the sixth scale degree since $6/10 = 0.6$ is closer to $\log_2(3/2)$ than $5/10 = 0.5$. However, this gives a perfect fifth of $2^{6/10} \approx 1.5157$, which is a little more than 18 cents sharp of a true $3:2$ perfect fifth. This error, which is just shy of the Pythagorean and syntonic commas, is too large a gap to accept, since it will negatively impact the other important intervals. The resulting perfect fourth (in this case scale degree 4) will be the same distance away from a true $4:3$ perfect fourth, except it is 18 cents flat (see Exercise 2). Still worse, the new major third (scale degree 3) ends up being more than 26 cents flat of a just $5:4$ major third. The combination of these errors renders the 10-tone equally tempered scale unusable.

In Table 4.8 we show the results of dividing the octave into n equal parts, focusing on the error with the just $3:2$ perfect fifth. It is clear from the table that 12-part division gives the best approximation to the just perfect fifth for $10 \leq n \leq 23$. Note that $n = 41$ and $n = 53$ give better approximations, as these divisions correspond to the denominators of the next two best rational approximations to $\log_2(3/2)$ given by the sequence of convergents in (4.21). Other reasonable approximations are $n = 17, 19, 22, 31$, and 43. It is possible to hear these different scales by altering the number of notes in the *Pythagorean, Meantone, and Equal Temperament Musical Scales* demonstration tool.

4.6.2 Meantone scales

In many ways we have only grazed the subject of tuning and temperament. The history of the discovery of different tuning systems is long and extensive. The reader is encouraged to explore the detailed text of Barbour [1951], as well as the excellent summary and analysis provided in Benson [2007]. Before the widespread use of equal temperament, numerous musicians, scientists, and mathematicians strove to create their own scales intent on using their own type of just intonation. The list of innovators includes many well-known historical figures, including Bartolomé Ramos de Pareja, Vincenzo Galilei (Galileo's father), Johannes Kepler, Marin Mersenne, Friedrich Wilhelm Marpurg, and the great Leonhard Euler.

During the Renaissance, *meantone* scales were in common use. These scales are created by adjusting the fifths by a fraction of the syntonic comma, in order to make the major thirds sound better. For example, in the classical meantone scale, sometimes called *quarter-comma meantone*, the major third between scale degrees 1 and 3, 4 and 6, and 5 and 7 is established in the just $5:4$ ratio. The notes in between these major thirds, scale degrees 2, 5, and 6, respectively, are then placed directly in the "middle" of their surrounding

n = number of steps in octave	m = number of steps in P5	New P5 (cents)	Error (cents)	$2^{m/n}$	% Error
10	6	720.0	18.0	1.5157	1.048
11	6	654.5	−47.4	1.4595	−2.701
12	**7**	**700.0**	**−2.0**	**1.4983**	**−0.113**
13	8	738.5	36.5	1.5320	2.131
14	8	685.7	−16.2	1.4860	−0.934
15	9	720.0	18.0	1.5157	1.048
16	9	675.0	−27.0	1.4768	−1.545
17	10	705.9	3.9	1.5034	0.227
18	11	733.3	31.4	1.5274	1.829
19	11	694.7	−7.2	1.4938	−0.416
20	12	720.0	18.0	1.5157	1.048
21	12	685.7	−16.2	1.4860	−0.934
22	13	709.1	7.1	1.5062	0.413
23	13	678.3	−23.7	1.4796	−1.359
24	14	700.0	−2.0	1.4983	−0.113
31	18	696.8	−5.2	1.4955	−0.299
41	24	702.4	0.5	1.5004	0.028
43	25	697.7	−4.3	1.4963	−0.247
53	31	701.9	−0.1	1.500	−0.004

TABLE 4.8. The results of dividing the octave into n equal parts for various values of n. Here m is the best possible choice (when compared to the just $3:2$ perfect fifth) for the number of steps for the new "perfect fifth." The fourth and sixth columns measure the error when compared with a true $3:2$ perfect fifth. Positive (negative) error means that the approximate fifth is sharper (flatter, respectively) than the true one.

notes, a type of averaging. Since we are interested in ratios when comparing frequencies, "middle" means the square root. In other words, the whole step between scale degrees 1 and 2 equals the whole step between 2 and 3. Together they make a $5:4$ major third, so we find that

$$W^2 = \frac{5}{4}, \quad \text{or} \quad W = \frac{\sqrt{5}}{2}.$$

All five whole steps of the major scale are then determined by this irrational ratio. The two half steps in the scale are also taken to be equivalent; a little algebra shows that $H = 8:5\sqrt[4]{5}$ or $H = 8:5^{5/4}$. This results in a perfect fifth of $\sqrt[4]{5} \approx 1.49535$, which translates into 696.6 cents, just over 5 cents flat of the true $3:2$ perfect fifth. This discrepancy is clearly audible in the *Pythagorean, Meantone, and Equal Temperament Musical Scales* demonstration tool. Note that even though quarter-comma meantone loses the $3:2$ perfect fifth, it has a consistent whole step (unlike just intonation) and half step, along with just major thirds.

The problem with the meantone scales is that while some keys sound great, others are terrible. The issue of changing keys within a piece can be problematic. During the Baroque period, more complicated systems of averaging were used, known as *well-tempered* systems. In these tuning systems, all keys could be utilized, but different keys took on different moods. For example, C major was "completely pure," D major was the "key of triumph," and E♭ major was the "key of love"; however, A♭ major was the "key of the grave."[8]

As the title indicates, Bach's *Well-Tempered Clavier* was intended to be played on a keyboard that was well tempered, despite the daunting challenge of playing preludes and fugues in all 24 major and minor keys. Many have mistakingly assumed that Bach wrote the massive work to champion equal temperament as a superior tuning system. There is ample evidence to suggest that this was not the case. Some modern organs possess the ability to play in these older tuning systems in order to provide a historically accurate performance. The Charles Fisk organ at Stanford University has a lever that can change the organ between meantone and well-tempered systems. The lever does not change the white keys, but the black keys are altered from one system to the other.

4.6.3 Other equally tempered scales

Table 4.8 suggests that other equal divisions of the octave can lead to novel and appealing musical scales. For example, an equally tempered scale with 19 parts will have a new perfect fifth (scale degree 11) that is 7.2 cents flat from a just perfect fifth and a new major third (scale degree 6) that is 7.4 cents flat from a just $5:4$ major third. These are calculated via

$$1200 \log_2 \left(2^{11/19}\right) - 1200 \log_2 \left(\frac{3}{2}\right) = 1200 \left(\frac{11}{19} - \log_2 \left(\frac{3}{2}\right)\right) \approx -7.218 \text{ cents}$$

and

$$1200 \log_2 \left(2^{6/19}\right) - 1200 \log_2 \left(\frac{5}{4}\right) = 1200 \left(\frac{6}{19} - \log_2 \left(\frac{5}{4}\right)\right) \approx -7.366 \text{ cents,}$$

respectively. When compared with the corresponding intervals in the 12-note equally tempered scale, the fifth is more out of tune, but the major third is now half the distance (7 cents as opposed to 14) from a just major third. The seventeenth-century Dutch mathematician and scientist Christiaan Huygens was one of the first to use the 19-tone equally tempered scale.

Many composers have made use of the 24-tone equally tempered scale. This scale is typically known as the *quarter-tone scale* because the usual $2^{1/12}$ half step has been split in half again (take the square root) to become $(2^{1/12})^{1/2} = 2^{1/24}$. In other words, a quarter tone (50 cents) corresponds to a musical interval that is half the size of our usual half step (100 cents). Since 24 is just $2 \cdot 12$, all the frequency ratios of equal temperament are preserved (e.g., a perfect fourth is now the 10th scale degree, but the multiplier is unchanged since $2^{10/24} = 2^{5/12}$). The quarter-tone scale has an additional 12 notes to

---

[8]Schubart, *Ideen zu einer Ästhetik der Tonkunst.*

choose from while maintaining the same notes from our usual chromatic scale. Composers who have used the quarter-tone scale include Pierre Boulez, Charles Ives, Alois Hába, Bela Bartók, Valentino Bucchi, Krzysztof Penderecki, and Ivan Wyschnegradsky. In 1912, Andrzej Milaszewski (1861–1940) went so far as to patent a quarter-tone piano in Vienna.

The 53-note equally tempered scale has a particularly good perfect fifth *and* major third (see Exercise 5). In this scale, one "step" is obtained via multiplication by $2^{1/53}$ and is equal to approximately 22.6 cents. This is very close to the Pythagorean comma (23.5 cents) and syntonic comma (21.5 cents). The scale actually works out quite well when compared with our usual major scale. A traditional whole step will be made up of nine little steps, while the usual half step contains four little steps. This all fits nicely together into the octave, since $5 \cdot 9 + 2 \cdot 4 = 53$. The scale was appealing enough for Robert Bosanquet, who in 1876 constructed a "generalized keyboard harmonium" featuring 53 notes in an octave (see Figure 4.7). Music theorists have employed the 53-note equally tempered scale as an effective theoretical system for studying Turkish classical music.

FIGURE 4.7. Bosanquet's enharmonic harmonium with its 53 keys per octave.

Exercises for Section 4.6

1. Using proof by contradiction and the Fundamental Theorem of Arithmetic, show that $\log_2(3)$ is irrational. Since $\log_2(3/2) = \log_2(3) - \log_2(2) = \log_2(3) - 1$, conclude that $\log_2(3/2)$ is also irrational.

2. Suppose that the octave is divided into 10 equal parts, so that a "half step" corresponds to multiplication by $2^{1/10}$. Show that the new perfect fifth (six half steps) is the same number of cents sharp of a true $3:2$ perfect fifth as the new perfect fourth (four half steps) is flat from a true $4:3$ perfect fourth.

3. Suppose that the octave is divided into n equal parts, where m_1 steps approximate the perfect fourth and m_2 steps approximate the perfect fifth. Let e_1 be the number of cents difference between the new perfect fourth and a true $4:3$ perfect fourth, and let e_2 be the number of cents difference between the new perfect fifth and a true $3:2$ perfect fifth. Show that if $m_1 + m_2 = n$, then $e_1 + e_2 = 0$. In other words, if the new fourth and fifth are inverse musical intervals (they combine to create an octave), then each is the same distance from its "true" value, with one sharp and one flat.

4. Recall that in the quarter-comma meantone scale, all five whole steps in the major scale are equivalent, given by the irrational multiplier $W = \sqrt{5}/2$. The two half steps are also identical.

   a) Given that the octave is in a $2:1$ ratio, show that a half step in this meantone scale corresponds to the multiplier $H = 8:5^{5/4}$. This is the ratio that, when multiplied by a frequency, raises the pitch by one half step.

   b) Show that a perfect fifth in quarter-comma meantone is given by the multiplier $\sqrt[4]{5} = 5^{1/4}$.

5. Consider the 53-tone equally tempered scale, where the distance between consecutive scale degrees is $2^{1/53}$. For this scale, Table 4.8 indicates that 31 steps of the scale give an excellent approximation to a just perfect fifth.[9]

   a) Show that going up by 12 perfect fifths and then down by seven octaves is equivalent to going up by one scale degree (one step) in the scale. In essence, this means that the circle of fifths does not close, and the gap (or comma) is $2^{1/53}$, approximately 22.6 cents.

   b) Which scale degree corresponds to a major third? In other words, how many steps in the scale make up a major third? Find the number of cents in the new major third and compute the error (in cents) with the just $5:4$ major third.

References for Chapter 4

Barbour, J. M.: 1951, *Tuning and Temperament: A Historical Survey*, Michigan State College Press.

Barbour, J. M.: 1957, A Geometrical Approximation to the Roots of Numbers, *American Mathematical Monthly* **64**(1), 1–9.

Benson, D. J.: 2007, *Music: A Mathematical Offering*, Cambridge University Press.

Bibby, N.: 2003, Tuning and Temperament: Closing the Spiral, *in* J. Fauvel, R. Flood and R. Wilson (eds), *Music and Mathematics: From Pythagoras to Fractals*, Oxford University Press, pp. 12–27.

[9]This problem is based on Section 6.3 of Benson, *Music*.

Blatner, D.: 1997, *The Joy of π*, Walker and Company, New York.

Cohen, H. F.: 2007–2014, Stevin, Simon, *Grove Music Online*, Oxford University Press.

DeBoer, L.: March 2010, The Indiana Pi Bill, 1897. `http://www.agecon.purdue.edu/crd/localgov/Topics/Essays/Pi_Bill_Indiana_1897.htm`.

Duffin, R. W.: 2007, *How Equal Temperament Ruined Harmony (and Why You Should Care)*, W. W. Norton & Company, Inc.

Dunne, E. and McConnell, M.: 1999, Pianos and Continued Fractions, *Mathematics Magazine* **72**(2), 104–115.

Hutchinson, L.: May 2002, An Interview with James Taylor, *Performing Songwriter* **61**.

Lam, J. S. C.: 2007–2014, Zhu Zaiyu, *Grove Music Online*, Oxford University Press.

Lindley, M.: 2007–2014, Just [Pure] Intonation, *Grove Music Online*, Oxford University Press.

Lindley, M., Campbell, M. and Greated, C.: 2007–2014, Interval, *Grove Music Online*, Oxford University Press.

Schubart, C.: 1806, *Ideen zu einer Ästhetik der Tonkunst*, Degen. Translated by Rita Steblin in *A History of Key Characteristics in the 18th and Early 19th Centuries*, UMI Research Press, 1983.

Silverman, J. H.: 2006, *A Friendly Introduction to Number Theory*, 3rd edition, Pearson Prentice Hall.

Stewart, I.: 2003, Faggot's Fretful Fiasco, *in* J. Fauvel, R. Flood and R. Wilson (eds), *Music and Mathematics: From Pythagoras to Fractals*, Oxford University Press, pp. 60–75.

Strähle, D. P.: 1743, Nytt Påfund, Til at Finna Temperaturen, i Stämningen för Thonerne på Claveret Ock Dylika Instrumenter, *Proceedings of the Swedish Academy* **IV**, 281–291.

Chapter 5

Musical Group Theory

Quaerendo invenietis (Seek and ye shall find)
> — inscription given by Johann Sebastian Bach in his *Musical Offering*

This chapter focuses on the many ways composers have used symmetry to build complex pieces out of simple musical themes and motifs. Emphasis is placed on the mathematical transformations of translation, vertical and horizontal reflection, and rotation. In musical terms, these are transposition, retrograde, inversion, and retrograde-inversion, respectively. Several examples are presented from composers such as Bach, Barber, Bartók, Beethoven, Haydn, Hindemith, Liszt, and Sousa.[1] Tonal inversions, such as the type frequently used by Bach, are contrasted with the exact inversions favored by more modern composers.

The life and music of the famous Hungarian composer Béla Bartók, and his supposed use of Fibonacci numbers and the golden ratio, are investigated in detail, with a particular focus on his *Music for Strings, Percussion and Celesta*. Is there clear evidence that Bartók used the Fibonacci numbers and the golden section to help structure his music? If so, was this use intentional or a fortuitous coincidence overemphasized by theorists? There is little agreement among scholars on the answer to these questions. Our intention here is to present the reader with enough information and insight in order to form their own opinions on the matter. As part of the development, we discuss the Fibonacci numbers, their surprising appearance in nature, and their connection to the golden ratio.

The chapter closes with an introduction to group theory, culminating in a detailed look at the dihedral group of degree 4 (the symmetries of the square), which contains the musical symmetries mentioned above as a special subgroup. The development of group theory here lays the groundwork for later applications in Chapter 6 (change ringing) and Chapter 7 (12-tone music).

5.1 Symmetry in Music

Symmetry is prevalent in art, architecture, nature, and music. It is one of those aesthetic qualities that is inherently appealing, providing balance and representing equality. When a piece of music, or work of literature, starts and ends with the same theme, the closure

---

[1]Many of the examples in Section 5.1 come from two excellent sources: Harkleroad, *Math behind the Music*; and Hodges, *Geometry of Music*.

achieved provides us with a sense of satisfaction. The A-B-A ternary form, commonly found in music, is a symmetric structure where part A presents the main theme, part B contains contrasting material, and then part A returns to recap the main idea. With vocal works, a symmetric musical line can be used to underscore the beauty of the subject appearing in the text of the voice. Composers have used symmetry to help structure their works and to reuse musical motifs in new ways to construct more complicated pieces.

5.1.1 Symmetric transformations

Mathematically speaking, an object has symmetry if, after applying a certain geometric transformation, the appearance of the object is unchanged. For example, a circle has rotational symmetry since any rotation of the circle about its center preserves the shape of the circle. The transformations we focus on here are reflections and rotations. In addition, we will consider translations because they also maintain the shape of an object and are commonly used in music. A *translation* is simply a shift in some direction. We focus on vertical translations because they are equivalent to transpositions in music.

We will refer to an object as having *vertical symmetry* if it is symmetric under a vertical reflection about a central axis, called the *axis of symmetry*. Likewise, an object has *horizontal symmetry* if it is symmetric with respect to a reflection about a horizontal axis of symmetry. For instance, the letters A, M, and T have vertical symmetry since their shape is unchanged upon reflection about a vertical axis through the center of each letter. However, they do not possess horizontal symmetry because they change to ∀, ꟽ, and ⊥ after a reflection about a central horizontal axis. In contrast, the letters C, E, and K possess horizontal symmetry, but not vertical symmetry, since they are altered to Ɔ, Ǝ, and ꓘ under a vertical reflection. The letters H, O, and X have both vertical and horizontal symmetry. Some objects with vertical or horizontal symmetry (or both) are shown in Figure 5.1.

Objects can also have *rotational symmetry*, where the object is unchanged by a rotation through a particular angle. We will let R_θ denote a rotation by θ degrees in the clockwise direction about a central point. For example, R_{90} is a 90-degree rotation, R_{180} is a 180-degree rotation, and R_{-90} is a rotation by 90 degrees in the counterclockwise direction. The rotation R_0 does not rotate the object at all. Such a transformation is called the *identity map* since it preserves the identity of the figure regardless of whether it has symmetry or not. Note that since there are 360 degrees in one full rotation of the circle, we have equivalences such as $R_{90} = R_{450} = R_{-270}$ and $R_{180} = R_{-180}$. In general, two rotations

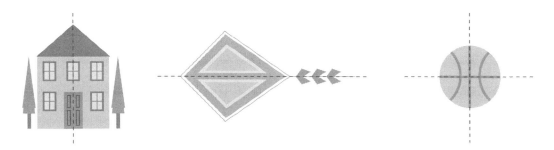

FIGURE 5.1. Some figures with vertical symmetry (*left*), horizontal symmetry (*center*), or both (*right*).

are equivalent if the difference between their angles is a multiple of 360. Mathematically speaking, we have

$$R_\theta = R_\phi \text{ if and only if } \theta - \phi = 360k \text{ for some integer } k. \qquad (5.1)$$

Equation (5.1) is an example of an *equivalence relation*, something we have already encountered in music in our discussion of music theory in Chapter 2. For instance, two notes an octave apart receive the same letter name, or two chromatic scales are considered to be equivalent because the total set of pitches is the same. These musical equivalencies use the same concept as a mathematical equivalence relation.

An equilateral triangle will have three rotational symmetries: $R_0, R_{120},$ and R_{240}. A regular hexagon (a six-sided polygon with equal sides and equal angles between adjacent vertices) will have six rotational symmetries: $R_0, R_{60}, R_{120}, R_{180}, R_{240},$ and R_{300}. A circle has the strongest rotational symmetry, since it is unaltered by a rotation through *any* angle. The letters N, S, and Z each have the rotational symmetry R_{180}.

In music, there are four standard symmetry transformations, known as *transposition, inversion, retrograde,* and *retrograde-inversion*. Their mathematical counterparts are displayed in Table 5.1. Composers frequently apply these transformations to a melodic *theme* or short *motif* to generate additional music that builds on the qualities of the original. As we shall see in our ensuing discussion, some of these transformations are more common than others, as it requires a great deal of skill to create a successful retrograde or retrograde-inversion.

Musical Term	**Mathematical Transformation**	**Example**		
Transposition	Vertical Translation			
Inversion	Horizontal Reflection	C	E	D
Retrograde	Vertical Reflection	A	M	T
Retrograde-Inversion	180° Rotation	N	S	Z

TABLE 5.1. The mathematical transformations corresponding to the four different types of musical symmetries. The examples are shapes that are invariant (remain the same) under the given transformation.

We introduced the idea of transposing a melody or theme in Section 2.4.4. Recall that a *transposition* is simply a vertical translation of all notes on the staff by some fixed interval. Technically, the musical interval that each note is shifted by should be identical for all notes, a task that key signatures easily ensure is achieved. However, in some of the examples we present, we will relax this strict definition to some degree.

Transposition can be used to great effect to reinforce a particular rhythmic idea or to remind the listener of a key musical phrase. Even repetition is a form of transposition (a vertical translation by zero units—the identity map). Sports enthusiasts will recognize the technique of transposition as it is often used to generate excitement and anticipation among the fans in a stadium. Simply raising the pitch of a simple melody, sometimes called a *riff*, successively by a half step can generate enthusiasm amidst the crowd (see Figure 5.2 for an example).

FIGURE 5.2. A typical sports chant at a Worcester Sharks hockey game demonstrating the use of transposition.

An *inversion* is a horizontal reflection about a particular note on the staff, sometimes called the *pivot note*. We will distinguish between two types of inversions, *tonal* or *exact*. A tonal inversion remains within the given key and may not preserve the precise interval relationships before and after the inversion. For example, a rising minor third may be inverted to a falling major third. An exact inversion preserves all of the musical intervals precisely. This usually means leaving the given key or having no previously specified key. The different types of inversions are discussed in Section 5.1.2.

A *retrograde* is playing the theme or motif backward, mathematically akin to a vertical reflection. A simple example is the opening notes in the melody of the song *Lean on Me* (1972) by Bill Withers, where the four-note ascending pattern from the first to fourth scale degree is immediately sung backward. In a similar vein is the melodic line accompanying the text "for the Lord God Omnipotent reigneth," in the triumphant *Hallelujah chorus* from George Frideric Handel's *Messiah* (1741). As with the melody from *Lean on Me*, this motif also rises from scale degree 1 to 4 and then returns, a musical palindrome given by the notes D E F♯ G (G) G F♯ E D. The parenthetical G is an octave lower than its surrounding Gs. Although the rhythm is not quite a perfect retrograde, Handel is using vertical symmetry to proclaim the brilliance and balance of the "Lord God." To emphasize this significance, the motif appears repeatedly, often in different voices, throughout the piece.

Our last kind of musical symmetry is a *retrograde-inversion*, the rarest of transformations, which combines the two operations of a retrograde and inversion. Here the melody is inverted and then the result is written backward, a process identical to rotating the music by 180°. It is worth noticing that a retrograde-inversion may also be created by first writing the motif backward and then inverting. Regardless of the order of operations, the same music will be obtained.[2] Mathematically speaking, this is a consequence of the fact that horizontal (H) and vertical (V) reflections commute. Try drawing a figure on a piece of paper and applying a horizontal and vertical reflection to it in different orders. The result should always be the same, as $H * V = V * H$. The $*$ symbol means to compose the two transformations. For example, $H * V$ means to first do a horizontal reflection and then follow with a vertical reflection.

The three techniques of inversion, retrograde, and retrograde-inversion are demonstrated on a simple melody in Figure 5.3. Here we have used a tonal inversion that is a reflection about the middle of the staff (the note B), which is the starting note of the melody.

---

[2]This assumes that in each case the inversion is accomplished with respect to a fixed position on the staff. A different definition of inversion may not commute with a retrograde, as discussed with the 12-tone technique in Section 7.1.1.

FIGURE 5.3. A simple melody along with its inversion, retrograde, and retrograde-inversion.

5.1.2 Inversions

Inversions are the musical equivalent of a horizontal reflection. If the notes in a melody jump up in pitch, then in the inversion, those same notes now move down in pitch, and vice versa. Often a musical inversion is not a true reflection in the strictest mathematical sense. Composers may desire the effect of an inversion but prefer that the music remain in the given key.

For example, suppose we are writing a melody in C major and we begin on the tonic C, followed by a major third up to E. If we want to invert our melody about the tonic, then we would begin on C and go down a major third. But this lands us on A♭, which is *not* in the key of C major. If we wanted to stay in the given key and still go down by a third, we would land on A, which is a minor third below C. Thus, choosing the note A inverts a rising major third to a falling minor third. We call such an inversion a *tonal inversion* because the reflection is not mathematically precise (four half steps up is reflected to three half steps down). An inversion where the interval relationships are preserved exactly is called an *exact inversion*. If the melody goes up by *n* half steps, then an exact inversion of the melody will go down by precisely *n* half steps. This relationship may or may not be preserved in a tonal inversion.

In Figure 5.4 we demonstrate the difference between a tonal and exact inversion on a short motif in the key of C major. Each inversion reflects about C, the second space from the top of the staff and the opening note of the melody. (We call C the pivot note.) The successive musical intervals for the original melody and each transformation are shown in Table 5.2. Notice that the arrows for the inversions are flipped from the original melody, and that two intervals in the tonal inversion are adjusted from minor thirds to major thirds to remain in the key of C major. The exact inversion introduces two new notes, B♭ and A♭, which are foreign to C major. Depending on your preference, the key for the exact inversion could be taken as F minor, A♭ major, or even C Phrygian mode. Notice the symmetry in the lower-right excerpt, where both the original melody and the exact inversion are shown together. Here it is easy to see the precise horizontal reflection about an axis of symmetry through C.

One composer who made frequent use of inversions was the great Johann Sebastian Bach (1685–1750). Bach was a master at using tonal inversions, and occasionally exact inversions, in his fugues and cannons. A *fugue* is a style of music that begins by announcing a fundamental theme, called the *subject*, which is then repeated in later measures, called the *answer*, sometimes exactly, but often in an altered form. The subject, ensuing answers, and a possible *countersubject*, a contrasting melody often played while the answers to the subject are presented, are cleverly woven together in a rich style of composition called *counterpoint*.

FIGURE 5.4. A simple melody along with its tonal and exact inversions. Here the horizontal reflection pivots about C, as can be viewed clearly in the lower-right excerpt.

Excerpt	Sequence of Intervals
Original Melody	↑ P4, ↓ m3, ↑ whole step, ↑ m3, ↓ P5
Exact Inversion	↓ P4, ↑ m3, ↓ whole step, ↓ m3, ↑ P5
Tonal Inversion	↓ P4, ↑ M3, ↓ whole step, ↓ M3, ↑ P5

TABLE 5.2. The interval sequences for the excerpts in Figure 5.4.

Bach was a master of counterpoint, highly skilled at modifying his musical subjects. In addition to using inversions, transpositions (often to the dominant scale degree) were commonly used, as well as techniques such as *augmentation*, *diminution*, and *stretto*. Augmentation occurs when the rhythms of the theme are lengthened but the notes remain the same. For example, a quarter note becomes a half note, an eighth note becomes a quarter note, and so forth. Diminution is the opposite process, where the durations are shortened rather than lengthened. If the theme is quickly imitated in succession by different voices, then we have a stretto. The most elaborate transformations of the subject typically occurred in the developmental section of a piece (the middle section in ternary form).

Bach's *Fugue No. 8 in D-sharp Minor*

Figure 5.5 shows the opening subject of Bach's *Fugue No. 8 in D-sharp minor* from the *Well-Tempered Clavier*, vol. I, as well as two inverted subjects, one beginning in measure 45 and the other starting in bar 30. The horizontal reflection symmetry between the subject and its inversions is readily apparent. It is instructive to compare the two inversions. The interval between the opening note of the subject (D♯) and the first note of the inversion in measure 45 (A♯) is a perfect fifth, or seven half steps. Technically, the symmetry axis would correspond to a note 3.5 half steps between the two notes, which is a note halfway between F♯ and G. Instead, it is easier to regard the inversion as being about the space on the bottom of the staff. Based on a note-by-note comparison, the inversion is tonal. Although the intervals in the first measure match up exactly, the interval crossing from the first measure into the second measure of the subject is a descending major second

FIGURE 5.5. The subject and two different inversions of the subject in Bach's *Fugue No. 8 in D-sharp minor* from the *Well-Tempered Clavier*, vol. I.

(G♯ to F♯), but the corresponding interval moving from measure 45 to 46 is an ascending *minor* second (E♯ = F to F♯—don't forget those six sharps in the key signature). Moreover, the subject jumps up a P4 between beats 3 and 4 of the second measure, while the inverted subject jumps down a P5 in the same spot.

It does not make musical sense to think of the second inversion in Figure 5.5 as taking place about some point in the staff, as the interval between starting notes is a minor 10th, or 15 half steps. Thus, the pivot point would be located 7.5 half steps from each pitch, which is the note halfway between A♯ and B. It is better to just focus on the intervals between successive notes in each melodic line. As with the first example, this inversion is also tonal, and Bach takes liberties with his choice of intervals. For example, the subject opens with an ascending P5, but the inversion in measure 30 opens with a descending P4, two half steps off. Although the symmetry in both inversions is not exact, each transformation contains enough of a reflection of the subject to be clearly recognizable by the listener.

Figure 5.6 demonstrates an augmentation of the main subject of the fugue, which occurs in the left hand beginning in measure 62. Except for the very last note, each note of the subject has been doubled in length. Notice also that the opening interval of the augmentation is a P4 (dominant to tonic) as opposed to the P5 that opens the subject (tonic to dominant).

FIGURE 5.6. The augmented subject in measure 62 of Bach's *Fugue No. 8 in D-sharp minor* from the *Well-Tempered Clavier*, vol. I.

FIGURE 5.7. An exact inversion at the start of the aptly named *Subject and Reflection*, no. 141, vol. 6 of *Mikrokosmos* by Béla Bartók. The pivot note is B♭. Even the number of the piece, 141, is symmetric.

Bartók's *Subject and Reflection*

Modern twentieth-century composers, who were not restricted by a major-minor tonal system, could employ exact inversions more freely. One such artist was the great Hungarian composer Béla Bartók (1881–1945).

An excellent example of an exact reflection comes from Bartók's *Subject and Reflection*, no. 141, vol. 6 in his series of piano pieces *Mikrokosmos* (see Figure 5.7). The entire *Mikrokosmos* contains 153 pieces for piano ranging in difficulty from beginner to advanced and features some noteworthy musical traits, such as the use of whole-tone scales, Hungarian folk tunes, syncopation, and modal harmonies. In no. 141, where even the number of the piece is symmetric, Bartók writes the left-hand part as an exact inversion of the right-hand melody about B♭, with the former transposed down an octave. Unlike the previous piece by Bach, the inversion happens *simultaneously*, as both the original and reflected parts are played together. In a fugue, when the subject is inverted, it nearly always happens several measures after the main theme has been introduced.

To see the symmetry more clearly, Figure 5.8 shows the two parts in the same clef with the phrasing and inner voice parts removed. In each measure of the excerpt, every pair of notes is precisely the same musical distance from B♭. This can also be viewed in terms of successive intervals. The opening five intervals in the upper part are up an M2, up an m3, down an m2, up an m3, and down an M2. One can check that the same interval sequence is used in the lower part, with the "ups" and "downs" reversed. This is also clearly audible when listening to the piece.

Bartók maintains the horizontal reflection throughout the short piece (it runs about 1.25 minutes in length). However, in some measures he deletes a note from one of the parts, while in others, such as the section marked Vivacissimo, the inversion is delayed by half a beat in the left hand. Mathematically, this is akin to a *glide reflection*, where the reflected objects are slightly offset. When the main opening melody returns, at the

FIGURE 5.8. Reduction of *Subject and Reflection* to one staff, demonstrating the precise horizontal symmetry about the middle of the staff.

section marked Tempo I, the inverted left-hand part now enters a full beat behind the right hand, and the inversion is less exact, as a few notes are shifted a half step away from their previous locations. The different types of inversions utilized by Bartók as the piece proceeds create an interesting musical effect, one that is not rigidly mathematical. The syncopated delay when the melody returns near the end of the piece is quite reminiscent of Bach's fugal counterpoint.

Back to Bach

Although most of Bach's inversions consist of reflected answers entering after the original subject has been heard, there is at least one example where the inversion occurs simultaneously with its subject. We present it here to demonstrate the similarity with Bartók's *Subject and Reflection*. This special inversion appears in the three-part fugue *Contrapunctus XI*, from Bach's *The Art of the Fugue*, a lengthy but unfinished work that Bach composed during the last decade of his life. On the second beat of measure 158, the subject and an inverted subject a perfect 11th higher are played together *simultaneously*, a stunning musical moment well ahead of its time (see Figure 5.9). Writer and keyboardist David Schulenberg described the inversion as a "splendid moment, perhaps marking the culmination of the whole main series of eleven contrapuncti."[3]

FIGURE 5.9. The simultaneous occurrence of the subject and inverted subject in measures 158–162 and 164–168 in Bach's *Contrapunctus XI*.

Both inversions in Figure 5.9 are tonal, although the example in the bass clef is nearly exact. This inversion takes place about A (the top line of the staff), as can easily be seen on beat 3 of measure 165 and on beat 4 of measures 166 and 167, where the two parts come together on a shared A. If each B♭ in the bottom example were changed to a B♮, then the inversion would become exact.

5.1.3   Other examples

Transposition

The idea of applying a vertical translation to an entire melodic line or to some short motif has been used by nearly every composer. It is a simple and effective way to generate more music, and it can be used to increase dramatic tension during a piece. As with inversions, a transposition can remain in the given key or can shift into a new tonality. In Ella Fitzgerald's classic rendition of *Mack the Knife* (1960; recorded live in Berlin), where she

<hr>

[3]Schulenberg, *Keyboard Music of J.S. Bach*, p. 361.

FIGURE 5.10. The opening measures, with the key voices condensed to one staff, of Beethoven's famous 5th Symphony. The mark over the half notes in the second and fifth measures is called a *fermata* and indicates that the note should be held beyond its usual duration. Notice the different ways (e.g., transposition, inversion) that Beethoven employs his legendary four-note motif.

famously forgot the lyrics midway through the song, each successive verse is transposed up a half step into a new key. This standard translation is a great method of keeping the listener interested, particularly if other aspects of the song are falling apart.[4]

One of the greatest classical examples to feature repeated translations of a short musical idea is the first movement of Ludwig van Beethoven's *Symphony No. 5 in C minor* (1807–1808). This famous symphony took the concept of developing a simple motif to new heights. The opening "da da da dum" four-note motif is transposed, inverted, elongated, altered, and continually repeated throughout the piece to create the bulk of the music (see Figure 5.10 for the opening measures of the movement). The entire first minute of the symphony consists solely of this dramatic motif. Beethoven works variations of the motif into other movements of the symphony, bringing it back unexpectedly in the final movement.

Another well-known classical example involving repeated transpositions is the *Adagio for Strings* (1936) by Samuel Barber. The composer reworked the slow movement from his *String Quartet in B minor*, op. 11 (1936) into this glorious piece for string orchestra. Some readers may be familiar with this work from its frequent use in television and cinema, most notably as the background for Oliver Stone's Vietnam War film *Platoon*. The piece drips with luscious harmonies and a tantalizingly slow and somber melody that appears to be continuously climbing up the scale. As with Beethoven's 5th Symphony, Barber uses repetition and transposition to develop a simple motif consisting of three quarter notes into a primary theme that generates most of the music (see Figure 5.11). The initial quarter-note pattern, consisting of the notes A, B♭, C, is repeated once and

FIGURE 5.11. The hauntingly beautiful opening melody to Barber's *Adagio for Strings*.

---

[4]Sometimes a little memory loss can work in your favor. Without any lyrics, Ella did what many great jazz musicians do—she improvised. Her memorable rendition earned her two Grammy awards in 1961.

then transposed up a half step to the notes B♭, C, D♭. These new notes are also repeated once before being raised up a whole step to C, D♭, E♭. To keep the three-note motif within the key of B♭ minor (the key of the piece), the first transposition of the motif is tonal, while the second one is exact (up a minor third).

Haydn's *Menuetto al Rovescio*

An excellent example of a retrograde is the *Menuetto al Rovescio* (Minuet in Reverse) from Franz Joseph Haydn's *Piano Sonata in A major* (Hob. XVI/26 or Landon 41), written in 1773 and dedicated to Prince Nicolaus Esterházy. The magical second movement of this sonata consists of a minuet and a trio, each of which are exact retrogrades. The music first appeared under a different guise as part of the third movement of Haydn's *Symphony No. 47 in G major*, "The Palindrome," written a year earlier.

The complete minuet is shown in Figure 5.12. Notice the vertical reflection at the end of measure 10: both the right and left hands proceed to play their parts backward. If we stretched this musical palindrome so that all 20 measures fit on one long line, the music would possess a perfect vertical symmetry about the end of bar 10. To elongate the effect of the retrograde, Haydn uses repeat signs at the end of each 10-bar phrase (forward and backward). The recording of the sonata by pianist John McCabe features a trill (a quick flourish of extraneous notes) in measures 8 and 9 which is played in retrograde on the return trip [Haydn, 1974–1977]. The 12-bar trio that follows the minuet is also a perfect retrograde, and it too is repeated for effect. The movement ends with one more pass through the minuet.

FIGURE 5.12. The wonderful retrograde (at the end of measure 10) in Haydn's *Menuetto al Rovescio* from his *Piano Sonata in A major* (Hob. XVI/26 or Landon 41).

The $\frac{3}{4}$ meter seems crucial to making the retrograde work harmonically, as Haydn often emphasizes the key notes of the underlying chord in beats 1 and 3 of each measure. For instance, the first and last notes of measures 1, 3, and 5 use only the notes of an A major chord (the tonic). The same beats in bars 2, 4, and 10 feature the notes of an E dominant seventh chord. Thus, on playing each measure in reverse, the chord and corresponding harmony remain unchanged.

Bach's *Musical Offering*

In 1747, three years before his death, Bach visited the palace of Frederick the Great in Potsdam, Germany. Having heard of the prodigious skills of Bach as composer and key-boardist, the king of Prussia, a music lover, flutist, and composer himself, made several requests to Bach's second son, Carl Philipp Emanuel, expressing a desire to meet and hear the great musical talent. Upon arrival, Bach was presented with a theme of music, the "Royal theme," for which he was asked to improvise a three-part fugue on the spot. After accomplishing this, the King pushed Bach further, asking for a six-part fugue. This was too much of a demand even for Bach, although he stunned the court audience by improvising and playing a six-part fugue on a theme of his own choosing. After return-ing home, Bach composed a six-part fugue, then called a *ricercar*, as well as several other pieces based on the Royal theme, and sent the musical collection to the king as a gift, his *Musical Offering*.

The work contains some excellent examples of symmetry transformations used to gen-erate and develop one specific theme, in this case, the Royal theme (see Figure 5.13). Writ-ten in the key of C minor, the theme begins with the notes of a C minor triad. A measure later it descends down the chromatic scale from the dominant G to the leading tone B (using the harmonic version of the minor scale), before finding its way back to the tonic on C.

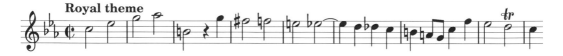

FIGURE 5.13. The Royal theme presented to Bach by Frederick the Great, upon which the *Musical Offering* is based.

There are 13 pieces in the *Musical Offering*: two fugues (one in three parts based on Bach's improvisation in front of the king, and the other in six parts composed later); a trio sonata for flute, violin, and keyboard; and 10 *canons*. A canon is a sophisticated version of a round (e.g., *Row, Row, Row Your Boat*), where one melodic line is imitated in some form and played after the main melody has begun. The imitations of the main theme can come in a variety of forms, including direct repetition; repetition at a particular interval above or below the original (e.g., a transposition by a perfect fifth); inversion; and augmentation, where one voice sounds the theme while the other plays it out over twice the duration. It is even possible to use a retrograde, where the theme and its response move in opposite directions at the same time. This is known as a *crab canon*, presumably because crabs possess the ability to quickly move from side to side.

FIGURE 5.14. The marvelous *Crab Canon* from Bach's *Musical Offering,* a perfect musical palindrome. Note the precise retrograde (vertical reflection) at the end of measure 9.

Remarkably, Bach uses all of these forms in his 10 canons based on the Royal theme. Four of the canons use inversion to imitate the theme, one of which also employs augmentation. The inversions are tonal, often reflecting about the third scale degree (either E♭ or E♮). For instance, in piece no. 4, *Canon in Contrary Motion*, the notes G F E♭ D C are inverted to C D E♭ F G, a tonal inversion that stays in the key of C minor.

Perhaps the most impressive of the canons is piece no. 9, *Crab Canon*, which features a remarkable retrograde of the Royal theme (see Figure 5.14). The canon is for two parts (or voices), and each part is an exact retrograde of the other. The top voice announces the Royal theme in the first nine measures and then follows with an elaborate eighth-note countermelody over the next nine measures. Not only does this entire 18-measure part sound great when played backward, but it works forward and backward at the same time! While the top voice plays its part, the bottom voice is playing the same part in retrograde. Even though they are traveling in opposite directions, the two voices blend beautifully.

This type of retrograde is more advanced than Haydn's *Menuetto al Rovescio*. Here we have a vertical reflection occurring at the end of measure 9, and each part moves in retrograde, except that the two parts are *interchanged*; the first part plays the second part backward and vice versa. Mathematically speaking, this type of retrograde can be visualized on a *Möbius strip*. Write the primary 18-bar phrase as one long line of music on a thin strip of paper. Cut the paper in half, at the end of measure 9. Glue the two parts together, but make a twist before gluing the parts at the end of measure 9. Each player now travels in opposite directions around the Möbius strip, with the vertical reflection taking place when the two parts pass each other after one "loop." The twist represents the interchanging of the parts.[5]

Interestingly, the canons in the *Musical Offering* are written as a kind of musical puzzle, as the full parts were not transcribed by Bach in the original manuscript. All of the canons except for one were given in abbreviated form. Three of the canons, including the crab canon, had no title or instructions as to how and when the imitations of the Royal theme were to be played. Instead, the inscription *Quaerendo invenietis* (Seek and ye shall find) was provided. Cunningly, Bach left clues as performance directions, using clefs in odd positions to indicate the opposing parts. For example, at the end of the main melodic line in the crab canon, Bach wrote the opening clef, time signature, and key signature backward (a vertical reflection).[6] The "puzzle" of Bach's *Musical Offering* was solved and first published by his student Johann Philipp Kirnberger.

The *Musical Offering* aptly demonstrates Bach's clear interest in applying symmetry to music in a variety of contexts. Even the general structure of the work displays symmetry, as the fugues open and close the work, and the trio sonata, the longest piece in the collection, lies at the center, with five canons to either side (see Figure 5.15). A masterful use of building intricate and evolved music from just one theme, the *Musical Offering* is a testament to Bach's musical (and mathematical) genius.

FIGURE 5.15. The overarching symmetrical structure of Bach's *Musical Offering*.

Music with multiple symmetries

Some works, such as Bach's *Musical Offering*, strive to incorporate more than one symmetry type. Another such example is Paul Hindemith's *Ludus Tonalis* (Game of Tones; 1942), a delightful piano piece modeled after Bach's *Well-Tempered Clavier*. As with Bach's collection of preludes and fugues, *Ludus Tonalis* uses all 24 major and minor keys. One of its more striking features is that the closing Postludium (excluding the final C major chord) is a retrograde-inversion of the opening Praeludium (see Figure 5.16). As a twentieth-century composer, Hindemith was less restricted by the rules of tonality and could therefore meet the challenge of constructing an entire movement that works on its own as well as after a rotation by 180°.

---

[5]For a fantastic rendering of Bach's *Crab Canon* on a Möbius strip see the video created by Jos Leys at https://www.youtube.com/watch?v=xUHQ2ybTejU.

[6]Actually, in the Bach-Gesellschaft edition of the *Musical Offering* [Bach, 1992], the time signature ℂ (common time) is rotated by 180 degrees to ↄ, which is different from applying only a vertical reflection.

FIGURE 5.16. The opening and closing measures of Hindemith's *Ludus Tonalis* (excluding the final chord) are in retrograde-inversion. The entire final movement is equivalent to the first movement rotated by 180°. © Copyright 1942 by Schott Music GmbH & Co. KG. © Copyright renewed. All Rights Reserved. Used by permission of European American Music Distributors Company, sole U.S. and Canadian agent for Schott Music GmbH & Co. KG, Mainz, Germany.

Technically speaking, the inversion shown in Figure 5.16, which is taken about middle C, is not exact. For example, the fourth note in the first measure of the Praeludium, an E, should be inverted to an A♭. Instead, Hindemith uses an A in the corresponding location in the bass clef in the Postludium (four notes from the end of the last measure). This discrepancy is primarily an artifact of musical notation; an exact inversion using the staff often requires the introduction of accidentals. Hindemith focused on applying a 180° rotation that included the accidentals, rather than inverting each musical interval precisely.

Ludus Tonalis features 12 three-part fugues that utilize many different kinds of symmetry, including inversion, retrograde, augmentation, and diminution. The piece was presented to Hindemith's wife on her birthday, and for fun the composer sketched a different type of lion at each entrance of the subject in the 12 fugues. Taken as a whole, *Ludus Tonalis* displays all four of the symmetry transformations discussed in this section.

For a musical excerpt that possesses reflection and rotational symmetry all at once, consider György Kurtág's *Játékok for Piano I, Hommage à Eőtvős Péter*. *Játékok* (Games; begun in 1973) is an ongoing series of piano pieces meant to capture the simple childlike joy of playing the piano. The *Hommage à Eőtvős Péter* contains a measure that has vertical, horizontal, and rotational symmetry, all present together. In other words, applying any of these three transformations to the music will return the exact same music.

Another Hungarian composer to combine multiple symmetries in the same piece was the great Franz Liszt (1811–1886). During the developmental section of Liszt's popular piano piece *Hungarian Rhapsody No. 2* (1847), there is a challenging set of measures that utilize transpositions, inversions, and retrogrades. The use of these symmetries is fairly easy to hear and see. The full rhapsody is one of the most popular piano solos around and was used in many animated cartoons, including the *Bugs Bunny* short "Rhapsody Rabbit" (Warner Brothers) and the *Tom and Jerry* episode "The Cat Concerto" (MGM), which won the 1946 Academy Award for best cartoon.

As a final observation, we note that many 12-tone pieces or 12-tone inspired works contain exact inversions, retrogrades, and retrograde-inversions of their primary tone row (sometimes called the basic set). This is a standard feature of the 12-tone technique of composing, a technique established and developed by Arnold Schoenberg and adopted by members of his Second Viennese School, such as Anton Webern and Alban Berg. We will discuss the four symmetry transformations, their use, and some of the special properties they possess in a detailed study of the 12-tone technique in Chapter 7.

Exercises for Section 5.1

1. Describe the type(s) of symmetry present (vertical or horizontal reflection, rotational, none) in each of the following letters or symbols: **B**, **Λ**, **L**, **V**, **Θ**, **Ǝ**, and ∞.

2. Using staff paper, write out the retrograde, inversion (tonal; pivot about A), and retrograde-inversion forms of the musical subject below.

3. Write out the *exact* inversion of the following melody reflecting about the note B♭.

4. Notate each musical interval between consecutive notes in the opening seven bars of Bartók's *Subject and Reflection* (use Figure 5.8). Do this for both the upper and lower parts in order to confirm the horizontal symmetry.

5. Figure 5.17 shows an inversion in the opening of the march *The Thunderer* (1889) by John Philip Sousa.[7] List the intervals between adjacent notes in both the left- and right-hand parts. Is the inversion tonal or exact? In terms of the letter names of the notes, what other symmetry is present?

FIGURE 5.17. The opening measures of the march *The Thunderer*, by John Philip Sousa.

[7]Example from Harkleroad, *Math behind the Music*.

6. Figure 5.18 shows the music and lyrics for *Mary Had a Little Lamb*.

   a. Write out the first eight measures of *Mary Had a Little Lamb* after applying a *tonal* inversion about B♭, staying in the key of B♭. Use the treble clef with the correct key signature. (You do not need to include or invert the text.)

   b. Write out the first eight measures of *Mary Had a Little Lamb* after applying an *exact* inversion about B♭. Use the treble clef.

   c. Write out the first eight measures of *Mary Had a Little Lamb* after applying a retrograde-inversion about B♭, staying in the key of B♭. Use the treble clef with the correct key signature.

   d. Play the original and each of the three transformed melodies from the previous exercises. How do the new melodies sound? Which do you prefer and why?

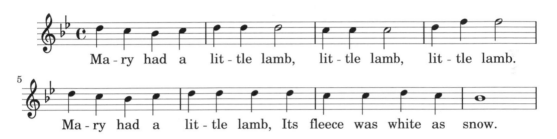

FIGURE 5.18. The musical setting of the nursery rhyme *Mary Had a Little Lamb*, words by Sarah Hale (1830) and music by Lowell Mason.

7. As discussed in the text, an exact inversion usually takes the original theme out of its original key, as accidentals are required to make the inversion a true horizontal reflection. In this exercise we learn how to find the new major key after performing an exact inversion.

   a. Take an ascending C major scale beginning on C and invert it exactly about the starting C. Write down the notes of this new descending scale using a diatonic spelling. For instance, the first note is C and the second note is a B♭ because we go down by a whole step instead of up. The diatonic spelling is B♭, not A♯, because we always use consecutive letters of the musical alphabet. How many flats are in the new descending scale? The set of notes in the descending scale correspond to which major key?

   b. Repeat the instructions for part **a.** on an ascending A major scale. How many flats are in the new scale? Which major key does it correspond to?

   c. Going in the other direction, what major scale, when inverted exactly about its tonic, gives precisely the notes in the C major scale?

   d. State a general rule for finding the new key when a major scale is inverted exactly about its tonic. How can you find the new key quickly? *Hint:* The circle of fifths may be helpful here.

8. Identify the type of symmetry present in each of the following two examples. Be as precise as possible (e.g., tonal inversion about C).

a.

b.

5.2   The Bartók Controversy

Let my music speak for itself, I lay no claim to any explanation of my works!

— Béla Bartók[8]

Typically, when a mathematical process is being employed by a composer, it is readily identifiable upon careful analysis. It is clear that Bartók's *Subject and Reflection* is a work about horizontal symmetry. The brilliance of Bach's *Musical Offering* has to do with the multiple applications of symmetry to elaborate and develop the Royal theme. Hindemith clearly intended to use rotational symmetry in *Ludus Tonalis*. In all of these cases, mathematical ideas are at play. We will explore some other inherently and intentionally mathematical compositions in the following three chapters.

In contrast, there are some musical examples where the mathematical meaning or intent is less clear. While some analysts delight in finding hidden mathematical connections, others cry foul and accuse their brethren of overreaching by making exaggerated claims without conclusive evidence. One controversial example worth a closer look is Bartók's *Music for Strings, Percussion and Celesta* (1936).

Bartók's life

Our story begins with the young Bartók (1881–1945), who could play 40 songs on the piano by the age of four. He wrote his first piece of music at age six and quickly rose to become a chapel organist, accomplished pianist, and respected composer. Bartók studied at the Catholic Gymnasium (high school) in Pozsony, where he excelled in math and physics, in addition to music. He entered the Academy of Music in Budapest in 1899. Bartók, who spent the last five years of his life in New York, is generally considered to be one of Hungary's greatest composers.

---

[8]Lendvai, *Béla Bartók*, p. 96.

Bartók was an avid collector of folk music, particularly Hungarian, Romanian, Slo-vakian, and Turkish. He traveled the countryside with fellow Hungarian and composer Zoltán Kodály collecting thousands of folk songs that would later exert a strong influence on his music. He drew inspiration from contemporaries such as Debussy and Kodály, as well as past masters such as J. S. Bach.

One important trait of Bartók, particularly relevant to this story, was his deep interest in nature. He built an impressive collection of plants, insects, and minerals and was reportedly fond of sunflowers and fir cones. In his article "A Népzene Forrásainál" (At the Sources of Folk Music; 1925), Bartók writes, "We follow nature in composition ... folk music is a phenomenon of nature. Its formations developed as spontaneously as other living natural organisms: the flowers, animals, etc."[9] His interest in folk music and respect for nature are well established and can be discerned in his music.

The controversy began in 1955, after Bartók's death, when the Hungarian musical an-alyst Ernő Lendvai started to publish his findings claiming the existence of the Fibonacci numbers and the golden ratio in many of Bartók's pieces. Lendvai drew connections between Bartók's compositional traits and his love of nature and "organic" folk music. He took a broad view, examining form (structure of pieces, where climaxes occur, phras-ing, etc.) as well as tonality (modes and intervals), in discerning a substantial use of the golden ratio and the Fibonacci numbers. While some found Lendvai's work fascinating and proceeded to build on his initial ideas, others found errors in his analysis and began to discredit him. Lendvai soon became a controversial figure in the study of Bartók's music.

5.2.1   The Fibonacci numbers and nature

Before proceeding with our story, we pause to define the golden ratio and discuss a fascinating connection between the Fibonacci numbers and nature. Recall from Section 1.4 that the Fibonacci numbers are the sequence

$$1, \ 1, \ 2, \ 3, \ 5, \ 8, \ 13, \ 21, \ 34, \ 55, \ 89, \ \ldots,$$

a list also studied by the Indian scholar Hemachandra.

Remarkably, these special numbers show up in a variety of contexts in nature. The number of petals in many flowers is often a Fibonacci number. For example, we find three-leaf clovers, buttercups with five petals, black-eyed Susans with 13, and daisies with 21 or 34 petals (see Figure 5.19 for some additional examples). The number of spirals in bracts of a pine cone or pineapple, in opposing directions, are typically consecutive Fibonacci numbers. The same holds true for the number of spirals in the seed heads on daisy and sunflower plants.

In 1994, botanist Roger Jean conducted a survey of the literature encompassing 650 species and 12,500 specimens. He estimated that among plants displaying spiral or multijugate phyllotaxis ("leaf arrangement"), roughly 92% demonstrated *Fibonacci phyl-lotaxis* [Jean, 1994]. In other words, the numbers of leaves counted were highly likely to be Fibonacci numbers. These surprising connections to nature actually have a biological and mathematical basis. Some of the facts about spirals in seed heads can be explained using continued fractions and the golden ratio.

---

[9]Translation from Lendvai, *Béla Bartók*, p. 29.

FIGURE 5.19. Fibonacci flowers (from left to right): red columbine (five petals); black-eyed Susan (13 petals); common chicory (21 petals); sunflower (34 petals). Photo credits (from left to right): © ljhimages/iStock/Thinkstock; © herreid/iStock/Thinkstock; © ArminStautBerlin/iStock/Thinkstock; © Racide/iStock/Thinkstock.

5.2.2 The golden ratio

Definition 5.2.1. *Consider two quantities a and b and suppose that a > b. If the ratio of the larger quantity to the smaller, a:b, is the same as the ratio of the total sum to the larger, (a + b):a, then the ratio of the two quantities is called the* golden ratio. *The golden ratio is denoted by ϕ (pronounced "fai" or "fee") and has the value* $\phi = \dfrac{1 + \sqrt{5}}{2} \approx 1.61803398875$.

Figure 5.20 shows a segment divided into two lengths based on the golden ratio. It might seem that this special ratio depends on the values of a and b. This is incorrect; there is only one golden ratio, and it comes from simplifying the defining equation

$$\frac{a}{b} = \frac{a + b}{a} . \tag{5.2}$$

If we divide the top and bottom of the fraction on the right-hand side of Equation (5.2) by b, we obtain

$$\frac{a}{b} = \frac{\frac{a}{b} + 1}{\frac{a}{b}} . \tag{5.3}$$

This new equation is now written solely in terms of the quantity a/b, the quantity we desire. Setting $\phi = a/b$ in Equation (5.3) yields an equation in one variable,

$$\phi = \frac{\phi + 1}{\phi} . \tag{5.4}$$

This last equation is equivalent to the quadratic equation $\phi^2 - \phi - 1 = 0$. Using the quadratic formula, the positive solution of this equation is $\phi = \dfrac{1 + \sqrt{5}}{2}$, the golden ratio.

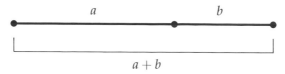

FIGURE 5.20. The ratio $a:b$ equals the ratio $(a + b):a$, called the golden ratio.

The golden ratio is also known as the *golden mean*, the *golden section*, and even the *divine proportion*. Many artists and architects consider figures whose proportions are in the golden ratio to be the most aesthetically pleasing, hence the "divine" characterization. In his book *Math and the Mona Lisa: The Art and Science of Leonardo da Vinci*, author Bülent Atalay argues for the existence of the golden ratio in Leonardo da Vinci's *Mona Lisa*. In fact, the ratio of the height to the width in this famous painting equals the golden ratio.

The Fibonacci numbers are intimately related to the golden ratio. Taking the ratios of consecutive Fibonacci numbers gives a sequence of numbers which limits on (or "converges to," using more fancy mathematical language) the golden ratio. Rounding to six decimal places, the ratios of consecutive Fibonacci numbers are

$$1, 2, 1.5, 1.666667, 1.6, 1.625, 1.615385, 1.619048, 1.617647, 1.618182, \ldots,$$

a sequence alternating above and below ϕ. We will verify that this sequence converges to the golden ratio in Exercise 5. Since $\sqrt{5}$ is irrational, so is ϕ. If we compute the continued fraction expansion of the golden ratio, we find that $\phi = [1; 1, 1, 1, 1, 1, 1, \ldots]$. This means that ϕ is the least "rational-like" irrational number. Moreover, the convergents of ϕ, those rational numbers that are the best approximations to ϕ, are the ratios of consecutive Fibonacci numbers. (See Section 4.5.2 and Equation (4.15) for a review of continued fractions and convergents.)

We can now provide some justification as to why the Fibonacci numbers arise in flowers. In order to propagate the species, a flower strives to optimize the number of seeds in its seed head. Starting at the center, each successive seed occurs at a particular angle to the previous one, on a circle slightly larger in radius. This angle needs to be an irrational multiple of 2π; otherwise, there is wasted space. But it also needs to be poorly approximated by rationals; otherwise, there is still wasted space. Thus, the best angle is one that divides the circumference of the circle using the golden ratio ϕ. This special angle, which is approximately $137.5°$, is called the *golden angle*. Since we must have an integer number of seeds, the convergents to ϕ will be the best approximations to this angle, and this is why we see Fibonacci numbers. Since the petals of flowers are formed at the extremities of the seed spirals, we also see Fibonacci numbers in the number of flower petals. In some sense, mother nature knows her math.[10]

5.2.3 *Music for Strings, Percussion and Celesta*

One of the pieces studied in detail by Lendvai was Bartók's *Music for Strings, Percussion and Celesta*. Among other items, he pointed to some structural traits of the first movement: 89 measures in length, a dynamic climax at the end of bar 55, string mutes removed in measure 34, and an exposition that ends after 21 bars. Voilà: all four of these special measures correspond to Fibonacci numbers.

Unfortunately, Lendvai made some questionable assumptions, cherry-picked favorable data, and even fudged some of his numbers. For example, the first movement of *Music for Strings, Percussion and Celesta* is actually 88 bars long, not 89. Lendvai includes a footnote explaining the discrepancy, stating that "[t]he 88 bars of the score must be completed by a whole-bar rest, in accordance with the Bülow analyses of Beethoven."[11] This is a dubious

---

[10]Smith College manages a very informative and interactive website focusing on Fibonacci phyllotaxis: http://www.math.smith.edu/phyllo/index.html .

[11]Lendvai, *Béla Bartók*, p. 28.

assumption. Some of Lendvai's conclusions are incomplete, such as the claim that string mutes are removed in measure 34. In fact, the second, third, and fourth violins are the *only* string parts to remove their mutes in measure 34. The viola mutes come off at the end of bar 33, the mutes on the first violins and cellos are removed at the start of measure 35, and the bass mutes are withdrawn at the end of measure 37. Thus, the claim that string mutes are removed in a "Fibonacci measure" is simply false.

Howat's analysis

Lendvai's errors are addressed in the excellent and thorough paper "Bartók, Lendvai and the Principles of Proportional Analysis" by analyst Roy Howat [1983a]. Figure 5.21 shows Lendvai's "ideal proportions" versus the actual values computed by Howat, which can easily be confirmed by examining the score. While some of Lendvai's observations are accurate (e.g., the dynamic climax of the piece is clearly at the end of bar 55), others are off by a measure, presumably to provide a better fit with the Fibonacci numbers. In addition, some key moments of the first movement are neglected by Lendvai, such as the first entrance of the celesta in bar 77. Given that the title of the work includes the celesta, this entrance surely warrants as much attention as the removal of the mutes in the strings.

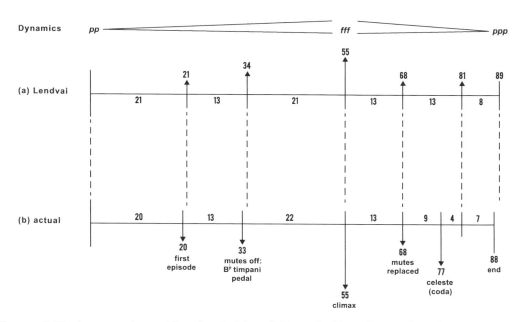

FIGURE 5.21. A comparison of Lendvai's (a) and Howat's (b) analyses of the first movement of *Music for Strings, Percussion and Celesta*.

Despite Lendvai's mistakes, he does raise some interesting questions about structure and form in the work. Howat pushes the analysis further, providing a fair and in-depth investigation into all facets of the complete work (particularly movements 1 and 3), one that is not predisposed toward a "Fibonacci bias." In addition to structural elements such as dynamic climaxes and instrumentation, Howat carefully examines the tonal structure in the work. His analysis of the first movement leads to some interesting conclusions.

The first movement is a fugue (à la Bach), with the opening subject played by the viola starting on A. This chromatic theme only ranges a tritone in distance (six half steps) to E♭. The successive answers to the main theme are transposed, alternating between ascending perfect fifths (A → E → B → F♯, etc.) and descending perfect fifths (A → D → G → C, etc.). This serves to keep each subject musically close, as demonstrated by the circle of fifths. The two different responses to the subject eventually arrive to the same starting note of E♭, which is halfway around the circle of fifths. Thus, the harmonic progressions of the responses to the subject mirror the actual theme itself. This important *tonal climax* occurs at measure 44, exactly halfway through the piece.

In the second half of the movement, after the dynamic climax in measure 55, the subject is inverted exactly and moves back around the circle of fifths to return to the opening A. Often, only the opening five notes are used (e.g., measure 65). The original five-note opening of the subject returns in measure 82, dividing the coda (defined by the entrance of the celesta) into $4 + 7$ bars. The first stretto in the fugue, where the initial subject is interrupted by another entrance of the subject before completing, occurs just before the end of measure 26 on the pitches F♯ and then C. These are precisely halfway around the circle of fifths. These return in inversion in the second half of the movement, ending in measure 68, giving a subdivision in the ratio 42:26, not Fibonacci numbers, but an excellent approximation to the golden ratio.

The very end of the movement contains a critical musical event that may give the best clues as to the composer's intentions. As the dynamic level becomes very soft, a magnificent exact inversion based on the scale of the main theme occurs between the first and second violins (all other instruments are silent; see Figure 5.22). The inversion is an exact horizontal reflection about A, reaffirming it as the tonal center of the movement.

FIGURE 5.22. The exact inversion about A in the violins at the end of the first movement of Bartók's *Music for Strings, Percussion and Celesta*. The first violins play the top part, while the seconds play the bottom. The motion from unison A to E♭ and back to A recaps the tonal structure of the fugue. © Copyright 1937 by Universal Edition A.G., Wien/UE 34129.

The inversion starts and ends on A (the two parts in unison) and rises and falls a tritone to E♭ in both voices, a wonderfully clever recap of the harmonic structure of the entire piece. Not only does the inversion highlight the tonal trip around the circle of fifths taken during the fugue, but it also serves to connect the two halves of the piece, as the second half features the inverted subject of the first. By closing this dramatic opening movement with a perfect inversion, Bartók pays tribute to the master of inversions and fugues, J. S. Bach. The final four notes of the first violins, C B♮ B♭ A, which translates in German (Bach's native tongue) to C H B A, also point to the great composer.

According to Howat's analysis, the third movement is the better example for discerning Fibonacci elements. Horror movie fans may recognize the chilling music from this movement (particularly memorable are the glissandi in the timpani and strings), as it was featured prominently in Stanley Kubrick's *The Shining*. This movement really does consist of 89 measures (a Fibonacci number), when counted in common time. The piece opens with a xylophone solo that follows an obvious Fibonacci rhythmic pattern of

$$1, 1, 2, 3, 5, 8, 5, 3, 2, 1, 1,$$

with a crescendo (<) followed by a decrescendo (>), and a dynamic climax at the middle of the sequence (see Figure 5.23). Based on Bartók's clear opening rhythmic statement, Howat opts to divide the movement using quarter-note beats rather than measure numbers as the metric for his analysis. Thus, a measure in $\frac{3}{2}$ counts as 1.5 and a measure in $\frac{2}{4}$ counts as 0.5. With this convention, Howat finds structural aspects of the movement that involve the Fibonacci numbers and the golden ratio.

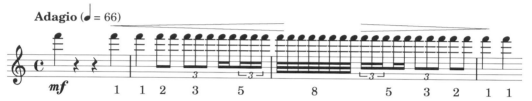

FIGURE 5.23. The clear Fibonacci pattern in the xylophone solo that opens the erie third movement of *Music for Strings, Percussion and Celesta*. © Copyright 1937 by Universal Edition A.G., Wien/UE 34129.

Influence of other composers

It is worth pointing out that some of the works of Bartók's contemporaries, such as Zoltán Kodály (1882-1967) and Claude Debussy (1862-1918), also show signs of using the golden ratio to highlight key structural moments in their works. These pieces were published before *Music for Strings, Percussion and Celesta*, so it stands to reason that Bartók was influenced by their work.

One such example is Kodály's *Méditation sur un Motif de Claude Debussy* (1907). Just as with the fugue from *Music for Strings, Percussion and Celesta*, the piece opens *pp* and ends *ppp*, with a central climax marked *fff*. Counting by quarter notes rather than measures, there are 508 beats in the piece. The golden ratio of 508 is 314 (to the nearest integer), and this just happens to be smack in the middle of the two climatic bars at *fff*.

Debussy, whose work Kodály was bringing to the attention of Bartók, published a collection of three piano pieces in 1905 called *Images* which soon became a part of Bartók's piano repertoire. The three pieces, *Reflets dans l'eau*, *Mouvement*, and *Hommage à Rameau*, have forms that demonstrate the golden ratio. *Reflets* and *Mouvement* begin *pp* and finish *ppp* and *pp*, respectively. They also have main climaxes at *ff* and *fff*, respectively, located at measures that divide the total piece by the golden ratio. *Hommage à Rameau* has a similar structure dynamically and, according to Howat, has a clear connection with the Fibonacci numbers.

Final remarks

As we conclude our story, what are we to make of the arguments of Lendvai and the more thorough analysis of Howat? Did Bartók consciously use the Fibonacci numbers and the golden ratio in *Music for Strings, Percussion and Celesta*? Critics of this perceived mathematical-musical connection point to the fact that there is no written record that Bartók made any calculations illustrating Fibonacci or the golden section. Based on his training in mathematics, as well as his love of nature and folk music, it is clear that Bartók knew about the Fibonacci numbers, their presence in nature, and their connection to the golden ratio. But if he was so inclined to incorporate Fibonacci into his music, why are there no concrete written records of it in the sketches and notes of Bartók's scores? This is a fair question, one that certainly casts a shadow of doubt on the line of reasoning first begun by Lendvai.

There are two possible explanations for the lack of written evidence. First, Bartók was highly secretive about his works, even when conferring with his students. He rarely spoke about his compositional practices and secrets. Moreover, Bartók was already being criticized for writing music that was overly "cerebral." Identifying the mathematical patterns in structure and tonality in his music would have only added fuel to his critics' fire.

A second explanation is that researchers have been looking in the wrong place. While no written calculations pertaining to Fibonacci numbers exist in the manuscripts of his scores, there actually *is* written evidence that Bartók made these calculations in his analysis of folk music. As reported by Howat, manuscript 80FSS1 from the New York Bartók Archive contains early drafts of Bartók's transcriptions of Turkish folk songs. On one of the pages is a series of calculations involving, not Fibonacci, but the *Lucas numbers*, the sequence

$$1, 3, 4, 7, 11, 18, 29, 47, 76, 123, \ldots . \tag{5.5}$$

The numbers are named after Edouard Lucas, a nineteenth-century French mathematician who studied and generalized the Fibonacci numbers.

Like the Fibonacci and Hemachandra numbers, the Lucas numbers are generated using the same recursive formula, $L_n = L_{n-1} + L_{n-2}$, and differ only in the first two entries, 1 and 3. Moreover, the ratio of successive Lucas numbers also limits on the golden ratio, as this limit only relies on the recursive relation and is independent of the two numbers beginning the pattern (see Exercise 6). According to Bartók's own analysis of the folk song's meter, a subdivision in terms of the Lucas numbers is clearly discernible.

This hard evidence suggests that Bartók may have consciously decided to insert the Lucas numbers, in addition to Fibonacci and the golden ratio, into the structural features of his works. A quick glance at the numbers calculated by Howat for the first movement of *Music for Strings, Percussion and Celesta* shows a greater proclivity toward the Lucas numbers than Fibonacci. Regardless of where one sides in the Fibonacci/Bartók debate, the strength of the first movement, one that is inherently mathematical, lies with Bartók's use of symmetry. The tonal climax halfway through the piece, the inverted subject, the tonal symmetry around A, the mirrored trips around the circle of fifths, and the brilliant inversion in the violins at the end of the movement are key symmetric features that an attentive listener can hear and enjoy.

Exercises for Section 5.2

1. Find a flower somewhere on your campus (or in your neighborhood) and count the number of petals. Do you find a Fibonacci number? Given the season or climate, you may have to be creative where you look to locate a flower. State the location, type of flower, and number of petals counted.

2. How well would ϕ work as a musical interval? Regarding ϕ as a multiplier (as in Chapter 4), convert it into cents. Based on your answer, which musical interval is closest to the one determined by ϕ? Would it sound consonant or dissonant? Explain.

3. Suppose that a piece of music being composed is 144 measures long. Where should the climax of the piece be located in order to divide it into two parts whose proportion is in the golden ratio? There are two answers. Find both of them, rounding to the nearest measure. What is significant about these numbers? Complete the same problem for a piece of music that is 150 measures long. Why do you have to round off by more to get your answer with 150 than with 144? *Hint:* The last question has something to do with the properties of convergents and continued fractions (see Section 4.5.2).

4. Recall that the golden ratio is given by $\phi = (1 + \sqrt{5})/2$. Using only algebra, compute and simplify the following quantities:

   a. ϕ^2 **b.** $\dfrac{1}{\phi}$ **c.** $\phi^2 - \phi$ **d.** $\phi^2 - \dfrac{1}{\phi}$ **e.** $\phi^3 - \dfrac{2}{\phi}$

5. In this exercise, we will prove that

$$\lim_{n \to \infty} \frac{F_{n+1}}{F_n} = \phi = \frac{1 + \sqrt{5}}{2}.$$

In other words, the ratios of successive Fibonacci numbers approach the golden ratio; the further along in the sequence, the better the approximation. This key fact establishes an important connection between the Fibonacci numbers and the golden ratio.

   a. Beginning with the recursive formula $F_{n+1} = F_n + F_{n-1}$, divide both sides by F_n and simplify.

   b. We will assume that $\lim\limits_{n \to \infty} \dfrac{F_{n+1}}{F_n} = L$ exists. This requires a rigorous proof, but we will assume that the limit exists and denote it as L. What is $\lim\limits_{n \to \infty} \dfrac{F_n}{F_{n-1}}$?

   c. Rewrite the right-hand side of your simplified equation from part **a.** to include the ratio F_n/F_{n-1}. Now take the limit as $n \to \infty$ of both sides to get an equation solely in terms of L. You can apply such facts as "the limit of a sum equals the sum of the limits."

   d. Solve your equation for L from part **c.** and show that $L = \phi$.

6. The previous exercise only relies on the defining recursive relation $F_{n+1} = F_n + F_{n-1}$, which states that the next term in the sequence is found by summing together the previous two. Regardless of the starting numbers, the ratio of successive numbers always limits on the golden ratio.

   a. Check that the ratios of successive Lucas numbers (see Equation (5.5)) approach the golden ratio. Use a calculator to write out at least the first 10 ratios.

   b. Pick any two positive integers and then use the recursive formula $F_{n+1} = F_n + F_{n-1}$ to write out your own special Fibonacci-like sequence. Check that the ratios of successive numbers in your sequence approach the golden ratio. Use a calculator to write out at least the first 10 ratios.

5.3 Group Theory

As we discuss symmetry in music and look at the properties of transformations such as reflections and rotations, it becomes evident that there are some interesting mathematical ideas at work. These concepts are part of an important branch of mathematics known as *abstract algebra*.

Much of algebra is concerned with structure. What are the key properties that addition and multiplication possess that can be generalized in other settings? What are the inherent •structures? Which concepts should be defined and which can be deduced? After years of struggling with these questions, mathematicians came up with the fundamental concept, called a *group*. The subject of group theory will help us gain a fuller understanding of symmetry in music, as well as change ringing, to be explored in Chapter 6.

There are two features to a group, the members of the group, typically called its *elements*, and a group *operation*, denoted by $*$. The group operation explains how to combine two elements in a group to yield another element in the group. Since this and the notation are similar to the multiplication of two real numbers, we will often refer to $a * b$ as the *product* of a and b. We will use the standard notation of G for a group. The symbol $\in$ means "is an element of." For instance, $a \in G$ means that a is an element of the group G, and $\sqrt{2} \notin \mathbb{Q}$ means that $\sqrt{2}$ is not a rational number.

Definition 5.3.1. *The set G is a group under the operation $*$ if the following four properties are satisfied:*

(i) *Closure: If $a \in G$ and $b \in G$, then $a * b \in G$. This must be true for all elements a and b in the group G.*

(ii) *Associativity: $(a * b) * c = a * (b * c)$ for any elements $a, b, c \in G$.*

(iii) *Identity: There must exist an element $e \in G$ called the identity element such that $a * e = a$ and $e * a = a$.*

(iv) *Inverse: For every element $a \in G$, there must exist an element $a^{-1} \in G$ called the inverse of a, such that $a * a^{-1} = e$ and $a^{-1} * a = e$. Note that the inverse of each element must be in the group G.*

Let us clarify these important properties. Closure, the first property, ensures that elements from the group combine together to produce other elements still *inside* the group. If the group consisted of elements in a box, then pulling any two elements out of the box and computing their product would yield an element from the box—the box is "closed." The second property is about arrangements when multiplying. If you are multiplying three elements of a group together, it doesn't matter whether you multiply the first two together and then the third, or whether you multiply the last two together and multiply by the first. In other words, the notation $a * b * c$ is perfectly clear; there is no need for any parentheses.

What *is* important is the order in which group elements are multiplied. There are many examples of groups where $a * b \neq b * a$. We will study a few in this book. A group G for which $a * b = b * a$ for all a and b in the group is called a *commutative group*.

The identity element e is just like 0 for usual addition, or 1 for multiplication. These are numbers that have no effect on other numbers; they preserve the "identity" of the numbers they are adding or multiplying, respectively. The same trait applies to e, since $a * e = e * a = a$ preserves the identity of a.

Finally, inverses are the elements that return us to the identity, a bit like going backward. The product $a * a^{-1} = e$ means that a^{-1} "cancels out" what a is doing. Similarly, $a^{-1} * a = e$ means that a cancels out a^{-1}. In usual addition, the inverse of 4 is -4 since $4 + (-4) = 0$ or $-4 + 4 = 0$. For standard multiplication, the inverse of 7 is $1/7$ as $7 \cdot 1/7 = 1/7 \cdot 7 = 1$.

5.3.1   Some examples of groups

We now discuss some basic examples of groups.

Example 5.3.2. *Suppose that $G = \mathbb{Z}$ (the integers) with $* = +$, the usual addition of two numbers. For example, $3 * 5 = 8$, because $3 * 5$ really means $3 + 5$. We also have $-1 * 5 = 4$ and $9 * 0 = 9$ in this group. Closure is satisfied because the sum of two integers is another integer. Associativity follows because the addition of real numbers is associative. The identity element is $e = 0$, and $a^{-1} = -a$ since $a * a^{-1} = a + (-a) = 0 = e$. Note that a^{-1} is always in G because the negative of an integer is still an integer. Since $a + b = b + a$, this group is commutative.*

Example 5.3.3. *Let $G = \mathbb{R} - \{0\}$ (all real numbers except for 0) with $*$ equal to the usual multiplication of two real numbers. In this case, $3 * 5 = 15$ since $3 * 5 = 3 \cdot 5$. Closure is satisfied because the product of two nonzero real numbers is another nonzero real number. Associativity holds because multiplication of real numbers is associative. In this group, the identity element is $e = 1$, and $a^{-1} = 1/a$ since $a * a^{-1} = a \cdot 1/a = 1 = e$. Note that a^{-1} is defined (and in G) because we have excluded 0 from our group. Since multiplication is commutative, this group is commutative.*

We point out that the set $G = \mathbb{Z}$ (the integers) is *not* a group under multiplication.

Example 5.3.4. *Suppose that $G = \{0, 1, 2, \ldots, 10, 11\}$, and define $*$ to be <u>addition modulo 12</u>. In other words, when we add two elements together, if the result is greater than 11, we subtract a multiple of 12 until it is in G. For example,*

$$9 * 5 = 9 + 5 = 14 = 2 \pmod{12} \qquad (\text{since } 14 - 12 = 2),$$

or

$$8 * 4 = 8 + 4 = 12 = 0 \pmod{12} \qquad (\text{since } 12 - 12 = 0).$$

This commutative group is often denoted $\mathbb{Z}_{12}$ and called the <u>integers mod 12</u>. The identity element is $e = 0$ and $a^{-1} = 12 - a$ since $a + (12 - a) = 12 = 0 \pmod{12}$. Although this group may seem strange, it is exactly how we tell time in countries that use 12-hour clocks. For example, 2 hours after 11:00 is 13:00 in Britain, but 1:00 in the United States, so $13:00 = 1:00 \pmod{12}$. Musically speaking, this group is akin to identifying the same note in different octaves. If the various instruments in an orchestra are all playing C in different registers, we still think of the note heard as just a "C," even if the frequencies are different. Identifying all integers (or notes) that are multiples of 12 (or octaves) apart from each other is called forming an <u>equivalence class</u>. This concept is a key principle in 12-tone music, discussed in Chapter 7.

Note that there is nothing special about the number 12 in the previous example. Modular arithmetic works with any natural number. For example, $\mathbb{Z}_3$, the integers mod 3, identifies every integer with either 0, 1, or 2. The group $\mathbb{Z}_2 = \{0, 1\}$ identifies every integer with 0 if it is even and 1 if it is odd. The fact that the sum of two odd numbers is even is captured by the fact that $1 + 1 = 0 \pmod 2$.

5.3.2 Multiplication tables

One of the best ways to understand a group and its inherent structure is to create a *multiplication table*, a square table that shows all of the possible products from within the group. This only makes practical sense for groups with a finite number of elements, known as *finite groups*. The groups defined using modular arithmetic are examples of finite groups. A group multiplication table is sometimes called a *Cayley table*, named in honor of the English mathematician Arthur Cayley (1821–1895).

Consider the group $\mathbb{Z}_4 = \{0, 1, 2, 3\}$, with $*$ equal to addition modulo 4. The multiplication table for $\mathbb{Z}_4$ is given by Table 5.3. Each entry is obtained by calculating $a * b$, where a is from the leftmost column and b from the top row, with the result shown in the corresponding row and column. For example, since $2 * 3 = 2 + 3 = 1 \pmod 4$, the entry in the row corresponding to 2 and column corresponding to 3 is 1. Similarly, we see from the table that $2 * 2 = 0$, which is correct since $4 = 0 \pmod 4$.

$*$	0	1	2	3
0	0	1	2	3
1	1	2	3	0
2	2	3	0	1
3	3	0	1	2

TABLE 5.3. The multiplication table for $\mathbb{Z}_4$.

There are several important points to make about Table 5.3. First, notice that every product is in $\mathbb{Z}_4$ (closure), and that each element of the group occurs exactly once in any row or column. Fans of *Sudoku* puzzles will recognize the latter property, as those puzzles require each digit from 1 to 9 to appear just once in each row and column. This "uniqueness" property will follow from the Cancellation Property discussed below. Second, note that the table is symmetric about the diagonal through $*$. This occurs because $\mathbb{Z}_4$ is commutative ($a * b = b * a$).

Third, there is a recognizable pattern to the placement of the entries. The first row is simply $0, 1, 2, 3$. The second row shifts these numbers to the left and moves the 0 to the end of the row. This is precisely the cyclic shift σ we discussed in Chapter 2 (see Definition 2.2.5). The third row is σ applied to the second row, or σ^2 applied to the first, and the last row is σ^3 applied to the first row. The shift σ^4 applied to the first row leaves it unchanged. This cycling through the numbers of the group is why $\mathbb{Z}_4$ is called a *cyclic group*. Finally, the inverse of each element in the group can be found by looking for the corresponding column element that produces a 0 in the given row. For instance, $1^{-1} = 3$ since $1 * 3 = 0$, and $2^{-1} = 2$ since $2 * 2 = 0$.

The fact that each row or column of Table 5.3 contains each element of $\mathbb{Z}_4$ exactly once is true for any group. This is a particularly useful property, one that helps ensure that a group multiplication table is constructed correctly. It follows from another useful fact about groups called the *Cancellation Property*. As we move forward in our discussion of groups, we will begin to drop the $*$ notation for convenience. Thus, it will be understood from here on that ab is equivalent to $a * b$.

Theorem 5.3.5 (The Cancellation Property). *For any elements a, b, c of a group G,*

$$\textbf{(i)} \; ab = ac \implies b = c, \qquad \textbf{(ii)} \; ba = ca \implies b = c.$$

Proof: The proof is straightforward and relies on the fact that a^{-1} exists. We carefully give the argument step by step. We have

$$
\begin{aligned}
ab = ac &\implies a^{-1} * (ab) = a^{-1} * (ac) && \text{(multiplication by } a^{-1} \text{ on the left)} \\
&\implies (a^{-1} * a) * b = (a^{-1} * a) * c && \text{(associative property)} \\
&\implies e * b = e * c && \text{(definition of inverse)} \\
&\implies b = c && \text{(definition of identity),}
\end{aligned}
$$

which proves item **(i)**. The proof of item **(ii)** is similar and is left as an exercise. □

Part **(i)** of the theorem is cancellation on the left, and part **(ii)** is cancellation on the right. While these may seem like common sense, notice that the proof used three of the four properties in the definition of a group. The Cancellation Property quickly explains why each element of a group appears precisely once in any row or column of a group's multiplication table. Suppose that a particular row had two equivalent entries. This would mean that there exists some group element a (corresponding to row a) and *different* group elements b and c such that $a * b = a * c$ or $ab = ac$. But then left cancellation quickly yields $b = c$, a contradiction. Thus, any particular row can never have repeated elements. Since there are as many entries in a row as there are elements in the group, every row has each element of the group exactly once. The argument for the columns is the same, except that now we have $ba = ca$ for different elements b and c (corresponding to column a), and right cancellation yields the desired contradiction.

Example 5.3.6. *Recall the complex number $i = \sqrt{-1}$, defined via the equation $i^2 = -1$. Let $G = \{1, i, -1, -i\}$ and $*$ correspond to complex multiplication. Construct the multiplication table for G and confirm that G is a group.*

Solution: We start with a few sample multiplications in G. We have $i * -i = -i^2 = 1$ and $-1 * i = -i$. The complete multiplication table is shown in Table 5.4, demonstrating that G is indeed closed under complex multiplication. Associativity follows from the fact that complex multiplication, just like real multiplication, is associative. In this group, $e = 1$ is the identity element. Finally, the inverse of each element is easily discerned from Table 5.4, or can be found by $a^{-1} = 1/a$: $1^{-1} = 1$, $i^{-1} = -i$, $(-1)^{-1} = -1$, $(-i)^{-1} = i$. Since all four properties of a group are satisfied, we have shown that G is a group. Note the similarities between Table 5.3 and Table 5.4. G is also a commutative and cyclic group. □

$*$	1	i	-1	$-i$
1	1	i	-1	$-i$
i	i	-1	$-i$	1
-1	-1	$-i$	1	i
$-i$	$-i$	1	i	-1

TABLE 5.4. The multiplication table for $G = \{1, i, -1, -i\}$ under complex multiplication.

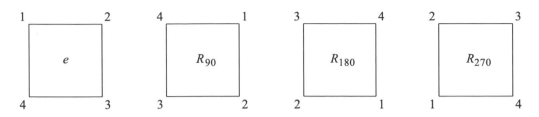

FIGURE 5.24. The original square along with clockwise rotations of $90°, 180°$, and $270°$.

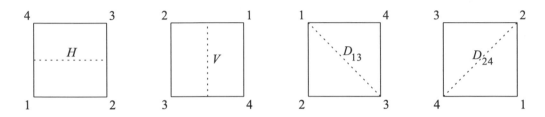

FIGURE 5.25. The four reflections of the original square through a horizontal, vertical, left-diagonal, and right-diagonal line of symmetry (shown dashed in each example).

5.3.3 Symmetries of the square

Groups can arise in many different arenas, not just in the realm of real or complex numbers. One group that has some interesting connections to music involves the symmetries of a square. We start with a square with vertices labeled $1, 2, 3, 4$ in clockwise order. A *symmetry of the square* is a geometric transformation that preserves the size and shape of the square.

For example, rotating the square $90°$ clockwise about its center returns the square to its original position, but with a different labeling of the vertices. As in Section 5.1.1, we denote this rotation as R_{90}. The square is also preserved under rotations of $180°$ (R_{180}) and $270°$ (R_{270}). A rotation of $360°$ returns the square to the same position with the same labeling, so this is really the identity element e (the transformation that doesn't move the square). Figure 5.24 shows the original square along with the three rotations just discussed. The numbering of the vertices helps us distinguish one transformation from another. Any other rotation either is equivalent to one of $\{e, R_{90}, R_{180}, R_{270}\}$ or doesn't return the square to its original position. For instance, a rotation of $45°$ still produces a square, but now it is a diamond balancing on one vertex, as opposed to a square resting on one side as shown in Figure 5.24.

In addition to rotations, it is possible to reflect the square about a line of symmetry and still preserve the original shape. There are four such lines: a horizontal line through the center (H), a vertical line through the center (V), a southeast diagonal between vertices 1 and 3 (D_{13}), and a northeast diagonal between vertices 2 and 4 (D_{24}). Figure 5.25 displays the four reflections along with the corresponding labels on the vertices. These are the only possible reflections that preserve the original position of the square.

While translations also preserve the shape of the square, the labeled vertices will not change, so we consider any translation equivalent to the identity element e. Thus, there are eight symmetries of the square, including the identity transformation. We claim that these eight operations form a group, with multiplication $*$ defined to be the composition of two transformations. This group is known as the *dihedral group of degree 4* and is denoted by D_4. To recap, the eight elements of D_4 are the symmetry transformations

$$D_4 = \{e, R_{90}, R_{180}, R_{270}, H, V, D_{13}, D_{24}\}.$$

Let us construct the multiplication table for D_4. We define the product $a * b$ to mean perform the transformation a on the original square first and then perform transformation b on the result. For example, $R_{90} * R_{180}$ means rotate the original square clockwise $90°$ and then rotate again clockwise by $180°$. The result is R_{270}. Similarly, $R_{180} * R_{270} = R_{90}$, because $180 + 270 = 450$ and R_{450} is equivalent to R_{90}.

Composing rotations and reflections is a bit more complicated. Using the labels on the vertices shown in Figures 5.24 and 5.25 helps determine the resulting product. It might also be illustrative to build a model and physically rotate and reflect the square. We will always list the ordering of a particular square by starting in the upper left corner and moving clockwise around the square. For instance, e corresponds to the ordering $1, 2, 3, 4$, while V has the ordering $2, 1, 4, 3$. When composing with D_{13} and D_{24}, keep in mind that these are always reflections about the southeast and northeast diagonals, respectively, regardless of the numbers at the vertices. In other words, D_{13} does *not* mean fix vertices 1 and 3 and interchange the other two. It just means reflect the square about the diagonal running from the upper left to the lower right corner.

Suppose that we want to compute $H * R_{90}$. Applying H first, we obtain a square with vertices ordered $4, 3, 2, 1$. Applying a clockwise $90°$ rotation then yields the ordering $1, 4, 3, 2$. Since $1, 4, 3, 2$ is the same ordering as D_{13}, we conclude that $H * R_{90} = D_{13}$ (see the top of Figure 5.26). On the other hand, $R_{90} * H = D_{24}$ (the order matters). To check this last product, we first apply R_{90} yielding the ordering $4, 1, 2, 3$. Then we apply a horizontal reflection, which interchanges the pair of vertices labeled 1 and 2, and also the pair of vertices labeled 4 and 3. The end result is a square with vertices ordered $3, 2, 1, 4$, corresponding to the group element D_{24} (see the bottom of Figure 5.26).

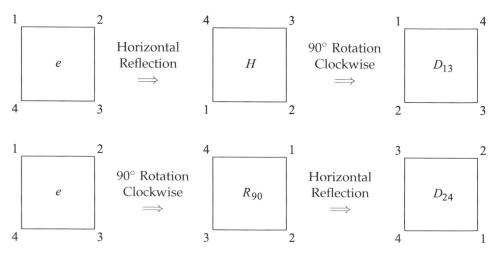

FIGURE 5.26. The product $H * R_{90} = D_{13}$, while $R_{90} * H = D_{24}$.

$*$	e	R_{90}	R_{180}	R_{270}	H	V	D_{13}	D_{24}
e	e	R_{90}	R_{180}	R_{270}	H	V	D_{13}	D_{24}
R_{90}	R_{90}	R_{180}	R_{270}	e	D_{24}	D_{13}	H	V
R_{180}	R_{180}	R_{270}	e	R_{90}	V	H	D_{24}	D_{13}
R_{270}	R_{270}	e	R_{90}	R_{180}	D_{13}	D_{24}	V	H
H	H	D_{13}	V	D_{24}	e	R_{180}	R_{90}	R_{270}
V	V	D_{24}	H	D_{13}	R_{180}	e	R_{270}	R_{90}
D_{13}	D_{13}	V	D_{24}	H	R_{270}	R_{90}	e	R_{180}
D_{24}	D_{24}	H	D_{13}	V	R_{90}	R_{270}	R_{180}	e

TABLE 5.5. The multiplication table for the eight symmetries of the square.

The complete multiplication table for D_4, the symmetries of the square, is given in Table 5.5. Notice that every element in the table belongs to D_4, which shows that the group is closed under composition. Moreover, every element in D_4 has an inverse, as indicated by the presence of e in each row and column. For example, since $R_{90} * R_{270} = R_{270} * R_{90} = e$, we know that R_{90} and R_{270} are inverses of each other. This makes sense geometrically: to undo a 90° rotation, we rotate backward 90°, which is equivalent to the rotation R_{270}. The inverses of each element in D_4 are

$$e^{-1} = e, \qquad (R_{90})^{-1} = R_{270}, \qquad (R_{180})^{-1} = R_{180}, \qquad (R_{270})^{-1} = R_{90},$$

and

$$H^{-1} = H, \qquad V^{-1} = V, \qquad (D_{13})^{-1} = D_{13}, \qquad (D_{24})^{-1} = D_{24}.$$

Note that except for R_{90} and R_{270}, all other elements of D_4 are their own inverse. In general, a group element that is its own inverse (i.e., satisfies the equation $a^2 = e$) is called an *involution*. A reflection is an involution because if it is applied twice in a row, then the original object is returned ($H^2 = e$, $V^2 = e$, etc.). Checking the other two group properties, associativity follows from the definition of the composition of functions, and the identity element e is contained in our set of symmetries. This finishes the argument that the elements in D_4 form a group under $* = $ composition. Since $a * b \neq b * a$ for several different products, D_4 is a noncommutative group.

5.3.4   The musical subgroup of D_4

Recall that the musical symmetries of inversion, retrograde, and retrograde-inversion correspond to the mathematical transformations of a horizontal reflection, a vertical reflection, and a 180° rotation, respectively. Transpositions correspond to translations, which we identify with the identity element e as we did with the symmetries of the square.

$*$	e	H	V	R_{180}
e	e	H	V	R_{180}
H	H	e	R_{180}	V
V	V	R_{180}	e	H
R_{180}	R_{180}	V	H	e

TABLE 5.6. The multiplication table for the four musical symmetries.

Consider the set of these four musical symmetries, $M = \{e, H, V, R_{180}\}$. There is something special about M. Suppose that we create a multiplication table for the four elements in M, where $*$ equals composition, just as with the symmetries of the square. The result is shown in Table 5.6.

We see from the multiplication table that all products of elements in M yield an element in M. Thus, M is closed under composition. Moreover, the identity element e is contained in M, and each element is its own inverse (an involution). This shows that M, the set of musical symmetries, is a group under composition. The special group M, which arises frequently in mathematics, is sometimes called the *Klein four-group*, named after the famous German mathematician Felix Klein (1849–1925).

Since M is a subset of D_4, we call M a *subgroup* of D_4. In general, a subgroup of a group is a nonempty subset that forms a group by itself, under the *same* group operation. Subgroups play an important role in group theory and help us understand the structure of the larger group. When checking whether a subset of a group is actually a subgroup, the hardest property to verify is closure. Associativity follows easily because it is true for the larger group. The identity element e must be included in any subgroup. Inverses must also be checked for inclusion in the subgroup (they exist in G, but they also have to be contained in the subgroup).[12]

One of the most useful facts about a subgroup has to do with its size in comparison with the larger group it belongs to. Suppose that G is a finite group containing n elements. We call the number of elements in G the *order* of G, denoted by $|G| = n$. For example, $|\mathbb{Z}_4| = 4$ and $|D_4| = 8$. Notice that $|M| = 4$ divides evenly into $|D_4| = 8$. This fundamental fact about subgroups is true in general, proven by the great Italian mathematician Joseph-Louis Lagrange (1736–1813).

Theorem 5.3.7 (Lagrange's Theorem). *Suppose that G is a finite group with order n and that M is a subgroup of G with order m. Then, n is a multiple of m, that is, m divides evenly into n.*[13]

Simply put, Lagrange's Theorem states that the number of elements in a subgroup must divide evenly into the number of elements in the larger group. Thus, a group of order 8, such as the symmetries of the square, can have subgroups of size 1, 2, 4, or 8, but no other size. A group of order 7 can only have subgroups of order 1 or 7, because 7 is a prime number. In fact, if G has a prime number of elements, then its two subgroups are

---

[12] Actually, if G is finite and M is a nonempty subset of G which is closed under the group operation $*$, then the inverse property automatically holds. In other words, nonempty subsets of finite groups that are closed under multiplication are always subgroups.

[13] For a proof of this theorem, see Herstein, *Abstract Algebra*, p. 70.

quite elementary. They are the subgroup with one element, $\{e\}$, and the group G itself. In general, these two subgroups are often called the *trivial subgroups*.

Example 5.3.8. *Recall from Example 5.3.6 that $G = \{1, i, -1, -i\}$ forms a group under complex multiplication. List all of the subgroups of G.*

Solution: Since $|G| = 4$, Lagrange's theorem implies that any subgroup of G will have order 1, 2, or 4. The trivial subgroups are $\{1\}$ (the only possible group with one element must consist of the identity element) and G itself. To find a subgroup of order 2, we start with the identity element 1 and then attempt to add an additional element while still satisfying closure and the inverse property.

Suppose we tried the subgroup $S = \{1, i\}$. This does not work for two reasons. First, $i * i = i^2 = -1$, but -1 is not contained in S, so closure fails. Moreover, $i^{-1} = -i$, so the inverse of i is not contained in S either. A similar argument shows that $S = \{1, -i\}$ also fails. On the other hand, $S = \{1, -1\}$ is a subgroup of G. It is closed since $-1 * -1 = 1 \in S$, and it satisfies the property of inverses since $(-1)^{-1} = -1$. In sum, there are three subgroups of G: $\{1\}, \{1, -1\}$, and G itself. $\square$

Exercises for Section 5.3

1. Here are some questions concerning closure.

   a. Which of the following sets are closed under addition? Explain.
      (i) $\mathbb{Z}$ (integers)   (ii) $\mathbb{Q}$ (rational numbers)   (iii) $\mathbb{R} - \mathbb{Q}$ (irrational numbers)

   b. Which of the following sets are closed under multiplication? Explain.
      (i) $\mathbb{Z}$ (integers)   (ii) $\mathbb{Q}$ (rational numbers)   (iii) $\mathbb{R} - \mathbb{Q}$ (irrational numbers)

2. Show by way of an example that the subtraction of integers does *not* satisfy the associative property.

3. Explain why the integers do *not* form a group under multiplication.

4. Finish the proof of the Cancellation Property by showing that for any elements a, b, c of a group G, if $ba = ca$, then $b = c$. This is known as "cancellation on the right."

5. Using the multiplication table for the symmetries of the square (Table 5.5), what type of transformation results when composing two rotations? or two reflections? or a rotation and a reflection? Your answer to each question should be one of the following: a rotation, a reflection, or neither. Assume that $e = R_0$ is a rotation.

6. Form the group multiplication table for $\mathbb{Z}_6 = \{0, 1, 2, 3, 4, 5\}$ with $*$ equal to addition modulo 6. State the inverse of each element.

7. Consider the "multiplicative" group $G = \{1, 3, 5, 7\}$, where $*$ is multiplication modulo 8. Recall that $a = b \pmod 8$ if a and b differ by a multiple of 8. For example, $3 * 5 = 7$ because $3 \cdot 5 = 15 = 7 \pmod 8$. Also, $7 * 7 = 7^2 = 1$ because $49 = 1 \pmod 8$. Construct the multiplication table for G and state the inverse of each element. Compare your table with the multiplication table for the Klein four-group (see Table 5.6). What do you notice?

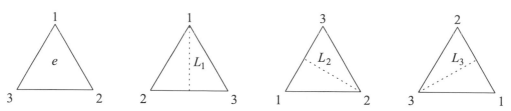

FIGURE 5.27. An equilateral triangle and its three reflections L_1, L_2, and L_3.

8. Consider the symmetries of an equilateral triangle, that is, transformations that leave the triangle unchanged. There are a total of six such transformations: three clockwise rotations (e, R_{120}, and R_{240}), and three reflections (L_1, L_2, and L_3) about the three altitudes of the triangle. We denote L_i as the reflection about the line of symmetry through vertex i (see Figure 5.27). The six transformations $\{e, R_{120}, R_{240}, L_1, L_2, L_3\}$ form a group with $*$ equal to composition of functions (just like the symmetries of the square). This group is called the *dihedral group of degree 3*, denoted D_3. Construct the multiplication table for this group, with the elements listed in the same order as above. Check that D_3 is indeed a group and state the inverse of each element. You may assume that $*$ is associative.

9. Suppose that a piece of music has two symmetries, a retrograde and an exact inversion about the middle line of the staff. What other type of symmetry must it necessarily possess? Explain using the musical group M. *Hint:* A retrograde corresponds to V, while an exact inversion is H. If the symmetries V and H are both present, then so is their composition $V * H$.

10. Generalizing the previous question, suppose that a piece of music has two of the following three symmetries: retrograde, inversion, and retrograde-inversion. Assume that both inversions are exact and about the middle line of the staff. Using group theory and the musical group M, explain why the third symmetry type must also be present. In other words, any piece with two of the three listed symmetries *automatically* has all three.

11. Recall that $\mathbb{Z}_4 = \{0, 1, 2, 3\}$ is a group with $*$ equal to addition modulo 4 (see Table 5.3). List all of the subgroups of $\mathbb{Z}_4$.

12. Find two subgroups of D_4, the symmetries of the square, that are order 2 (contain just two elements). Find a subgroup of D_4 of order 4 which is different from the musical group M.

13. Recall that $\mathbb{Z}$ (the integers) is a group under addition. Which of the following subsets S are subgroups of $\mathbb{Z}$ under addition? Explain carefully.

    a. $S = \{0, 1, -1\}$
    b. $S = \{0, 2, 4, 6, 8, \ldots\}$
    c. $S = \{0, \pm 2, \pm 4, \pm 6, \pm 8, \ldots\}$
    d. $S = \{0, \pm 1, \pm 3, \pm 5, \pm 7, \ldots\}$
    e. $S = \{0, \pm 10, \pm 20, \pm 30, \pm 40, \ldots\}$

References for Chapter 5

Atalay, B.: 2006, *Math and the Mona Lisa: The Art and Science of Leonardo da Vinci*, Smithsonian Books.

Bach, J. S.: 1992, *The Art of the Fugue and A Musical Offering*, Dover Publications, Inc., New York. Edited by Alfred Dörffel and Wolfgang Graeser.

Barber, S.: 1939, *Adagio for Strings*, G. Schirmer, Inc.

Bartók, B.: 1937, *Music for Strings, Percussion and Celesta*, Boosey & Hawkes Music Publishers Ltd.

Bartók, B.: 1966, A Népzene Forrásainál (At the Sources of Folk Music), *in* A. Szőllősy (ed.), *Bartók Béla Összegyűjtött Írásai (Selected Writings of Béla Bartók)*, Zeneműkiadó, Budapest, pp. 576–578.

Bartók, B.: 1987, *Mikrokosmos: 153 Progressive Piano Pieces*, Vol. 6, Nos. 140–153, Boosey & Hawkes Music Publishers Ltd.

Beethoven, L.: 1989, *Symphonies Nos. 5, 6 and 7* (in full score), Dover Publications, Inc., New York.

David, H. T.: 1945, *J. S. Bach's Musical Offering: History, Interpretation and Analysis*, Dover Publications, Inc., New York.

Goldberg, L.: 1995, *The Well Tempered Clavier of J.S. Bach: A Handbook for Keyboard Teachers and Performers*, Music Sources, Center for Historically Informed Performances, Inc.

Harkleroad, L.: 2006, *The Math behind the Music*, Cambridge University Press.

Haydn, J.: 1974–1977, *The Piano Sonatas*, Decca Record Company Limited, London. Performed by John McCabe. CD.

Herstein, I. N.: 1990, *Abstract Algebra*, 2nd edition, Macmillan Publishing Company.

Hindemith, P.: 1943, *Ludus Tonalis*, edition 3964, B. Schott's Söhne, Mainz.

Hodges, W.: 2003, The Geometry of Music, *in* J. Fauvel, R. Flood and R. Wilson (eds), *Music and Mathematics: From Pythagoras to Fractals*, Oxford University Press, pp. 90–111.

Hofstadter, D. R.: 1979, *Gödel, Escher, Bach: An Eternal Golden Braid*, Vintage Books, New York.

Howat, R.: 1983a, Bartók, Lendvai and the Principles of Proportional Analysis, *Music Analysis* 2(1), 69–95.

Howat, R.: 1983b, *Debussy in Proportion: A Musical Analysis*, Cambridge University Press.

Jean, R.: 1994, *Phyllotaxis: A Systemic Study in Plant Morphogenesis*, Cambridge University Press.

Kamien, R. (ed.): 1984, *The Norton Scores: An Anthology for Listening*, 4th edition, W. W. Norton & Company.

Lendvai, E.: 1955, *Bartók Stilusa (Bartók's Style)*, Zeneműkiadó, Budapest.

Lendvai, E.: 1971, *Béla Bartók: An Analysis of His Music*, Kahn & Averill, London.

Schubert, G.: 2007–2014, Hindemith, Paul, *Grove Music Online*, Oxford University Press.

Schulenberg, D.: 1992, *The Keyboard Music of J.S. Bach*, Schirmer Books.

Sousa, J. P.: ca. 1912, *The Thunderer*, Carl Fischer, New York.

Chapter 6

Change Ringing

Change ringing is a non-competitive and non-violent team activity that is highly stim-ulating intellectually and mildly demanding physically, and makes a beautiful sound. It develops mental and physical skills in a context of communal effort. The intense concentration required brings euphoric detachment that cleanses the mind of the day's petty demands and frustrations.

— North American Guild of Change Ringers[1]

In this chapter we discuss the mathematics behind the intriguing activity of change ringing. Beginning in England and developing over the course of many centuries, the sophisticated and highly structured practice of team bell ringing has evolved into a special blend of music, mathematics, and precision. The specific rules and requirements of change ringing are best understood in the mathematical language of permutations. In fact, the first change ringing composers were actually doing some group theory before group theory had even been mathematically defined. Specific examples on three and four bells are analyzed in detail, with the dihedral group making a surprise appearance.[2]

6.1   Basic Theory, Practice, and Examples

Large and booming bells are rung in the belfry of towers across the world to herald important events such as weddings, funerals, and coronations or to summon worshipers to service. A unique type of bell ringing, called *change ringing*, or sometimes known as campanology, began in England around the year 1610. It originated in part out of a desire to have a more amplified sound from the bell, as well as increased control over the precise timing of a ring. The technological advancement that facilitated these aims was to mount the bell on a large wheel, enabling it to swing full circle and be rung in the mouth-up position (bell facing upward), thereby allowing for greater dispersement of the sound (see Figure 6.1). Bell ringers pull on a large rope hanging overhead and attached to the wheel

---

[1]http://www.nagcr.org/pamphlet.html

[2]Much of the material for this chapter is based on the groundbreaking work of mathematician and composer Arthur White, who introduced the author to this subject many years ago as part of a splendid, undergraduate winter-term project at Western Michigan University.

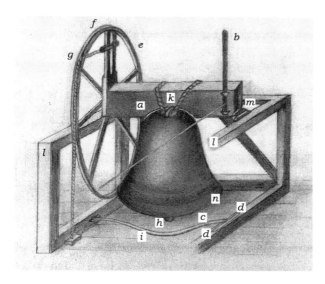

a. Stock
b. Stay
c. Slider
d. Blocks
e. Wheel
f. Groove of Wheel
g. Fillet
h. Ball of Clapper
i. Flight of Clapper
k. Cannons
l. Timber of Cage
m. Gudgeons
n. Lip of Bell

FIGURE 6.1. A large bell and the apparatus
required to make it ring.

to make the bell sound. The slider and the stay provide a great degree of control over the timing of the ring, but after a bell sounds, it can take up to 2 seconds before sounding again.

The bells in question are quite large, ranging in size from 200 lbs to several tons. There are typically anywhere from 4 to 12 bells in a particular tower, with some towers containing as many as 16 bells. Due to the overwhelming size and sophisticated ringing mechanics, it is difficult, if not impossible, to play a traditional "melody" with the bells. Instead, what is commonly (and historically) practiced is to ring the bells in some prescribed arrangement, where each bell is rung exactly once. The order of the bells can change from one arrangement to the next. How these "changes" are determined and subsequently carried out by the bell ringers is the essence of change ringing.

Change ringing is primarily practiced in England, where there are nearly 5000 towers serving around 40,000 ringers. There are about 45 North American towers that perform change ringing, with a similar number in Australia, New Zealand, Ireland, Wales, and Scotland. A smaller number of towers also exist in South Africa. During the 2012 London Olympic Games, change ringing occurred across the countryside, highlighted by the piece *Five Rings Triples*, specifically composed for the occasion by Howard Skempton. Remarkably, before the Olympics were held and in honor of the Queen's Diamond Jubilee, a floating belfry with eight bells rang a quarter peal (about 45 minutes in length) while traveling down the Thames River, leading a huge contingent of celebratory boats (see Figure 6.2).

6.1.1 Nomenclature

Suppose that there are n bells (n ringers) numbered consecutively as $1, 2, 3, \ldots, n$. Bells are rung as a team, with ringers standing in a circle, each assigned to one bell. Musically, the notes assigned to each bell decrease in pitch with bell 1 (the *treble*) being the highest and bell n (the *tenor*) being the lowest.

FIGURE 6.2. The Bell Tower (*left*) in Perth, Western Australia, home to the Swan Bells. The floating belfry (*right*) on the Thames River, London, during the Queen's Diamond Jubilee in June 2012.

A *change* is a specific arrangement of the bells so that each bell is rung exactly once, a rather democratic approach. For example, the four arrangements

$$(1\ 2\ 3\ 4\ 5),\quad (1\ 3\ 5\ 2\ 4),\quad (5\ 4\ 3\ 2\ 1),\quad \text{and}\quad (3\ 2\ 4\ 5\ 1)$$

are all valid changes on $n = 5$ bells, since each bell appears precisely once in each example. On the other hand,

$$(1\ 1\ 1\ 1\ 1),\quad (1\ 3\ 5\ 3\ 5),\quad (5\ 4\ 4\ 2\ 2),\quad \text{and}\quad (3\ 2\ 4\ 5\ 5)$$

are not valid changes because, in each case, at least one bell is repeated before all the bells have sounded. For a change such as $(3\ 2\ 4\ 5\ 1)$, it is understood that bell 3 is rung first, bell 2 goes second, bell 4 is third, bell 5 is next, and bell 1 is last.

Mathematically speaking, a change is simply a *permutation*, that is, a reordering of the numbers $1, 2, 3, \ldots, n$. Thus, valid changes correspond precisely to permutations. The special permutation $(1\ 2\ 3\ \cdots\ n)$, where the bells are rung in numerical order, is called *rounds*. This particular change is very significant, as all pieces must begin and end on rounds. Often, the ringers will repeat rounds several times to indicate that a composition is about to commence.

6.1.2 Rules of an extent

One of the primary goals of change ringing is to successfully complete an *extent*, a sequence of changes that conforms to a particular set of rules. Rules 1–3 are mandatory for any extent, while Rules 4–6 are considered optional, although they are often prized by change ringers.

Rules of an Extent

Mandatory:

1. The first and last changes must be rounds, $(1 \ 2 \ 3 \ \cdots \ n)$.

2. Other than rounds, all of the other possible changes on n bells must occur exactly once.

3. Between successive changes, no bell moves more than one position.

Optional:

4. No bell rests for more than two (sometimes relaxed further to four) consecutive changes.

5. Each *working bell* should do the same amount of "work" (obey the same overall pattern).

6. The list of moves used to help create the extent should form a palindrome (a retrograde symmetry). This is called the *palindrome property*.

Rules 2 and 3 will be explained in greater detail below. Rules 5 and 6 will be explored in a particular example in Section 6.1.5.

Rule 2: The importance of n!

Rule 2 yields one of the most interesting and challenging features of change ringing: no permutation (except for rounds) may be repeated during a piece. This ensures music that is constantly changing while also demanding great concentration on the part of the ringer, who must be mindful of his or her current location and ensuing position. Moreover, Rule 2 states that *all* of the possible permutations of $(1 \ 2 \ 3 \ \cdots \ n)$ must occur in the extent.

How many permutations of n bells are there? The answer is $n!$, read as "n factorial."

Definition 6.1.1. *Suppose that n is a positive integer. Then $\underline{n \text{ factorial}}$, denoted $n!$, is the product of n and all positive integers smaller than n. In other words,*

$$n! \ = \ n \cdot (n-1) \cdot (n-2) \cdots 3 \cdot 2 \cdot 1.$$

For example, $5! = 120$ since $5 \cdot 4 \cdot 3 \cdot 2 \cdot 1 = 120$, and $3! = 6$ since $3 \cdot 2 \cdot 1 = 6$. We define $0! = 1$ in mathematics to extend many useful formulas involving factorials. Note that $n! = n \cdot (n-1)!$, which is useful when computing. For example, $6! = 6 \cdot 5! = 6 \cdot 120 = 720$.

To understand why $n!$ arises in change ringing, we invoke an important branch of mathematics focused on counting, called *combinatorics*. Consider placing each bell in one of n slots. We can choose any bell we want for the first position, so there are n choices. But once we make a choice for the first bell, there are now $n-1$ remaining bells to choose from for the second slot, because we cannot repeat any numbers when creating a valid change. Continuing along, there are $n-2$ choices for the third slot because we have to eliminate the first two bells from contention. The counting continues in this fashion, decreasing by one as we move to the right, until we have two choices in the penultimate slot, followed by just one choice in the final slot (see Figure 6.3). It follows that the number of possible

n	$n-1$	$n-2$	$\cdots$	3	2	1

FIGURE 6.3. Counting the number of possible changes. The number of choices in a given slot depends on how far to the right that slot is positioned. The farther to the right, the less choices there are because numbers cannot be repeated.

permutations of n bells is $n!$, because we simply multiply the number of choices in each slot, $n \cdot (n-1) \cdot (n-2) \cdots 2 \cdot 1 = n!$. (Why we multiply these numbers together instead of adding requires a little bit of thought.)

As a quick check of our formula, note that when $n = 2$ there are only $2! = 2$ possible changes, (1 2) or (2 1). If $n = 3$, there are $3! = 6$ possible changes, given by

$$(1\ 2\ 3),\ (1\ 3\ 2),\ (2\ 1\ 3),\ (2\ 3\ 1),\ (3\ 1\ 2),\ \text{and}\ (3\ 2\ 1). \tag{6.1}$$

There are $4! = 4 \cdot 3! = 24$ possible permutations when $n = 4$. One way to write them down in a methodical fashion is to take all six changes for $n = 3$ given in Equation (6.1) and place a 4 in front of them. Then, for the remaining possibilities, we could replace any particular number in the changes in Equation (6.1) by a 4, and put the number being replaced at the front, creating six new changes. For example, if we pick 1 as the number to be replaced by 4, the six new changes are

$$(1\ 4\ 2\ 3),\ (1\ 4\ 3\ 2),\ (1\ 2\ 4\ 3),\ (1\ 2\ 3\ 4),\ (1\ 3\ 4\ 2),\ \text{and}\ (1\ 3\ 2\ 4).$$

These are all of the possible permutations for $n = 4$ that start with a 1. Since this procedure can be accomplished for each of the three numbers 1, 2, or 3, there are a total of $6 + 3 \cdot 6 = 24$ changes on four bells, as predicted by the $n!$ formula.

There is a rather daunting problem in composing an extent that satisfies Rule 2: the value of $n!$ grows exceedingly fast. In fact, $n!$ grows faster than any exponential function. To play all $n!$ changes in an extent requires not only a great deal of skill and planning but also a great deal of time. For each n, Table 6.1 displays the value of $n!$, the approximate length of time required to play all $n!$ changes, and the usual titles appended to pieces composed for the corresponding number of bells.

An important term in change ringing is a *peal*, which consists of at least 5040 changes for seven or fewer bells, or at least 5000 changes when ringing eight or more bells. Since the times required to play a complete extent on n bells are so massive, it is typical in practice to perform just a portion of the full extent. For example, weddings or major events are often announced by ringing a *quarter peal*, which means ringing at least 1/4 of the changes in a peal. Quarter peals take about 40–45 minutes to perform.

Rule 3: No jumping

Next, consider Rule 3, which states that no bell can move more than one position between successive changes. This rule, which can be interpreted as the "no jumping" rule, is necessitated by the size of the bells and the method with which they are rung. It also helps the ringer keep track of the path of their particular bell. Since bells cannot hop over one another, Rule 3 limits the number of allowable maneuvers on n bells. The only possible moves are for adjacent bells to interchange positions. Although this sounds rather limiting, the repeated application of allowable moves can generate a wide variety of compositions.

n	$n!$	Approximate Duration	Name
3	6	15 seconds	*Singles*
4	24	1 minute	*Minimus*
5	120	5 minutes	*Doubles*
6	720	30 minutes	*Minor*
7	5040	3 hours	*Triples*
8	40,320	24 hours	*Major*
9	362,880	9 days	*Caters*
10	3,628,800	3 months	*Royal*
11	39,916,800	3 years	*Cinques*
12	479,001,600	36 years	*Maximus*

TABLE 6.1. The approximate durations to ring an extent on n bells covering all $n!$ changes. The titles given to extents on n bells are also displayed. For example, *Plain Bob Minimus* is a composition written for four bells, while *Grandshire Triples* is an extent on seven bells.

Below are two possible sequences of changes for six bells. The parentheses around each change have been dropped to improve readability, a practice we will continue to use when listing a sequence of changes.

```
1 3 5 2 4 6            1 3 5 2 4 6
3 1 5 4 2 6            5 3 1 6 4 2
3 5 1 4 6 2            3 1 5 4 2 6
```

The sequence on the left is valid because no bell moves more than one position between successive changes. On the other hand, the sequence on the right violates Rule 3 because bells 5 and 6 move two positions to the left between changes 1 and 2, and two positions to the right between changes 2 and 3.

6.1.3 Three bells

We now consider the simple case of ringing an extent on three bells. Given Rule 3, there are only two possible moves allowed between changes: either the first two bells can interchange positions, or the last two bells can interchange. Any other maneuver will have one bell moving more than one position, violating Rule 3. For example, we cannot have the consecutive changes (1 2 3), (2 3 1), because bell 1 has moved two positions.

Denote by a the move that interchanges the bells in the first two positions and by b the move interchanging the bells in the last two positions. We write $a = (1\ 2)$ and $b = (2\ 3)$, to indicate the positions of the bells that are being interchanged. Note that it is the *positions* being interchanged, not the numbers of the bells. For instance, the move a always interchanges the first two bells, which is different from interchanging bells numbered 1 and 2.

Beginning with rounds, $(1\ 2\ 3)$, suppose that we apply the move a and interchange the first two bells. This leads to the change $(2\ 1\ 3)$. Next, we must apply move b, since applying a brings us right back to rounds (too early). We must apply some move because we are not allowed to repeat any changes. After interchanging the last two bells, we arrive at the third change in our sequence, $(2\ 3\ 1)$. If we were to apply b again, we would return to the previous change, which violates Rule 2. Thus, we apply a once more, interchanging the first two bells, obtaining $(3\ 2\ 1)$ as our fourth row. The process continues by alternating between moves a and b until we magically return to rounds.

A	**B**
1 2 3	**1** 2 3
2 **1** 3	1 3 **2**
2 3 **1**	3 **1** 2
3 2 **1**	3 2 **1**
3 **1** 2	2 3 **1**
1 3 2	**2** 1 3
1 2 3	**1** 2 3

FIGURE 6.4. The only two possible extents on three bells. Extent A is generated by the sequence of moves $[a\,b]^3$, while extent B arises from $[b\,a]^3$. In each extent, bell 1 is plain hunting up, while bell 3 is plain hunting down.

This extent (labeled A) is shown in Figure 6.4. Note that it satisfies the first three rules in the definition of an extent. It begins and ends with rounds, all six permutations are used exactly once (except for rounds), and no bell moves more than one position between rows. It is created by alternately applying moves a and b three times: $a\,b\,a\,b\,a\,b = [a\,b]^3$. (Note that $[a\,b]^3 \neq a^3 b^3$ in the usual fashion—the order in which transformations are composed matters.) One important observation is that once a has been chosen as the opening move, the remaining moves are completely determined in order to satisfy Rule 2. The only other possible extent (labeled B in Figure 6.4) begins with move b and proceeds to alternate between moves b and a. Again, the changes are completely determined once b opens the extent, following the pattern $b\,a\,b\,a\,b\,a = [b\,a]^3$. Any other choice would return to a previous permutation, violating Rule 2.

Note the simple zigzag pattern (in boldface) that bell 1 follows in extent A, sweeping quickly from position 1 to position 3 and then back again. In change ringing parlance, bell 1 is *plain hunting*. It only needs to do this once to complete the extent. Since bell 1 initially moves from the first to last position, we say that it is *hunting up*. Bell 3 is also hunting in extent A, although it is *hunting down* starting with the second change. In extent B, bell 1 also hunts up, beginning with the second change, while bell 3 hunts down right from the start.

6.1.4   The number of permissible moves

Before considering extents on more than three bells, we focus on the number of allowable moves for n bells that are in accordance with Rule 3. To be clear, a permissible "move" is a transformation that takes us from one change to the next without moving any bell more than one position. It follows that the only types of permissible moves are those that interchange adjacent bells.

For $n = 2$ bells, there is only one possible move, (1 2), where the two bells interchange positions. As we saw in the previous section, there are two permissible transformations for $n = 3$ bells, (1 2) and (2 3). After some thought, there are four possible moves for $n = 4$ bells. They are (1 2) (interchange the first two bells), (2 3) (interchange the middle pair of bells), (3 4) (interchange the last pair of bells), and (1 2)(3 4) (interchange both the first pair and last pair of bells). Unlike the case $n = 3$, with four bells we can either restrict a transformation to just one pair or move both pairs simultaneously.

To recap, the number of permissible transformations for $n = 2, 3$, and 4 bells is $1, 2$, and 4, respectively. Noticing the doubling pattern, it seems likely that the number of legitimate moves on $n = 5$ bells is eight. However, this is incorrect; there are only seven possible moves for $n = 5$ bells. To see this, we consider what happens to the number of permissible transformations when a bell is added. We will divide the set of possible moves into two groups, one containing those moves that keep the new bell fixed, and one for those that do not. The key observation is that there is only one way to move the new bell: it must be interchanged with the adjacent bell, which is in the penultimate position. In this case, that corresponds to (4 5).

We list the seven permissible transformations on $n = 5$ bells below, organizing them into three distinct groups:

$$(1\ 2), (2\ 3), (3\ 4), \qquad (1\ 2)(4\ 5), \qquad \underbrace{(4\ 5)}_{\text{new move}}.$$
$$\underbrace{(1\ 2)(3\ 4),}_{\text{moves from } n = 4} \qquad \underbrace{(2\ 3)(4\ 5),}_{\text{moves from } n = 3}$$

The group of four on the left are the only allowable transformations that do not move the bell in position 5 (the last bell). These are just the moves we found in the case $n = 4$. The remaining three transformations are the ones that move the last bell. The middle set contains the two moves from the case $n = 3$, with the new move (4 5) appended, while the last element on the right is simply the new allowable transformation (4 5). Thus, the number of permissible moves for $n = 5$ bells is $4 + 2 + 1 = 7$.

We now generalize the previous argument. Denote by C_n the number of allowable transformations on n bells. Thus, $C_2 = 1, C_3 = 2, C_4 = 4$, and $C_5 = 7$. As we did with the case $n = 5$, we break C_n into two subsets depending on whether the transformation keeps the last bell fixed or not. The number of allowable moves that keep the last bell fixed is given by C_{n-1}, since this number represents all the different ways to move the first $n - 1$ bells without moving the new bell. The number of permissible transformations that move the last bell is $C_{n-2} + 1$, since we can append $(n - 1\ n)$ to each of the moves counted by C_{n-2}. The +1 is needed for the new, simple move $(n - 1\ n)$. Thus, we have discovered the formula

$$C_n = C_{n-1} + C_{n-2} + 1. \tag{6.2}$$

Equation (6.2) is reminiscent of an equation we have seen several times before. Except for the +1, it is the same recursive equation that defines the Fibonacci, Hemachandra, and

n	# of allowable moves	# of allowable moves $+1$
2	1	2
3	2	3
4	4	5
5	7	8
6	12	13
7	20	21
8	33	34
9	54	55
10	88	89
11	143	144
12	232	233

TABLE 6.2. The number of allowable moves for n bells, where no bell moves more than one position, as a function of n. The third column contains the famous Hemachandra–Fibonacci numbers.

Lucas numbers discussed in Sections 1.4 and 5.2.1. We can use this recursive formula to determine the number of permissible moves for different n, as shown in Table 6.2. Here an entry in the second column is obtained by adding 1 to the sum of the entries in the previous two rows.

Note that if we add 1 to each element in the second column of Table 6.2, the resulting sequence is the Hemachandra–Fibonacci numbers. Technically speaking, the Fibonacci numbers begin $1, 1, 2, 3, 5, 8, \ldots$, so if F_n denotes the nth Fibonacci number, we have shown that $1 + C_n = F_{n+1}$. For example, if $n = 5$, then $C_5 = 7$ and $F_6 = 8$, so the formula holds. We have found a rather surprising connection between change ringing and the Fibonacci numbers:

The # of permissible moves on n bells is one less than the $(n+1)$th Fibonacci number.

6.1.5 Example: *Plain Bob Minimus*

We now examine two extents on $n = 4$ bells. These will be useful for explaining the optional Rules 4, 5, and 6. Figure 6.5 shows the 25 changes of the extent *Plain Bob Minimus*, which are displayed in three columns, called *leads*. The changes are read by moving down each lead and then hopping over to the top of the next lead. Note that bell 1 (the treble) goes plain hunting, requiring three round trips from position 1 to position 4, before finishing the extent. When the treble bell is plain hunting, the extent is called a *method*; otherwise, it is known as a *principle*.

```
1 2 3 4          1 3 4 2          1 4 2 3
2 1 4 3          3 1 2 4          4 1 3 2
2 4 1 3          3 2 1 4          4 3 1 2
4 2 3 1          2 3 4 1          3 4 2 1
4 3 2 1          2 4 3 1          3 2 4 1
3 4 1 2          4 2 1 3          2 3 1 4
3 1 4 2          4 1 2 3          2 1 3 4
1 3 2 4          1 4 3 2          1 2 4 3
                                  1 2 3 4
```

FIGURE 6.5. The changes for the extent *Plain Bob Minimus*, displayed with its three leads (columns). Bell 1 is highlighted to show that it goes plain hunting. Note that bells 2, 3, and 4 hunt down in leads 2, 3, and 1, respectively. This extent factors as $[(a\,b)^3\,a\,c]^3$.

It is worth verifying the first three rules of an extent. The hardest rule to check is that all 24 permutations of (1 2 3 4) are included and never repeated (except for rounds). Notice that for each of the numbers 1, 2, 3, or 4, there are six rows that begin with that number, none of which are the same. Since $4 \times 6 = 24$, we know that all 4! permutations have been included and none repeated, satisfying Rule 2.

To check Rule 3, it helps to write *Plain Bob* in terms of its defining transformations. We call this *factoring the extent*, something we have already accomplished for the two extents on three bells. Recall that there are four allowable transformations on four bells. Let $a = (1\ 2)(3\ 4)$, $b = (2\ 3)$, $c = (3\ 4)$, and $d = (1\ 2)$. This labeling may seem inconsistent with the labeling used for $n = 3$, but its meaning will become apparent later. Note that a is the only transformation that moves all of the bells.

The first lead (column) alternates between moves a and b, with a going first. It can be written as $a\,b\,a\,b\,a\,b\,a = (a\,b)^3\,a$. Then, the transformation that brings us from the bottom of the first lead to the top of the second is c. Now something interesting happens: the remaining two leads follow the exact same set of transformations (check this). It follows that the full extent factors as $[(a\,b)^3\,a\,c]^3$.

The fact that each transformation in *Plain Bob* is one of our permissible moves verifies Rule 3. Rule 4, that no bell rests (remains in the same position) for more than two consecutive changes, is satisfied here because the transformation a is used to generate every other row (as can be seen from the factorization). Since a moves all of the bells, no bell ever remains in the same position for more than two consecutive changes.

To understand Rule 5, we say that bells 2, 3, and 4 are all *working*, but bell 1, since it is plain hunting (regarded as a relatively simple pattern), is *not* considered to be a "working" bell. The idea behind Rule 5 is to check whether the working bells are all following the same path. Focus on the path traced out by bell 2 in the first lead of *Plain Bob*. This same path is followed by bell 3, but in the second lead. Then, the path that bell 2 follows in the second lead is precisely the same traced out by bell 3 in the last lead. Finally, the path of bell 2 in the last lead is equivalent to the path of bell 3 in the first lead. If the extent is repeated multiple times, then bell 2 and bell 3 are following the same patterns, with bell 3 one lead behind that of bell 2. Thus, bells 2 and 3 are doing the same amount of "work."

Next consider bell 4. It follows the path of bell 3, but one lead behind. In other words, the path of bell 4 in the second (third) lead is equal to the path of bell 3 in the first (second) lead. We see that all three working bells do the same amount of work, so that Rule 5 is satisfied. This is akin to a round, where bell 4 follows bell 3, which in turn follows bell 2. One of the useful aspects of Rule 5 is that the ringer can perform on any of the working bells. Once a ringer has learned the path of a working bell, he or she may play any of those bells, as long as they are attentive to the particular lead being rung.

All of this is easily recognized mathematically from the factorization of the extent. Since *Plain Bob* factors as something *cubed*, the working bells all follow the same, but shifted, path. Each lead is essentially the same except that the working bells have been permuted (note that bell 1 is always in the same spot—position 1—at the start of each lead). Since the leads are each generated by the same sequence of moves, each bell will follow the path determined by its predecessor in the opening lead. In sum, an extent that factors as a power of transformations from the opening lead will always satisfy Rule 5.

The final rule to be considered is Rule 6, the *palindrome property*. To check this rule, it is necessary to factor the extent. If the moves that generate the opening lead form a palindrome, then Rule 6 is satisfied. The first lead of *Plain Bob* is created by the moves $a\,b\,a\,b\,a\,b\,a$, which is a palindrome (the same "word" forward and backward) in the letters a and b. The move c, which brings us to the top of the next lead, is excluded from the list. Thus, Rule 6 is verified. Part of the significance of possessing the palindrome property, in addition to the aesthetic desire to have symmetry, is that it helps the ringers memorize their parts.

6.1.6   Example: *Canterbury Minimus*

1 2 3 4	1 3 4 2	1 4 2 3
2 1 4 3	3 1 2 4	4 1 3 2
2 4 1 3	3 2 1 4	4 3 1 2
2 4 3 1	3 2 4 1	4 3 2 1
4 2 3 1	2 3 4 1	3 4 2 1
4 2 1 3	2 3 1 4	3 4 1 2
4 1 2 3	2 1 3 4	3 1 4 2
1 4 3 2	1 2 4 3	1 3 2 4
		1 2 3 4

FIGURE 6.6. The changes for *Canterbury Minimus*.

Example 6.1.2. *Consider the changes for Canterbury Minimus shown in Figure 6.6. Is this a legitimate extent? In other words, does it satisfy the first three rules? Factor the set of changes and determine whether Rules 4, 5, and 6 are satisfied.*

Solution: Yes, this is a legitimate extent. It begins and ends on rounds, and all 24 permutations of (1 2 3 4) are sounded precisely once (except for rounds). The extent factors as $[a\,b\,c\,d\,c\,b\,a\,b]^3$. This means that Rule 3 is satisfied because all changes are generated using one of the four allowable transformations. In the strict sense, Rule 4 is not satisfied because the working bells (bells 2, 3, and 4) remain in the first position for three consecutive changes. This happens twice for each bell during the extent. However, if Rule 4 is relaxed to prohibiting four consecutive changes as opposed to two, then it is satisfied. Rule 5 holds since the extent can be factored into a perfect cube, and Rule 6 is satisfied because $a\,b\,c\,d\,c\,b\,a$ is a palindrome.   □

It is interesting to note the similarities and differences between *Plain Bob* and *Canterbury Minimus*. Their overall structure is very similar. In each extent bell 1 goes plain hunting, there are three leads with eight changes, and each lead begins with the same permutation of rounds. Assuming that Rule 4 is relaxed, each extent satisfies all six rules. The main difference between the two extents is that *Canterbury* features all four possible transformations, while *Plain Bob* has only three. Also, none of the working bells ever hunt down in *Canterbury* as they do in *Plain Bob*.

Exercises for Section 6.1

1. How many different ways can 10 numbered bells be arranged so that the first bell is always bell 1 and the last bell is always bell 10, but all the other bells are arbitrarily arranged?

2. How many different ways can 12 numbered bells be arranged assuming that the even bells (bells 2, 4, 6, etc.) are located in the even positions (positions 2, 4, 6, etc.), while the odd bells are located in the odd positions? The location within their respective groups is arbitrary.

3. Suppose that n is a positive integer. Simplify the expression $\dfrac{(n+2)!}{n!}$.
   Hint: How do you express $(n+2)!$ in terms of $n!$?

4. List the 12 permissible moves on $n = 6$ bells. List them in the form used in the text (e.g., (1 2) or (1 2)(5 6)).

5. List the 20 permissible moves on $n = 7$ bells. List them in the form used in the text (e.g., (1 2) or (1 2)(5 6)), and break them into two groups according to whether the bell in the seventh position moves or not.

6. How many permissible moves exist for $n = 15$ bells? What about $n = 20$ bells?

For Exercises 7–11, let $a = (1\ 2)(3\ 4)$, $b = (2\ 3)$, $c = (3\ 4)$, and $d = (1\ 2)$ represent the four permissible transformations for four bells.

7. Consider the sequence of eight moves $[a\,b\,c\,d]^2$.

   a. Beginning with rounds, write out the next eight changes determined by this sequence of moves.

   b. Why couldn't this sequence be used for the start of a legitimate extent?

   c. Does bell 1 go plain hunting in this sequence?

8. Write out the 25 changes (in three leads) of the extent on four bells determined by $[d\,b\,a\,d\,a\,b\,d\,c]^3$. This extent is called *St. Nicholas Minimus*. Does bell 1 go plain hunting? Which of the six rules for an extent are satisfied? Explain.

9. Write out the 25 changes (in four leads) of the extent on four bells determined by $[(d\,b)^2\,d\,a]^4$. This extent is named *Erin Minimus*. Does bell 1 go plain hunting? Which of the six rules for an extent are satisfied? Explain. What is the connection between each lead and "extent A" for three bells?

10. Examine the sequence of 25 changes in Figure 6.7. Is this a legitimate extent? Factor the sequence of changes in terms of the moves a, b, c, and d. Determine whether Rules 4, 5, and 6 are satisfied.

1 2 3 4	1 3 4 2	1 4 2 3
2 1 3 4	3 1 4 2	4 1 2 3
2 3 1 4	3 4 1 2	4 2 1 3
3 2 4 1	4 3 2 1	2 4 3 1
3 4 2 1	4 2 3 1	2 3 4 1
4 3 1 2	2 4 1 3	3 2 1 4
4 1 3 2	2 1 4 3	3 1 2 4
1 4 3 2	1 2 4 3	1 3 2 4
		1 2 3 4

FIGURE 6.7. A possible extent on four bells?

11. You are the composer. Make up your own extent, satisfying at least the first three rules, on four bells which is different from *Plain Bob Minimus*, *Canterbury Minimus*, *St. Nicholas Minimus*, *Erin Minimus*, and the changes in the previous exercise. List all 25 rows, as well as the factored form of your extent. Try to make bell 1 go plain hunting in your composition. Does your extent satisfy Rules 4, 5, and 6?

12. The first lead (24 changes) of *Oxford Treble Bob Minor* is shown in Figure 6.8. Let $a = (1\ 2)(3\ 4)(5\ 6)$, $b = (2\ 3)(4\ 5)$, $c = (3\ 4)(5\ 6)$, and $d = (1\ 2)(5\ 6)$ be four permissible transformations on six bells. Factor the lead in terms of a, b, c, and d. Does this extent satisfy the palindrome property?

1 2 3 4 5 6	2 5 6 4 3 1
2 1 4 3 6 5	5 2 4 6 1 3
1 2 4 3 5 6	5 2 6 4 3 1
2 1 3 4 6 5	2 5 4 6 1 3
2 3 1 6 4 5	2 4 5 1 6 3
3 2 6 1 5 4	4 2 1 5 3 6
3 2 1 6 4 5	4 2 5 1 6 3
2 3 6 1 5 4	2 4 1 5 3 6
2 6 3 5 1 4	2 1 4 3 5 6
6 2 5 3 4 1	1 2 3 4 6 5
6 2 3 5 1 4	2 1 3 4 5 6
2 6 5 3 4 1	1 2 4 3 6 5

FIGURE 6.8. The first lead of *Oxford Treble Bob Minor*.

6.2 Group Theory Revisited

As we study the extents on four bells discussed in Section 6.1, we begin to appreciate the rich mathematical structure inherent in each composition. Perhaps you struggled to compose your own valid extent in Exercise 11 of the previous section. How is it possible to construct an extent, other than by trial and error? Why do some sequences of transformations help create legitimate extents, while others return us to a previous change, thereby violating Rule 2. Surprisingly, these questions can best be answered using group theory.

6.2.1 The symmetric group S_n

Recall that a group G is a set with an associative binary operation $*$ that contains the identity element e, as well as inverses for each element. It turns out that the set of all permutations of $(1\ 2\ 3\ \cdots\ n)$ is a group with the operation $*$ equal to the product of two permutations. This group is denoted S_n and is often called the *symmetric group on n symbols*. To understand this important group, we first need to know how to multiply two permutations.

Suppose that $s = (2\ 3\ 1\ 5\ 4)$ and $t = (4\ 5\ 1\ 2\ 3)$. What is meant by $s * t$? The key is to view a given permutation in S_n as a function that assigns each number in rounds to a new number. For example, $s = (2\ 3\ 1\ 5\ 4)$ is really a function that maps $1 \mapsto 2, 2 \mapsto 3, 3 \mapsto 1, 4 \mapsto 5$, and $5 \mapsto 4$. This function has a domain and range equal to $\{1, 2, 3, 4, 5\}$ and can be visualized as

$$
\begin{array}{ccccc}
1 & 2 & 3 & 4 & 5 \\
\downarrow & \downarrow & \downarrow & \downarrow & \downarrow \\
2 & 3 & 1 & 5 & 4.
\end{array}
$$

Similarly, t is the function that looks like

$$
\begin{array}{ccccc}
1 & 2 & 3 & 4 & 5 \\
\downarrow & \downarrow & \downarrow & \downarrow & \downarrow \\
4 & 5 & 1 & 2 & 3.
\end{array}
$$

Each function is an example of a *bijection*, a function that is both one-to-one (each element in the range has a unique pre-image mapping to it from the domain) and onto (every element in the range has a pre-image in the domain).

Warning: We will treat every permutation in S_n as a function that maps each number in $\{1, 2, 3, \ldots, n\}$ to a number in $\{1, 2, 3, \ldots, n\}$. Here the focus is on the particular values of the numbers and where they are sent by the function. This is a *different* approach from the previous section, where, for example, we used notation such as $b = (2\ 3)$ to indicate that the bells in *positions* 2 and 3 should be interchanged, regardless of their numbers.

To find the product $s * t$, we compose s and t as functions. We will assume that $s * t$ means do s first, followed by t. (This is the opposite of traditional functional notation but is consistent with the notation we used in Section 5.3.) Therefore, $s * t$ can be represented as

$$
\begin{array}{cccccc}
 & 1 & 2 & 3 & 4 & 5 \\
s & \downarrow & \downarrow & \downarrow & \downarrow & \downarrow \\
 & 2 & 3 & 1 & 5 & 4 \\
t & \downarrow & \downarrow & \downarrow & \downarrow & \downarrow \\
 & 5 & 1 & 4 & 3 & 2,
\end{array}
$$

so that $s * t = (5\ 1\ 4\ 3\ 2)$. In other words, $s * t$ maps the top row $(1\ 2\ 3\ 4\ 5)$ to the bottom row $(5\ 1\ 4\ 3\ 2)$. For example, since $1 \mapsto 2$ under s and $2 \mapsto 5$ under t, the composition $s * t$ will have $1 \mapsto 5$. This is why the first entry in $s * t$ is a 5.

Example 6.2.1. *If $s = (2\ 3\ 1\ 5\ 4)$ and $t = (4\ 5\ 1\ 2\ 3)$, compute $t * s$ and s^2.*

Solution: To find $t * s$, we do the permutation t first, then s. This can be visualized via

$$
\begin{array}{cccccc}
 & 1 & 2 & 3 & 4 & 5 \\
t & \downarrow & \downarrow & \downarrow & \downarrow & \downarrow \\
 & 4 & 5 & 1 & 2 & 3 \\
s & \downarrow & \downarrow & \downarrow & \downarrow & \downarrow \\
 & 5 & 4 & 2 & 3 & 1,
\end{array}
$$

so that $t * s = (5\ 4\ 2\ 3\ 1)$. To find $s^2 = s * s$, we compose s with itself. This yields

$$
\begin{array}{cccccc}
 & 1 & 2 & 3 & 4 & 5 \\
s & \downarrow & \downarrow & \downarrow & \downarrow & \downarrow \\
 & 2 & 3 & 1 & 5 & 4 \\
s & \downarrow & \downarrow & \downarrow & \downarrow & \downarrow \\
 & 3 & 1 & 2 & 4 & 5,
\end{array}
$$

which means that $s^2 = (3\ 1\ 2\ 4\ 5)$. □

Note that $s * t \neq t * s$. This is crucial to remember: the order in which two permutations are composed matters. S_n is an example of a *noncommutative* group.

A key property of bijections is that when they are composed together, the result is another bijection. This means that multiplying two permutations of the same size (same number of bells) will always yield another permutation of the same size. This shows that the group S_n is closed under multiplication. The composition of functions is associative by definition. It doesn't matter how the functions are grouped as long as the same order is maintained. For our purposes, this means that $(s * t) * u = s * (t * u)$ for any permutations $s, t, u \in S_n$, and thus $*$ is associative.

The identity element in S_n is $e = (1\ 2\ 3\ \cdots\ n)$ (rounds), since this is the permutation that fixes every number. It is straightforward to check that $s * e = e * s = s$ for any permutation $s \in S_n$.

The remaining property required for S_n to be a group is that each permutation have an inverse that is also in S_n. The inverse of a permutation is found by reversing the direction of the arrows. In other words, if $3 \mapsto 1$ under s, then $1 \mapsto 3$ under s^{-1}. To construct the permutation s^{-1}, we simply reverse the direction of the arrows and then rearrange the pairs so that the domain is listed sequentially. Returning to our example, if $s = (2\ 3\ 1\ 5\ 4)$, then s^{-1} can be computed as

$$
s \begin{array}{ccccc} 1 & 2 & 3 & 4 & 5 \\ \downarrow & \downarrow & \downarrow & \downarrow & \downarrow \\ 2 & 3 & 1 & 5 & 4 \end{array} \implies s^{-1} \begin{array}{ccccc} 2 & 3 & 1 & 5 & 4 \\ \downarrow & \downarrow & \downarrow & \downarrow & \downarrow \\ 1 & 2 & 3 & 4 & 5 \end{array} \text{ or } s^{-1} \begin{array}{ccccc} 1 & 2 & 3 & 4 & 5 \\ \downarrow & \downarrow & \downarrow & \downarrow & \downarrow \\ 3 & 1 & 2 & 5 & 4, \end{array}
$$

which means that $s^{-1} = (3\ 1\ 2\ 5\ 4)$. One can check that $s^{-1} * s = e$ as well.

Theorem 6.2.2. *The set of permutations on* $(1\ 2\ 3 \cdots n)$, *together with the operation* $*$, *forms a noncommutative group with* $n!$ *elements. This group, denoted* S_n, *is called the* <u>symmetric group</u>.

6.2.2   The dihedral group revisited

Consider the first three rows of *Plain Bob Minimus*, $(1\ 2\ 3\ 4)$, $(2\ 1\ 4\ 3)$, and $(2\ 4\ 1\ 3)$, as permutations in S_4. The first row is just the identity element e. Let α and β denote the next two rows. It turns out that the remaining five changes in the first lead of *Plain Bob* can all be written in terms of α and β. For example, we have $\alpha * \beta = (4\ 2\ 3\ 1)$ and $\beta^2 = (4\ 3\ 2\ 1)$, which are the fourth and fifth changes, respectively, of the first lead. The entire lead can be described by α and β, as shown in Figure 6.9. Note that we have dropped the $*$ for ease of notation, so, for example, $\alpha\beta = \alpha * \beta$.

$$
\begin{array}{rcl}
e & = & 1\ 2\ 3\ 4 \\
\alpha & = & 2\ 1\ 4\ 3 \\
\beta & = & 2\ 4\ 1\ 3 \\
\alpha\beta & = & 4\ 2\ 3\ 1 \\
\beta^2 & = & 4\ 3\ 2\ 1 \\
\alpha\beta^2 & = & 3\ 4\ 1\ 2 \\
\beta^3 & = & 3\ 1\ 4\ 2 \\
\beta\alpha & = & 1\ 3\ 2\ 4
\end{array}
$$

FIGURE 6.9. The first lead of *Plain Bob Minimus* with each change described in terms of the permutations α and β.

Not only can each permutation be written in terms of α and β, but collectively, all eight permutations form a group. Let $K = \{e, \beta, \beta^2, \beta^3, \alpha, \alpha\beta^2, \alpha\beta, \beta\alpha\}$. We will show that K is a group, a subgroup of S_4. As discussed previously in Section 5.3.4, the key trait that needs to be confirmed is closure. In other words, we need to check that the product of any two elements in K remains in K. This can be accomplished by making a multiplication table and checking that all 64 products are indeed elements of K. We also need to find the inverses of each element and make sure those are also in K.

Instead of working every multiplication out by hand, there are some identities involving α and β which are particularly useful. In turn, these identities can be used to derive other useful relations. Here are some key identities:

$$\alpha^2 = e, \tag{6.3}$$
$$\beta^4 = e, \tag{6.4}$$
$$\beta\alpha\beta = \alpha. \tag{6.5}$$

Since α is the permutation corresponding to the move that interchanges both the first and last pair of bells, the relation $\alpha^2 = e$ confirms that doing this transformation twice will return the bells to their original position. The identity $\beta^4 = e$ is particularly useful: any time we compose four βs together, they cancel out. For example,

$$\beta^2 * \beta^3 = \beta^5 = \beta^4 * \beta = e * \beta = \beta.$$

Likewise, $\beta^3 * \beta^3 = \beta^6 = \beta^2$.

Suppose we wanted to find the product $\alpha * \beta^3 = \alpha\beta^3$ using the identities above. Begin by multiplying β^3 *on the right* of each side of Equation (6.5). This yields

$$\beta\alpha\beta * \beta^3 = \alpha * \beta^3.$$

The left-hand side of the previous equation simplifies to $\beta\alpha\beta^4 = \beta\alpha e = \beta\alpha$. Thus, $\alpha\beta^3 = \beta\alpha$. Note that it is critical to multiply in the same direction on each side of an equation since $*$ is not commutative.

By using the identities and arguments above, or by multiplying the permutations directly, we construct the multiplication table for the first lead of *Plain Bob Minimus* (see Table 6.3). To obtain each entry, we multiply an entry in the leftmost column by an element in the top row, taking the column entry first. For example, the product in the sixth row, third column is $\alpha * \beta = \alpha\beta$ (not $\beta * \alpha = \beta\alpha$).

There are several important things to notice about Table 6.3. First, every element in the table belongs to K. This shows that K is indeed closed under multiplication. Second, every element in K has an inverse in K, as indicated by the presence of e in each row and column. The inverses of each element in K are

$$e^{-1} = e, \qquad \beta^{-1} = \beta^3, \qquad (\beta^2)^{-1} = \beta^2, \qquad (\beta^3)^{-1} = \beta,$$

and

$$\alpha^{-1} = \alpha, \qquad (\alpha\beta^2)^{-1} = \alpha\beta^2 \qquad (\alpha\beta)^{-1} = \alpha\beta, \qquad (\beta\alpha)^{-1} = \beta\alpha.$$

Note that except for β and β^3, which are inverse of each other, all other elements of K are their own inverse. This completes the argument that the elements in K form a subgroup of S_4. Since $|K| = 8$ divides evenly into 24, the order of the larger group S_4, Lagrange's Theorem is satisfied.

But there is even more to say about this surprising connection to group theory. When algebraists find a group, the next question they ask is, "What group is it?" Consider the properties of K. It has eight elements and two *generators*, α and β, the first satisfying $\alpha^2 = e$ and the second satisfying $\beta^4 = e$. All the elements, except for β and β^3, are their own inverses. Does this sound familiar? If we treat β as a rotation and α as a reflection, then the structure inherent in Table 6.3 is identical to that of Table 5.5. For example, notice that

$*$	e	β	β^2	β^3	α	$\alpha\beta^2$	$\alpha\beta$	$\beta\alpha$
e	e	β	β^2	β^3	α	$\alpha\beta^2$	$\alpha\beta$	$\beta\alpha$
β	β	β^2	β^3	e	$\beta\alpha$	$\alpha\beta$	α	$\alpha\beta^2$
β^2	β^2	β^3	e	β	$\alpha\beta^2$	α	$\beta\alpha$	$\alpha\beta$
β^3	β^3	e	β	β^2	$\alpha\beta$	$\beta\alpha$	$\alpha\beta^2$	α
α	α	$\alpha\beta$	$\alpha\beta^2$	$\beta\alpha$	e	β^2	β	β^3
$\alpha\beta^2$	$\alpha\beta^2$	$\beta\alpha$	α	$\alpha\beta$	β^2	e	β^3	β
$\alpha\beta$	$\alpha\beta$	$\alpha\beta^2$	$\beta\alpha$	α	β^3	β	e	β^2
$\beta\alpha$	$\beta\alpha$	α	$\alpha\beta$	$\alpha\beta^2$	β	β^3	β^2	e

TABLE 6.3. The multiplication table for the eight permutations in the first lead of *Plain Bob Minimus*. Each entry is obtained by multiplying an entry in the leftmost column by an element in the top row, taking the column entry first.

the upper left and lower right blocks in Table 6.3 only consist of β and powers of β, just as the corresponding blocks in Table 5.5 consist only of a 90° degree rotation and multiples of R_{90}. Simply put, K is really the dihedral group D_4 in disguise!

To make this precise, consider the mapping ϕ that identifies elements of K with the symmetries of the square, as follows:

$$\begin{aligned}
\phi(\beta) &= R_{90} & \phi(\alpha) &= H \\
\phi(\beta^2) &= R_{180} & \phi(\alpha\beta^2) &= V \\
\phi(\beta^3) &= R_{270} & \phi(\alpha\beta) &= D_{13} \\
\phi(e) &= e & \phi(\beta\alpha) &= D_{24}.
\end{aligned}$$

The function ϕ is just a relabeling of the elements in K. The important feature here is that this relabeling converts the multiplication table for K into the multiplication table for the symmetries of the square. In essence, the groups are identical, or in group theory jargon, they are *isomorphic* ("iso" means *same*).

Theorem 6.2.3. *The first lead of Plain Bob Minimus is identical to the symmetries of the square. Mathematically speaking, K is isomorphic to D_4.*

The special mapping ϕ is an example of an *isomorphism* as it is a bijection between the two groups and satisfies the key property

$$\phi(a * b) = \phi(a) * \phi(b) \qquad \text{for all } a, b \in K. \tag{6.6}$$

We note that on the left-hand side of Equation (6.6), $*$ stands for multiplication in K, while on the right-hand side, $*$ represents composition of functions in D_4. In general, Equation (6.6) is the defining equation for ϕ to be an isomorphism. It states that corresponding elements in isomorphic groups will multiply together in the same manner. Or, in more basic terms, isomorphic finite groups have identical multiplication tables. Some other examples of isomorphic groups are discussed in the exercises.

6.2.3   Ringing the cosets

It is tempting to inquire about the other two leads in *Plain Bob Minimus*, to see whether they are also subgroups. However, neither of these contains rounds (the identity), so Property **(iii)** for groups fails to hold. However, it is easy to generate these leads from the subgroup formed by the first lead. Multiplication by the permutation (1 3 4 2) on the right takes the entire first column to the second column, and multiplication again by this same permutation takes the second column to the third. In group theory, the second and third leads are called *cosets*.[3]

Definition 6.2.4. *Suppose that H is a subgroup of a finite group G, and let a be an element of G that is not in H. Then the set $aH = \{a * h$, for each $h \in H\}$ is called a left coset of H in G, and the set $Ha = \{h * a$, for each $h \in H\}$ is called a right coset of H in G.*

Simply put, the left coset aH is obtained by multiplying a, on the *left*, by every element in the subgroup H. Likewise, the right coset Ha is obtained the same way except that a multiplies on the right in each product. It is crucial that $a \notin H$; otherwise, the fact that H is closed under $*$ would imply that aH and Ha reduce to H itself.

The key fact about left and right cosets of a subgroup H is that they are disjoint from H and have the same number of elements as H itself. This is relatively straightforward to show.

Theorem 6.2.5. *Suppose that H is a proper subgroup of a finite group G, and that $a \notin H$. Then aH and H are disjoint (no elements in common). Ha and H are also disjoint. Moreover, the number of elements in aH or Ha is the same as the number of elements in H.*

Proof: We use proof by contradiction. Suppose that aH and H have an element in common. This means that $a * h_1 = h_2$ for some $h_1, h_2 \in H$. Applying h_1^{-1} (which exists because H is a group on its own) on the right of each side of this equation yields $a = h_2 h_1^{-1} \in H$ (since H is closed). This contradicts the fact that $a \notin H$. Therefore, aH and H are disjoint. The proof that Ha and H are disjoint is similar and is left as an exercise.

To see that aH and H have the same number of elements, we construct a bijection between these two sets. Since a bijection is both onto (every element in the range gets counted) and one-to-one (no element in the range is counted more than once), the two sets will have the same size.

Consider the function $f : H \mapsto aH$ defined by $f(h) = ah$. By contradiction, suppose that f were *not* one-to-one. This would mean that there are two distinct elements of H, call them h_1 and h_2, such that $f(h_1) = f(h_2)$. But $f(h_1) = f(h_2)$ is equivalent to $ah_1 = ah_2$, and by the Cancellation Property, this implies that $h_1 = h_2$. Thus, we see that h_1 and h_2 are in fact not distinct. This contradiction proves that f is one-to-one.

To show that f is onto, pick any element $j \in aH$. By definition, $j = ah$ for some element $h \in H$. But then $f(h) = ah = j$, so h is the pre-image of j and f is onto. This shows that H and aH have the same number of elements. The proof that Ha and H have the same number of elements is similar and is left as an exercise. □

---

[3]The title of this section pays homage to the fine article of the same name by Arthur White [1987].

One of the useful features of cosets is that they *partition* a group into separate pieces, each with the same size. Recall from Lagrange's Theorem that the order of a subgroup divides evenly into the order of the group. Applying Theorem 6.2.5, it is possible to choose enough group elements $a_1, a_2, \ldots, a_i \notin H$ such that the cosets $a_1 H, a_2 H, \ldots, a_i H$ are all disjoint from each other and satisfy

$$G = H \cup a_1 H \cup a_2 H \cup \cdots \cup a_i H.$$

The number of elements in H and each coset is the same, so we have $(i+1) \cdot |H| = |G|$. For example, if G has 15 elements and H is a subgroup with three elements, then we need four distinct cosets of H to cover the whole group $(3 + 4 \cdot 3 = 15)$. This decomposition of the larger group into a subgroup and its cosets is precisely what occurs when change ringing composers create a full extent.

Let us revisit the three leads of *Plain Bob Minimus* and recast them into the language of group theory. Let $\omega = (1\ 3\ 4\ 2)$ represent the permutation at the top of the second lead. Note that $\omega * \omega = \omega^2 = (1\ 4\ 2\ 3)$ is the change at the top of the third and final lead. The first lead of *Plain Bob Minimus* is $K = \{e, \alpha, \beta, \alpha\beta, \ldots, \beta\alpha\}$, a subgroup of S_4 that is isomorphic to D_4, the symmetries of the square. (Note that we have rearranged the elements of the group so their order matches the order of the first lead shown in Figure 6.9.) The second lead is the right coset $K\omega = \{\omega, \alpha\omega, \beta\omega, \alpha\beta\omega, \ldots, \beta\alpha\omega\}$, and the third lead is the right coset $K\omega^2 = \{\omega^2, \alpha\omega^2, \beta\omega^2, \alpha\beta\omega^2, \ldots, \beta\alpha\omega^2\}$, which can be obtained from the first lead through multiplication on the right by ω^2, or from the second lead via multiplication on the right by ω.

Note the power of Theorem 6.2.5 in verifying Rule 2 for an extent. The subgroup K has no elements in common with either coset $K\omega$ or $K\omega^2$. Moreover, since $\omega^2 \notin K\omega$, the cosets $K\omega$ and $K\omega^2$ are also disjoint (see Exercise 11). Theorem 6.2.5 also implies that K, $K\omega$, and $K\omega^2$ each have the same number of elements. Since $8 + 8 + 8 = 24$, we are assured that all $4! = 24$ permutations of rounds have been rung. Mathematically speaking, the group S_4 has been partitioned into the dihedral group of degree 4 (represented by K) and two of its right cosets:

$$S_4 = K \cup K\omega \cup K\omega^2.$$

The structure of the right cosets also serves to verify Rule 5, that all working bells do the same amount of "work." In effect, multiplying on the right by the permutation $\omega = (1\ 3\ 4\ 2)$ fixes bell 1, replaces bell 2 with bell 3, replaces bell 3 with bell 4, and replaces bell 4 with bell 2. As a result of the coset structure, the path followed by a working bell in the opening lead will then be repeated in the second lead by the bell replacing it. Since ω^2 is the same as repeating the permutation ω twice, the same trait holds in the final lead as well, and Rule 5 is satisfied.

It should not be surprising to learn that $\omega^3 = e$ (rounds). Because of the cyclic structure of the extent, multiplying the final lead on the right by ω should return us to the first lead. In other words, $K\omega^3 = K$. The set $\{e, \omega, \omega^2\}$ is a subgroup of S_4 of order 3. It is called a *cyclic group* because it is produced by one element (ω) and its powers. We call ω the *generator* of the group. The adjective "cyclic" comes from the fact that as the generator is raised to higher and higher powers, the result cycles back to e and the lower powers. This new cyclic subgroup for *Plain Bob Minimus* is isomorphic to $\mathbb{Z}_3 = \{0, 1, 2\}$, where the group operation is addition modulo 3 (see Exercise 6).

Since $C = \{e, \omega, \omega^2\}$ is a subgroup of S_4, we can consider its cosets as well. While the right cosets of K form the columns of *Plain Bob Minimus*, the *left* cosets of C form the rows. The second row of the extent is given by the left coset $\alpha C = \{\alpha, \alpha\omega, \alpha\omega^2\}$, the third by $\beta C = \{\beta, \beta\omega, \beta\omega^2\}$, the fourth by $\alpha\beta C = \{\alpha\beta, \alpha\beta\omega, \alpha\beta\omega^2\}$, and so on. In this case, the subgroup C of order 3 and seven of its left cosets partitions the full group S_4:

$$S_4 = C \cup \alpha C \cup \beta C \cup \alpha\beta C \cup \cdots \cup \beta\alpha C.$$

There is clearly a great deal of group theory going on in the special construction of an extent. By "ringing the cosets," change ringers are assured of satisfying the key rules, and composers have a blueprint for creating valid extents. Interestingly, one of the "fathers of bell ringing" was the Englishman Fabian Stedman (1640–1713), who wrote a definitive text on change ringing which implicitly used group-theoretic concepts (see [White, 1996]) to explain and derive different extents such as *Plain Bob Minimus* and the future favorite *Stedman's Principle* (now called *Stedman Doubles*). His work, *Campanalogia*, was published in 1677, nearly a full century before group theory was established as a mathematical discipline [Stedman, 1677].

6.2.4   Example: *Plain Bob Doubles*

We end our discussion on change ringing by demonstrating how to generalize the construction of *Plain Bob* from four bells to five. We make bell 1 go plain hunting every lead, which means that we need 10 changes per lead, and 12 leads (since $12 \times 10 = 120 = 5!$).

First, we return to extent A on three bells and notice an interesting connection between it and *Plain Bob Minimus*. Take each change in extent A and add 1 to each number. Then, place a 1 at the front of each new change to obtain a permutation in S_4. The result is shown below:

1 2 3	$\Longrightarrow$	1 2 3 4
2 1 3		1 3 2 4
2 3 1	$\Longrightarrow$	1 3 4 2
3 2 1		1 4 3 2
3 1 2	$\Longrightarrow$	1 4 2 3
1 3 2		1 2 4 3.

The six new changes on four bells listed above are precisely the opening and closing changes of each lead in *Plain Bob Minimus*, in order. The reason this works is that to go from the opening change of a lead in *Plain Bob Minimus* to the bottom change in that lead (one cycle of bell 1 hunting up and down), we interchange the bells in positions 2 and 3. This is accomplished with the move $b = (2\ 3)$, which is equivalent to utilizing transformation $a = (1\ 2)$ every other move of extent A. Similarly, to move from the bottom of a lead to the top of the next lead in *Plain Bob*, we apply the move $c = (3\ 4)$, which is identical to employing $b = (2\ 3)$ every other move of extent A.

In addition, recall that the first five changes for extent A are $(ab)^2 a$, while the first seven changes for *Plain Bob Minimus* factor as $(ab)^3 a$. Here a changes meaning from $a = (1\ 2)$ in the case $n = 3$ to $a = (1\ 2)(3\ 4)$ for the case $n = 4$. In some sense, the new "a" is the natural extension of the old one when going from three to four bells.

We now generalize this argument to create *Plain Bob Doubles* using the rows of *Plain Bob Minimus*. The likely candidate to generate each lead is now $(ab)^4 a$, where the new transformation b is $b = (2\ 3)(4\ 5)$, while a remains the same as it was for the case $n = 4$.

As with *Plain Bob Minimus*, we use the transformation $c = (3\ 4)$ to move from the bottom of one lead to the top of the next. The first 40 changes of *Plain Bob Doubles* (four leads) are shown in Figure 6.10. These are known as the *Plain Course* of *Plain Bob Doubles*. It is worth checking that the permutations at the top and bottom of each lead (minus the 1 in front) are equivalent to adding 1 to each bell in every change of the first lead of *Plain Bob Minimus*. Notice that the working bells (bells 2, 3, 4, and 5) are also doing a version of plain hunting, except that they occasionally rest for four consecutive positions (at the bottom of one lead and then the top of the next) and must "dodge" bell 1.

1 2 3 4 5	1 3 5 2 4	1 5 4 3 2	1 4 2 5 3
2 1 4 3 5	3 1 2 5 4	5 1 3 4 2	4 1 5 2 3
2 4 1 5 3	3 2 1 4 5	5 3 1 2 4	4 5 1 3 2
4 2 5 1 3	2 3 4 1 5	3 5 2 1 4	5 4 3 1 2
4 5 2 3 1	2 4 3 5 1	3 2 5 4 1	5 3 4 2 1
5 4 3 2 1	4 2 5 3 1	2 3 4 5 1	3 5 2 4 1
5 3 4 1 2	4 5 2 1 3	2 4 3 1 5	3 2 5 1 4
3 5 1 4 2	5 4 1 2 3	4 2 1 3 5	2 3 1 5 4
3 1 5 2 4	5 1 4 3 2	4 1 2 5 3	2 1 3 4 5
1 3 2 5 4	1 5 3 4 2	1 4 5 2 3	1 2 4 3 5

FIGURE 6.10. The first four leads of the extent *Plain Bob Doubles*. Bell 1 is highlighted to show that it goes plain hunting. Each lead factors as $(a\,b)^4\,a$, with transformation c connecting the bottom of one lead to the top of the next.

Now something interesting happens. Notice that if we applied move $c = (3\ 4)$ to the 40th row (1 2 4 3 5), we would return to rounds prematurely. We cannot apply moves a or b either as they return us to the previous change or to the top of the lead, respectively. At this point a new transformation is required, called a *Bob*. The move needed is $d = (2\ 3)$, taking us to the change (1 4 2 3 5), which is new. At this point in the extent, the conductor of the band of ringers literally calls out "Bob," letting the ringers know that the special move d is to be performed. Note that this is also a generalization of *Plain Bob Minimus*, as the move $c = (3\ 4)$ is similar to a "Bob" at the bottom of each lead. In that case, had either of the previous two transformations been applied, the extent would violate Rule 2. After the "Bob," the next 40 changes return to the original pattern, until d is called for again at the 80th change and one last time as the final move to produce rounds. The final factorization of *Plain Bob Doubles* is therefore

$$\left[\left((a\,b)^4 a\,c\right)^3 \cdot \left((a\,b)^4 a\,d\right)\right]^3.$$

The 41st, 81st, and 120th changes are (1 4 2 3 5), (1 3 4 2 5), and (1 3 2 4 5), respectively.

Finally, we mention that the first lead of *Plain Bob* is always equivalent to the dihedral group D_n, where n equals the number of bells.

Exercises for Section 6.2

1. Suppose that $p = (1\ 3\ 5\ 2\ 4\ 6)$ and $q = (2\ 4\ 6\ 1\ 3\ 5)$ are two permutations in S_6. Compute the following:

      a. $p * q$ **b.** $q * p$ **c.** p^2

      d. p^4 **e.** p^{-1} **f.** q^{-1}

2. Suppose that $s = (2\ 4\ 5\ 1\ 3)$, $t = (3\ 4\ 2\ 5\ 1)$, and $u = (4\ 2\ 5\ 3\ 1)$. Compute $(s * t) * u$ and $s * (t * u)$ and check that they are equivalent.

3. Recall that $\alpha = (2\ 1\ 4\ 3)$ and $\beta = (2\ 4\ 1\ 3)$ are the second and third changes, respectively, of *Plain Bob Minimus*. Using Equations (6.3), (6.4), and (6.5), simplify each of the following into a power of β:

    a. $\alpha\beta\alpha$ **b.** $\beta^2(\beta\alpha)^2$ **c.** β^{201}

4. Consider the six changes of extent A (on three bells) as the group S_3 (see Figure 6.4). Let $e = (1\ 2\ 3)$, $\alpha = (2\ 1\ 3)$, and $\beta = (2\ 3\ 1)$.

    a. Write the other three changes in terms of α and β.

    b. Check that $\alpha^2 = e$, $\beta^3 = e$, and $\beta\alpha\beta = \alpha$.

    c. Form the group multiplication table for S_3, placing the elements in the order $e, \beta, \beta^2, \alpha, \alpha\beta, \beta\alpha$.

    d. Compare your table with the multiplication table for D_3, the dihedral group of degree 3 (see Exercise **8.** in Section 5.3). What do you notice? Make a bold mathematical statement about the groups S_3 and D_3.

5. Recall that $\mathbb{Z}_2 = \{0,1\}$, where the group operation is addition modulo 2. Let G be the group $\{+, -\}$ with the operation $*$ defined in terms of the usual rules for multiplying positive and negative numbers. For example, $+ * + = +$ and $+ * - = -$ because the product of two positive numbers is positive, and the product of a positive and negative number is negative. Make group multiplication tables for both $\mathbb{Z}_2$ and G, and show that the two groups are isomorphic by writing down a specific isomorphism ϕ between them.

6. Let $\omega = (1\ 3\ 4\ 2)$ be the permutation at the top of the second lead of *Plain Bob Minimus*, and let $G = \{e, \omega, \omega^2\}$. Recall that $\mathbb{Z}_3 = \{0,1,2\}$, where the group operation is addition modulo 3. Make multiplication tables for both $\mathbb{Z}_3$ and G, and show that they are isomorphic by constructing an isomorphism ϕ between the two groups.

7. Consider the subgroup $R = \{e, R_{90}, R_{180}, R_{270}\}$ of D_4, consisting of the symmetric rotations of the square, and the group $G = \{1, i, -1, -i\}$ under complex multiplication (see Example 5.3.6 in Section 5.3.2). Make a group multiplication table for R and show that the two groups are isomorphic by constructing an isomorphism ϕ from R to G.

8. Suppose that $(A, *)$ and $(B, \circ)$ are two isomorphic groups with group operations $*$ and $\circ$, respectively. This means that there exists a bijection ϕ, mapping from A to B, that satisfies

$$\phi(a * b) \;=\; \phi(a) \circ \phi(b). \tag{6.7}$$

   a. Show that $\phi(e_A) = e_B$, where e_A is the identity element for A and e_B is the identity element for B. In other words, ϕ maps the identity element of one group to the identity element of the other. *Hint:* Plug $b = e_A$ into Equation (6.7).

   b. Show that ϕ maps an inverse in one group to the inverse of the corresponding element in the other group. Specifically, show that $\phi(a^{-1}) = (\phi(a))^{-1}$ for any $a \in A$.

   Hint: Plug $b = a^{-1}$ into Equation (6.7), and then use the result from part **a.**

9. Recall that $\mathbb{Z}_4 = \{0,1,2,3\}$ is a group under addition modulo 4 and that $M = \{e, H, V, R_{180}\}$ is the Klein four-group of musical symmetries under composition. Show that these two groups are *not* isomorphic. This shows that groups with the same number of elements may not be isomorphic. *Hint:* Use proof by contradiction. If ϕ is an isomorphism from $\mathbb{Z}_4$ to M, what happens if you plug $a = b = 1$ into Equation (6.7)?

10. Suppose that H is a proper subgroup of the group G and that $a \notin H$. Following the proof of Theorem 6.2.5, show that Ha and H are disjoint.

11. Suppose that H is a proper subgroup of the group G and that $a, b \notin H$. In addition, suppose that $b \notin Ha$. Using proof by contradiction, show that the cosets Ha and Hb are disjoint, that is, they have no element in common. This explains why $K\omega$ and $K\omega^2$, representing the second and third leads of *Plain Bob Minimus*, respectively, are guaranteed to be disjoint.

12. Using four leads of 10 rows each, write out changes 81 through 120 for *Plain Bob Doubles*. Begin with the change (1 3 4 2 5) and end with (1 3 2 4 5). Compare your list of changes with the first four leads of *Plain Bob Doubles* shown in Figure 6.10.

13. Let $\alpha = $ (2 1 4 3 5) and $\beta = $ (2 4 1 5 3) be the second and third changes of *Plain Bob Doubles*, respectively.

   a. Write each of the remaining seven changes in the first lead of *Plain Bob Doubles* in terms of α and β (see Figure 6.10).

   b. Show that $\alpha^2 = e$, $\beta^5 = e$, and $(\alpha\beta)^2 = e$. Together with part **a.**, this shows that the first lead of *Plain Bob Doubles* is isomorphic to the dihedral group of degree 5.

14. Generalize the construction of the first lead in *Plain Bob* on four and five bells and write out the first lead (12 changes) of *Plain Bob Minor* on six bells.

References for Chapter 6

Gerstein, L. J.: 1987, *Discrete Mathematics and Algebraic Structures*, W. H. Freeman and Company.

North American Guild of Change Ringers: n.d., Change Ringing? What's That? `http://www.nagcr.org/pamphlet.html`.

Owen, P.: June 3, 2012, Ruling the Waves: Three Generations of Royals Join the Queen as She Sets Sail Down the Thames on Glorious Jubilee River Pageant, *Mail Online*. `http://www.dailymail.co.uk/news/article-2153969/`.

Philadelphia Guild of Change Ringers: n.d., What Is Change Ringing? `http://www.phillyringers.com`.

Stedman, F.: 1677, *Campanalogia*, W. Godbid, London.

White, A. T.: 1987, Ringing the Cosets, *American Mathematical Monthly* **94**(8), 721–746.

White, A. T.: 1993, Treble Dodging Minor Methods: Ringing the Cosets, on Six Bells, *Discrete Matematics* **122**, 307–323.

White, A. T.: 1996, Fabian Stedman: The First Group Theorist, *American Mathematical Monthly* **103**(9), 771–778.

White, A. T.: 2003–2014, Math and the Other Arts, Coursepack, Western Michigan University.

White, A. T. and Roaf, D.: 2003, Ringing the Changes: Bells and Mathematics, *in* J. Fauvel, R. Flood and R. Wilson (eds), *Music and Mathematics: From Pythagoras to Fractals*, Oxford University Press, pp. 112–129.

Chapter 7

Twelve-Tone Music

The main advantage of this method of composing with twelve tones is its unifying effect. In a very convincing way, I experienced the satisfaction of having been right about this when I once prepared the singers of my opera "Von Heute auf Morgen" for a performance. The technique, rhythm and intonation of all these parts were tremendously difficult for them, though they all possessed absolute pitch. But suddenly one of the singers came and told me that since he had become familiar with the basic set [the primary tone row], everything seemed easier for him. At short intervals all the other singers told me the same thing independently. I was very pleased with this . . .

— Arnold Schoenberg[1]

At the start of the twentieth century, as composers began to expand and experiment with different notions of harmony, the need for a tonal center in a piece of music became secondary. Music without a clearly defined key or a particular tonality is called, aptly enough, *atonal*. One of the early pioneers of this type of new music, often regarded as one of the most influential twentieth-century composers, was the Austro-Hungarian Arnold Schoenberg (1874–1951).

7.1   Schoenberg's Twelve-Tone Method of Composition

In order to de-emphasize a particular note or key, Schoenberg invented the *12-tone method* of composing, a technique that is inherently mathematical, relying on the musical symmetries discussed in Section 5.1. The fundamental property of 12-tone music is that each of the 12 notes in the chromatic scale is considered to be of equal importance. Instead of a tonic-dominant relationship that governs the harmonic musical structure, a particular ordering of all 12 notes is used which serves as the foundation for composition. The arrangement of the 12 notes of the chromatic scale in some particular order, with each note occurring precisely once, is called a *tone row* (also known as a *basic set*). Compositions using the 12-tone technique begin with a principal tone row and then apply different musical symmetries (transposition, inversion, retrograde, and retrograde-inversion) to this main

---

[1]Stein, *Style and Idea*, p. 244.

row in order to create new orderings of the 12 pitches. Different rows are then combined in some creative fashion (this is where the real composing occurs) to create the music. By restricting the notes to a particular tone row or one of its symmetric transformations, the equality of pitches and the lack of a primary tonality are guaranteed. Notice the structural similarities here to change ringing. Just as each bell in a change is to be rung exactly once, each note of the chromatic scale gets a unique place in any given tone row. In some ways, Schoenberg's technique of composing is a natural and musical extension of the British art of change ringing.

In the early 1920s, Schoenberg began to work out the ideas and structure to his 12-tone method in portions of works such as *Fünf Klavierstücke* (Five Pieces for Piano), op. 23 and the *Serenade*, op. 24. The first work to be completely organized around one primary tone row was his *Suite für Klavier* (Suite for Piano), op. 25, completed in 1923. We will examine this work in Section 7.2. Although initially music critics regarded the 12-tone technique as the compositional equivalent of "painting by numbers," many composers, such as Schoenberg's disciples Alban Berg (1885–1935) and Anton Webern (1883–1945), successfully developed the 12-tone technique into a more lyric musical form. Milton Babbitt (1916–2011) pushed the 12-tone technique to the limit by generalizing the 12-pitch concept to include rhythm, dynamics (not only loud versus soft, but also how the note is produced), and timbre. In this manner, Babbitt extended Schoenberg's 12-tone ideas to all facets of the music, imposing even more structural constraints on the composer. We will study a very similar compositional technique when we examine the use of magic squares by British composer Maxwell Davies (see Section 8.1).

7.1.1   Notation and terminology

We now explain some of the terminology and notation used in writing and analyzing 12-tone music. The first important point is that 12-tone music deals with *pitch classes* as opposed to focusing on the specific notes on the staff. Thus, all "Cs" on the musical staff (in any clef) are considered to be the same "C" in the tone row. The composer chooses which particular C on the staff is used at a given instant, but in terms of 12-tone analysis, all Cs are considered to be equivalent. Mathematically speaking, we have already explored this concept when discussing equivalence classes.

Given a tone row consisting of 12 distinct notes in some order, there are four symmetry operations that can be applied to the row. Since they correspond to mathematical transformations that preserve distance, each of these transformations will send the primary tone row to a different tone row that also contains 12 distinct notes. Musically, the transformations preserve the given intervals, so a newly created row cannot contain a repeated note. Mathematically, this is the concept of closure we encountered with group theory (see Section 5.3).

The four possible transformations come from our musical symmetries: transposition (translation), retrograde (vertical reflection), inversion (horizontal reflection), and retrograde-inversion (a 180° rotation). Each of the last three transformations can be followed by a transposition up a certain number of half steps. We will use the following notation to describe how a new tone row can be obtained from the primary one.

P_n = the **transposition** of the original tone row up n half steps.

R_n = the **retrograde** of P_n.

I_n = the **exact inversion** of P_n about its starting note. This is equivalent to transposing I_0 up n half steps.

RI_n = the **retrograde-inversion** of P_n, found by taking the retrograde of I_n.

The primary tone row, which typically appears in the first few measures of a piece or movement, is denoted P_0. If the row P_0 begins on a C, then the row P_1 begins on a C$\sharp$, while the row P_7 starts on G, a perfect fifth higher. The rows I_n and P_n always begin with the same pitch because the inversion is taken by reflecting about the starting note. Thus, if P_0 begins on a C, then so does I_0. The two different retrogrades, R_n and RI_n, are simply obtained by writing the rows P_n and I_n, respectively, backward.

The use of pitch classes implies that n need only range between 0 and 11. Specifically, we have the identities $P_{12} = P_0$, $R_{12} = R_0$, etc., since a transposition of 12 half steps is equivalent to changing each note by an octave, which does not alter the pitch class. Similarly, we also have $P_{13} = P_1$, $P_{14} = P_2$, $P_{-5} = P_7$, etc., equations reminiscent of arithmetic modulo 12. Thus, for each type of transformation (P_n, R_n, I_n, and RI_n), there are precisely 12 possible transpositions, $n \in \{0, 1, 2, \ldots, 11\}$, yielding a total of 48 possible variants of the original tone row. This gives the 12-tone composer a great deal of flexibility when writing music.

Figure 7.1 demonstrates some specific transformations on a primary tone row P_0. The row I_0 is the primary row inverted exactly about its starting pitch B. For example, the opening three notes of P_0 are B, C$\sharp$, and G$\sharp$, which is up a major second and then down by a perfect fourth. Consequently, the row I_0 begins B, A, D, which is *down* a major second, and then *up* by a perfect fourth. Alternatively, we can simply reflect each note about the middle of the staff, making sure the interval is maintained. For instance, the 8th note in the primary row is an F$\sharp$ on the top line of the staff, a perfect fifth above B. Therefore, the 8th note in the inverted row I_0 is E on the bottom line of the staff, a perfect fifth below B. In this case, since the inversion is taken about the middle of the staff, corresponding notes in the primary row and the inverted row are the same musical distance away from the center of the staff. Recall that notes an octave apart are considered equivalent in 12-tone music, so the intervals may have to be recomputed in a different octave. The row I_5 is I_0 transposed up by five half steps, or up a perfect fourth, while RI_5 is just the row I_5 played backward.

Warning: When writing out a tone row on the staff, we will consider the row to be one long measure. Therefore, it is necessary to include natural signs in order to cancel a prior accidental. For instance, in the row P_0 in Figure 7.1, the fifth note needs a natural sign to change it from a C$\sharp$ to a C; otherwise, the second and fifth notes would be identical. Always check your tone row to make sure the same note has not been accidentally repeated.

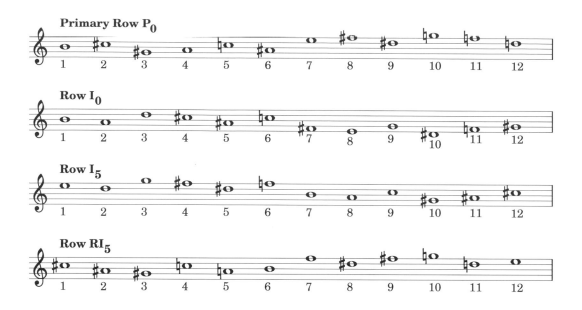

FIGURE 7.1. A primary tone row P_0, along with the transformations I_0 (inversion about the starting pitch B), I_5 (I_0 transposed up by five half steps), and RI_5 (I_5 backward). Note the use of natural signs in each row.

Regarding our notation, notice that the operations P and I commute, that is,

$$I_0 * P_n \; = \; P_n * I_0 \; = \; I_n. \tag{7.1}$$

The composition $I_0 * P_n$ means do I_0 first and then apply the operation P_n to the result. In other words, invert the row about its opening note and then raise the result up n half steps. This is equivalent to $P_n * I_0$, which first transposes the row up n half steps and then inverts about the *new* starting note. The first method is usually easiest because once the row I_0 is obtained, transposing it by n half steps is straightforward.

On the other hand, in our setup the operations R and I do *not* usually commute. Specifically, RI_n, the retrograde of I_n, is typically different from $R_n * I_0$, the inversion of R_n about its starting note. To understand why, compare the starting notes of the rows RI_n and $R_n * I_0$. By definition, the opening note of RI_n is the last note of row I_n. Meanwhile, the opening note of $R_n * I_0$ is the same as the starting note of R_n, which is equivalent to the last note of P_n. However, the last notes of the tone rows I_n and P_n are usually different, since an inversion only fixes the notes on the axis of symmetry (the starting note) and the note a tritone away ($+6 = -6 \pmod{12}$).

Therefore, as long as the last note of a tone row is not a tritone away from the opening note, the rows RI_n and $R_n * I_0$ will always be different. To avoid any confusion with our notation, the retrograde-inversion of a tone row, RI_n, should always be found by writing I_n backward. If, in fact, the first and last note of a tone row are a tritone apart, then and only then will $R_n * I_0$ be equivalent to RI_n (see Exercise 3 for an example).

7.1.2 The tone row matrix

One approach that is useful for composing 12-tone music, as well as analyzing existing compositions, is to form the *tone row matrix* associated with a primary row. This matrix is a 12×12 array featuring the primary tone row P_0 as the top row. The inverted row I_0 is then placed along the first column, with their shared starting note located in the top left corner of the matrix. Then, the primary tone row is transposed repeatedly to begin on each of the notes listed in the first column. Table 7.1 shows the tone row matrix for the primary row P_0 of Figure 7.1. The first four rows of the matrix, reading across from left to right, are P_0, P_{10}, P_3, and P_2. The retrogrades R_0, R_{10}, R_3, and R_2 are found by simply reading these same four rows from right to left. Next, the first four columns, reading from top to bottom, are the rows I_0, I_2, I_9, and I_{10}, while their retrogrades, RI_0, RI_2, RI_9, and RI_{10}, are obtained by reading the same four columns from bottom to top. Observe that the notes listed in column 7, which corresponds to rows I_5 and RI_5, agree with the music shown in Figure 7.1. Since the tone row matrix contains all 48 versions of the primary tone row, it is a very useful musical aid for creating or analyzing 12-tone music.

B	C♯	G♯	A	C	A♯	E	F♯	D♯	G	F	D
A	B	F♯	G	A♯	G♯	D	E	C♯	F	D♯	C
D	E	B	C	D♯	C♯	G	A	F♯	A♯	G♯	F
C♯	D♯	A♯	B	D	C	F♯	G♯	F	A	G	E
A♯	C	G	G♯	B	A	D♯	F	D	F♯	E	C♯
C	D	A	A♯	C♯	B	F	G	E	G♯	F♯	D♯
F♯	G♯	D♯	E	G	F	B	C♯	A♯	D	C	A
E	F♯	C♯	D	F	D♯	A	B	G♯	C	A♯	G
G	A	E	F	G♯	F♯	C	D	B	D♯	C♯	A♯
D♯	F	C	C♯	E	D	G♯	A♯	G	B	A	F♯
F	G	D	D♯	F♯	E	A♯	C	A	C♯	B	G♯
G♯	A♯	F	F♯	A	G	C♯	D♯	C	E	D	B

TABLE 7.1. The tone row matrix associated with the primary tone row P_0 in Figure 7.1. The "P" (transposition) tone rows are read from left to right, the "R" (retrograde) rows are read from right to left, the "I" (inversion) rows are read from top to bottom, and the "RI" (retrograde-inversion) rows are read from the bottom to the top.

Exercises for Section 7.1

1. Suppose that a primary tone row P_0 begins on a D and ends on a C.

   a. What is the first note of the row P_7?

   b. What is the first note of the row R_7?

   c. What is the first note of the row I_4?

   d. What is the first note of the row RI_4?

   e. List the names of the four tone rows that start on B♭.

2. Consider the primary tone row P_0 shown below:

   a. Copy the tone row P_0 onto staff paper and give each of the musical intervals between consecutive notes (m2, M2, P4, tritone, etc.).

   b. Form the entire tone row matrix for P_0 (see Table 7.1 for an example).

   c. In the treble clef, write out each of the following tone rows using correct accidentals: P_3, I_0, I_4, RI_4.

   d. Consider the following 12-tone excerpt based on the tone row P_0. Each of the four voice parts (soprano, alto, tenor, bass, arranged in order from the highest part to the lowest) follows a different tone row. For each part, identify the name of the tone row used.

3. Consider the primary tone row P_0 shown below:

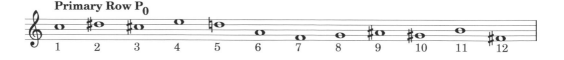

   a. Write out the tone rows RI_5 and R_5.

   b. Show that the rows RI_5 and $R_5 * I_0$ are identical. Note: RI_5 is the retrograde of row I_5, while $R_5 * I_0$ means to invert row R_5 about its starting pitch.

7.2   Schoenberg's *Suite für Klavier*, Op. 25

In the opening quote of this chapter, Schoenberg discusses the "unifying effect" of the 12-tone technique and notes the satisfaction that his singers attained once they understood the primary tone row and how it was being used in his opera. This underscores the importance of learning to analyze a 12-tone composition, particularly for musicians who perform these kinds of works. Musical analysis can help illuminate a composer's intent and provide a deeper meaning for the listener.

Let us examine some of the music from Schoenberg's *Suite für Klavier* (Suite for Piano; 1923), op. 25, his first piece to use the 12-tone method throughout. The suite consists of six parts, composed in a fashion similar to a Baroque suite. As is common with this compositional style, Schoenberg announces the primary tone row in the first three measures of the opening *Präludium*. The notes of the primary tone row P_0 are

$$E \quad F \quad G \quad D\flat \quad G\flat \quad E\flat \quad A\flat \quad D \quad B \quad C \quad A \quad B\flat,$$

as can be seen in the right-hand part (see Figure 7.2). The row P_0 generates the music for this piece, as well as the remaining five movements. Observe that the left-hand part does not open with the primary row (this would be too repetitive), but rather follows row P_6.

FIGURE 7.2. The opening measures of the *Präludium* from Schoenberg's *Suite für Klavier*, op. 25. The primary tone row P_0 is announced in the right hand while the left hand plays the row P_6. © Copyright 1925, 1952 by Universal Edition A.G., Wien/UE 7627.

The entire piano suite uses only P_0 and the rows P_6, I_6, I_0, R_6, and RI_6. These rows are shown in Figure 7.3 (note that the retrogrades can be obtained by reading the rows P_6 and I_6 backward, respectively). There are a few interesting features about the primary tone row and Schoenberg's particular choice of symmetry transformations. First, notice that the basic tone row starts on E and ends a tritone away, six half steps, on $B\flat$. Since Schoenberg only uses transformations with $n = 6$, all of the six rows utilized will begin and end on E or $B\flat$. This provides some continuity in the music and makes it easier to identify the tone rows in the piece.

Secondly, there are two more tritones within the primary tone row, between positions 3 and 4 (G to $D\flat$), and between positions 7 and 8 ($A\flat$ to D). By symmetry, there will also be tritones between these two positions in all six rows. Moreover, the notes G and $D\flat$ are equidistant from E (each a minor third away), and thus, an inversion about E will interchange their positions in the row. Again, since $n = 6$, the end result is that G and $D\flat$ always appear in the third and fourth positions of all six rows. The number 6 appears to be taking on increased significance as we conduct our analysis.

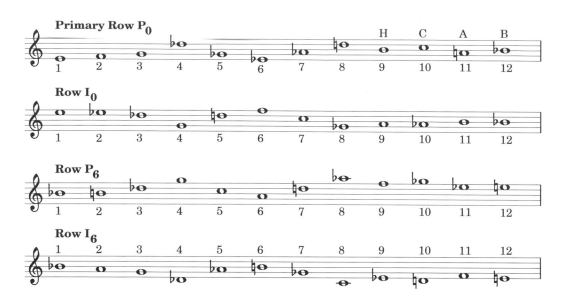

FIGURE 7.3. The primary tone row for Schoenberg's *Suite für Klavier*, op. 25, along with the key rows I_0, P_6, and I_6. Notice that each row begins and ends on either E or B♭, and that the tritone pair G, D♭ is always featured in the third and fourth positions of each row.

Thirdly, Schoenberg honors Bach at the end of his primary tone row. The last four notes of P_0 are B♮ C A B♭, which in German translate to H C A B. As the piece is in the Baroque style (Bach's time period), and as the very essence of 12-tone music is to apply symmetry to develop one "theme," namely, the row P_0, it is not at all surprising to see Schoenberg paying homage to the great master. In fact, in his 1941 article "Composition with Twelve Tones," Schoenberg describes another example where he intentionally used the B♭ A C B♮ "Bach pattern" in his music (see pp. 242–243 of [Stein, 1975]).

Figure 7.4 demonstrates another way to compose with a given tone row. Schoenberg displays some creativity in his use of the row R_6 in measure 5 of the *Präludium*. The row is not written linearly in one part; rather, it is divided into three different groups of four pitches each. As indicated in Figure 7.4, the bottom of the left-hand part contains the first four notes of the row R_6, the upper left-hand part contains the middle four notes, and the right-hand part completes the tone row.

The next piece in the suite is the *Gavotte*. As with the *Präludium*, Schoenberg declares the primary row P_0 in the opening bars with the right hand playing the first eight pitches and the left hand finishing the last four (see Figure 7.5). Then, in measures 2–3, the tone row I_6 is divided between both hands in a similar fashion (eight pitches in the right hand and four in the left). Note the reappearance of the Bach motif in the left hand, connecting the two rows.

As we see from this groundbreaking work, there is much more to the 12-tone method than a strict mathematical application of symmetry. The real challenge lies in turning the mathematical method into a creative and fluid compositional technique, a challenge certainly met by Arnold Schoenberg.

FIGURE 7.4. Measure 5 of the *Präludium* from Schoenberg's *Suite für Klavier*, op. 25. Here we see the row R_6 being subdivided into three pieces played simultaneously. © Copyright 1925, 1952 by Universal Edition A.G., Wien/UE 7627.

FIGURE 7.5. The opening three measures of the *Gavotte* from Schoenberg's *Suite für Klavier*, op. 25. Both the opening row P_0 and the ensuing row I_6 are divided between the two parts, with the Bach motif occurring in the left hand between the two rows. © Copyright 1925, 1952 by Universal Edition A.G., Wien/UE 7627.

Exercises for Section 7.2

1. What are some of the similarities between composing an extent in change ringing and writing a piece of music using the 12-tone method?

2. Figure 7.6 shows the first four measures of the *Gigue* from Schoenberg's *Suite für Klavier*, op. 25. Each measure contains a particular tone row based on the primary tone row P_0 shown in Figure 7.3. Identify the tone row used in each measure.

FIGURE 7.6. The opening four measures of the *Gigue* from Schoenberg's *Suite für Klavier*, op. 25. © Copyright 1925, 1952 by Universal Edition A.G., Wien/UE 7627.

7.3   Tone Row Invariance

How many possible tone rows exist? Given a particular tone row, are there always 48 *distinct* new rows arising from the four musical symmetries? What techniques have composers used to combine two tone rows simultaneously?

These questions can best be answered with a little combinatorics and some group theory. To count the total possible number of rows, we recall the concept of $n!$ from change ringing in Chapter 6. Since a tone row has no repeated notes, it can be viewed as a permutation of 12 numbers. As there are 12! possible permutations, we have a total of $12! = 479,001,600$ possible tone rows, a rather staggering number of possibilities. However, it is not the case that each tone row generates 48 new rows. It is possible that different operations may generate the same row, particularly when the prime row has some type of symmetry. We will call a tone row *invariant* if it is mapped to itself under one of the other 47 symmetry operations. The terminology comes from the idea that a transformation can leave a mathematical object unchanged (e.g., the eight symmetries of the square).

The chromatic scale revisited

For a simple example of invariance, take an ascending chromatic scale beginning on F as our primary tone row. Although this is musically trivial, it is easy to see why this particular row only generates 24 new tone rows, not the full 48. The key observation is that both an inversion and a retrograde of an ascending chromatic scale yield a descending chromatic scale. Specifically, I_0, the inversion of the original tone row, and R_1, the retrograde of the original row raised up by one half step, are both identical to a descending chromatic scale starting on F (see Figure 7.7). Since $I_0 = R_1$, we also have $I_1 = R_2$, just by raising each side of the equation up by one half step. In this case, the rows are each equivalent to a descending chromatic scale beginning on F$\sharp$. Continuing in this fashion,

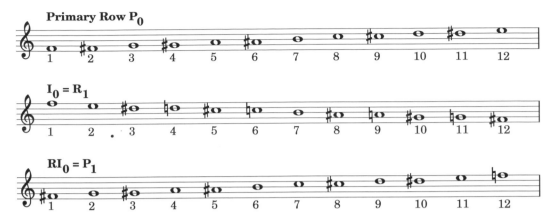

FIGURE 7.7. If the primary tone row P_0 is an ascending chromatic scale beginning on F, then the row I_0 is a descending chromatic scale beginning on F. Hence, row I_0 is equivalent to row R_1. Moreover, row RI_0 is an ascending chromatic scale beginning on F$\sharp$, which is identical to row P_1.

we have the following 12 tone-row equivalencies:

$$
\begin{array}{ll}
I_0 = R_1 & I_9 = R_{10} \\
I_1 = R_2 \quad \cdots & I_{10} = R_{11} \\
I_2 = R_3 & I_{11} = R_0,
\end{array}
$$

which represent the 12 possible descending chromatic scales.

Moreover, if we apply the retrograde transformation (vertical reflection) to both sides of the equation $I_0 = R_1$, we obtain the relation $RI_0 = P_1$, since a retrograde applied twice yields the original row (see Figure 7.7). In mathematical terms, we have $R_n * R_0 = P_n$. This yields an additional 12 equivalencies:

$$
\begin{array}{ll}
RI_0 = P_1 & RI_9 = P_{10} \\
RI_1 = P_2 \quad \cdots & RI_{10} = P_{11} \\
RI_2 = P_3 & RI_{11} = P_0,
\end{array}
$$

corresponding to the 12 possible ascending chromatic scales. The final equation, $P_0 = RI_{11}$, demonstrates the invariance of this particular tone row. In general, any identity of this sort, which involves the primary tone row, will always imply that only 24 new tone rows can be generated under the four symmetry operations. In the case of the chromatic scale, the 24 distinct rows are precisely the 24 different chromatic scales: 12 ascending and 12 descending.

Example 7.3.1. *Consider the tone row P_0 shown in Figure 7.8. It will only generate 24 distinct rows under the four symmetry transformations. Each inversion I_n is equivalent to some retrograde R_m, and each retrograde-inversion RI_n is equivalent to some transposition P_m. Find the specific equations relating the equivalent tone rows.*

Solution: First, notice the special numbers under each note in the row. Rather than indicate the position of the note in the row, these numbers reveal how many half steps

FIGURE 7.8. The primary tone row for Example 7.3.1, with numerical values beneath each note indicating the number of half steps above the starting pitch C.

the note lies above the starting pitch C. Thus, C is assigned a 0, C$\sharp$ is assigned a 1, D is given a 2, and so on. This numbering system will help create additional tone rows and also highlight the key symmetry underlying this example.

To form the row I_0, which is the original row inverted about the starting note C, we find the *inverse* of each integer modulo 12. Two integers in $\mathbb{Z}_{12}$ are inverses of each other if their sum equals 12, or more technically 0 (mod 12). For example, the second note of the primary row has a numerical value of 2, so its inverse would be $12 - 2 = 10$, or B$\flat$. The third note of P_0 has a value of 7, so its inverse is 5, which corresponds to an F. To determine the pitch that corresponds to a particular number, find the note above the given number in the primary tone row. In this fashion, we can quickly (and accurately) construct the row I_0.

Next, to determine whether I_0 is equivalent to a particular retrograde R_m, we transpose it up five half steps, so that the new starting note is an F, the last note of row P_0. This is easily accomplished by adding 5 (mod 12) to every number below the row I_0. Sure enough, the resulting row (I_5) is precisely the primary row backward. This shows that $I_5 = R_0$ (see Figure 7.9). Shifting both sides of this equation by successive half steps yields the full set of 12 equations connecting the inversions and retrogrades:

$$\begin{array}{lll} I_5 = R_0 & I_9 = R_4 & I_1 = R_8 \\ I_6 = R_1 & I_{10} = R_5 & I_2 = R_9 \\ I_7 = R_2 & I_{11} = R_6 & I_3 = R_{10} \\ I_8 = R_3 & I_0 = R_7 & I_4 = R_{11}. \end{array}$$

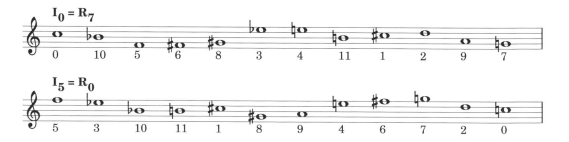

FIGURE 7.9. The tone rows I_5 and R_0 obtained from the primary row P_0 in Figure 7.8 are equivalent. It follows that the primary tone row P_0 is equivalent to RI_5, and that only 24 distinct rows are generated by the four symmetry transformations. Notice that I_5 is easily obtained from I_0 by adding 5 (mod 12) to each number beneath row I_0.

Finally, by applying the retrograde operation to both sides of $I_5 = R_0$, we obtain the relation $RI_5 = P_0$. Shifting this equation by successive half steps gives the 12 equations relating the transpositions to the retrograde-inversions:

$$
\begin{aligned}
RI_5 &= P_0 & RI_9 &= P_4 & RI_1 &= P_8 \\
RI_6 &= P_1 & RI_{10} &= P_5 & RI_2 &= P_9 \\
RI_7 &= P_2 & RI_{11} &= P_6 & RI_3 &= P_{10} \\
RI_8 &= P_3 & RI_0 &= P_7 & RI_4 &= P_{11}.
\end{aligned}
$$

This shows that the primary tone row P_0 is invariant, as it is equivalent to RI_5, and explains why it only generates 24 distinct new rows, not 48.   □

7.3.1 Using numbers instead of pitches

In Example 7.3.1 we introduced the idea of using numbers rather than notes to compute the inverse of the primary tone row. This is a very useful idea, as it simplifies the computation of other 12-tone rows and helps explain many of the key features of the technique. Instead of pitches and musical intervals, we work with numbers and arithmetic mod 12.

The basic technique is to denote the opening note of the primary row as 0, and then a note n half steps higher is assigned the number n. In this way, every number in the tone row is associated with an integer between 0 and 11. These numbers will be attached to their notes in all of the transformed tone rows. For example, if A♭ is represented by the number 9, then every A♭ in all of the subsequent tone rows will have a 9 underneath it.

One of the advantages of this numbering system is that the transformed rows are easy to construct, as long as addition is performed modulo 12. Thus, $6 + 10 = 4 \pmod{12}$, while $9 + 3 = 0 \pmod{12}$. Consider the primary tone row P_0 in Figure 7.8, which has the numerical representation 0 2 7 6 4 9 8 1 11 10 3 5. To find the row P_4, the transposition of the primary row up four half steps, just add 4 to each of these numbers, remembering to subtract 12 if the sum is 12 or greater. Hence, P_4 has the numerical representation 4 6 11 10 8 1 0 5 3 2 7 9.

As demonstrated in Example 7.3.1, the inverted tone rows are calculated by reflecting the numbers about the first value. When this value is 0, that is, if we are computing I_0, then the reflected value is simply the inverse of each number. If the note value is n, then the inverse is $-n$, which equals $12 - n \pmod{12}$. To invert P_4 about its starting note, we first determine how far each note value is from 4, and then add or subtract this number to 4 accordingly. For example, the number 11 is 7 units above 4, so it will reflect to $4 - 7 = -3 = 9 \pmod{12}$. On a clock, 11:00 am and 9:00 pm are each the same distance (5 hours) from 4:00 pm. The number 1 is 3 units below 4, so it will invert to $4 + 3 = 7$ (1:00 pm and 7:00 pm are each 3 hours from 4:00 pm). Continuing this process, we have

$$
\begin{aligned}
P_4 &= 4 \quad 6 \quad 11 \quad 10 \quad 8 \quad 1 \quad 0 \quad 5 \quad 3 \quad 2 \quad 7 \quad 9 \\
I_4 &= 4 \quad 2 \quad 9 \quad 10 \quad 0 \quad 7 \quad 8 \quad 3 \quad 5 \quad 6 \quad 1 \quad 11.
\end{aligned}
$$

One way to check the values for row I_4 is to observe that the sum of the numbers in the same position in rows P_4 and I_4 always totals 8 (mod 12). This follows from the horizontal symmetry. In general, the corresponding entries in rows P_n and I_n will always sum to $2n$ (mod 12). Also, since operations P and I commute, row I_4 can also be obtained by adding 4 to the numerical values of I_0 given in Figure 7.9. Finally, the two retrograde operations are easily found by writing the list of note values in the opposite direction:

$$R_4 \;=\; 9 \quad 7 \quad 2 \quad 3 \quad 5 \quad 0 \quad 1 \quad 8 \quad 10 \quad 11 \quad 6 \quad 4$$
$$RI_4 \;=\; 11 \quad 1 \quad 6 \quad 5 \quad 3 \quad 8 \quad 7 \quad 0 \quad 10 \quad 9 \quad 2 \quad 4$$

The tone row matrix of the special row in Example 7.3.1 is shown in Table 7.2, using numbers instead of pitches. Notice that the invariance of the row and the symmetries it produces are immediately discernible. Moreover, the equivalencies between the inversions and retrogrades and between the transpositions and retrograde-inversions are also readily apparent from the table.

	$I_0\downarrow$	$I_2\downarrow$	$I_7\downarrow$	$I_6\downarrow$	$I_4\downarrow$	$I_9\downarrow$	$I_8\downarrow$	$I_1\downarrow$	$I_{11}\downarrow$	$I_{10}\downarrow$	$I_3\downarrow$	$I_5\downarrow$	
$P_0 \rightarrow$	0	2	7	6	4	9	8	1	11	10	3	5	$\leftarrow R_0$
$P_{10} \rightarrow$	10	0	5	4	2	7	6	11	9	8	1	3	$\leftarrow R_{10}$
$P_5 \rightarrow$	5	7	0	11	9	2	1	6	4	3	8	10	$\leftarrow R_5$
$P_6 \rightarrow$	6	8	1	0	10	3	2	7	5	4	9	11	$\leftarrow R_6$
$P_8 \rightarrow$	8	10	3	2	0	5	4	9	7	6	11	1	$\leftarrow R_8$
$P_3 \rightarrow$	3	5	10	9	7	0	11	4	2	1	6	8	$\leftarrow R_3$
$P_4 \rightarrow$	4	6	11	10	8	1	0	5	3	2	7	9	$\leftarrow R_4$
$P_{11} \rightarrow$	11	1	6	5	3	8	7	0	10	9	2	4	$\leftarrow R_{11}$
$P_1 \rightarrow$	1	3	8	7	5	10	9	2	0	11	4	6	$\leftarrow R_1$
$P_2 \rightarrow$	2	4	9	8	6	11	10	3	1	0	5	7	$\leftarrow R_2$
$P_9 \rightarrow$	9	11	4	3	1	6	5	10	8	7	0	2	$\leftarrow R_9$
$P_7 \rightarrow$	7	9	2	1	11	4	3	8	6	5	10	0	$\leftarrow R_7$
	$RI_0\uparrow$	$RI_2\uparrow$	$RI_7\uparrow$	$RI_6\uparrow$	$RI_4\uparrow$	$RI_9\uparrow$	$RI_8\uparrow$	$RI_1\uparrow$	$RI_{11}\uparrow$	$RI_{10}\uparrow$	$RI_3\uparrow$	$RI_5\uparrow$	

TABLE 7.2. The tone row matrix, with numbers rather than pitches, for the invariant tone row P_0 from Example 7.3.1. Note that $R_0 = I_5$ and $P_0 = RI_5$. The matrix is symmetric about the diagonal between the upper right (5) and lower left (7) corners, demonstrating the invariance of P_0 and the fact that this special primary row only produces 24 distinct rows under the usual four transformations.

7.3.2 Further analysis: The symmetric interval property

How common are tone rows such as the one in Example 7.3.1, where the primary row only generates 24 new rows, rather than the expected 48? What traits do these special rows possess, and how can they be created (or avoided)?

The numbering system described above helps answer these questions. Consider the tone row P_0 in Table 7.2. If we sum the first and last entries of the row, 0 and 5, we obtain 5. The sum of the second and penultimate entries, 2 and 3, is also 5, as is the sum of the third and third-to-last entries, 7 and 10, given that we take the sum modulo 12. Continuing with this method of grouping, we find that the numbers in position m and position $13 - m$ always sum to 5. This key feature makes the row invariant. In fact, any tone row that has this grouping pattern, where each pair sums to 5, will be invariant and satisfy the relation $I_5 = R_0$.

Complimentary pairs

We will call a pair of two entries in a tone row *complimentary* if they correspond to entries in positions m and $13 - m$. In other words, complimentary pairs occur in positions 1 and 12, 2 and 11, 3 and 10, 4 and 9, 5 and 8, and 6 and 7 (see Figure 7.10). In this fashion, each entry from the first six notes of the tone row is paired with an entry from the last six notes, but in the opposite order. Notice that the retrograde operation will interchange the positions within each complimentary pair (e.g., position 3 becomes position 10, and vice versa), but not alter the pairs themselves.

Theorem 7.3.2. *Suppose that P_0 is a primary tone row such that every complimentary pair sums to n (mod 12). Then P_0 is invariant and satisfies the relation $I_n = R_0$. To be a legitimate tone row, n must be odd.*

Proof: Using the numerical method of notating tone rows, let a and $n - a$ represent an arbitrary complimentary pair in the tone row P_0. To find row I_0, we simply negate each entry, as this corresponds to a reflection about the number 0 (the first entry in row P_0). We then add n to each entry to obtain row I_n, as shown below:

$$
\begin{array}{ccccccc}
P_0 & = & 0 & a & \cdots & n - a & n \\
I_0 & = & 0 & -a & \cdots & -(n - a) & -n \\
I_n & = & n & n - a & \cdots & a & 0 & = R_0.
\end{array}
$$

Notice that the complimentary pair $(a, n - a)$ becomes $(n - a, a)$ under the transformation I_n. Since every complimentary pair sums to n, it follows that all complimentary pairs

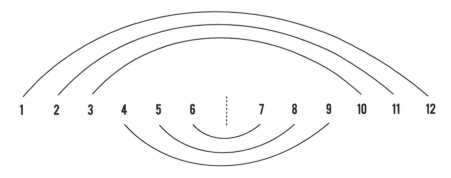

FIGURE 7.10. Complimentary pairs in a tone row combine the entries in the outermost positions and then proceed inward until reaching the pair between the 6th and 7th positions.

in the original row will be sent to themselves under the transformation I_n, but in the reverse order. This proves that $I_n = R_0$.

If n is even, then the numbers $n/2$ and $n - n/2$ are equivalent, and one of the complimentary pairs will contain the same notes. For instance, if $n = 8$ and $a = 4$, then the complimentary pair $(a, n - a)$ reduces to $(4, 4)$. But this violates the main property of a tone row, as no note can be repeated. Consequently, n must be odd, that is, $n \in \{1, 3, 5, 7, 9, 11\}$. Then, each complimentary pair consists of an even and an odd number, and the tone row P_0 can actually be constructed.   □

It is worth mentioning that the only other example of an invariant tone row we have seen thus far, the ascending chromatic scale, satisfies the hypothesis of Theorem 7.3.2. Since the scale ascends by consecutive half steps, the tone row is simply 0 1 2 3 4 5 6 7 8 9 10 11, a row whose complimentary pairs all sum to 11. By the theorem, this row is invariant and $I_{11} = R_0$, as discovered earlier.

From a musical perspective, a tone row constructed with each complimentary pair summing to the same odd number will always contain a vertical symmetry (a palindrome) in the list of musical intervals between consecutive notes. If we let "u" represent an interval rising up in pitch and "d" denote an interval moving down in pitch, then the intervals of the tone row P_0 from Example 7.3.1 are

$$\text{uM2, uP4, dm2, dM2, uP4, } \textbf{dm2} \text{, uP4, dM2, dm2, uP4, uM2.}$$

This is a symmetric list, including the direction, about the middle interval (dm2). Note that we have regarded the sixth note in the row, A, as being an octave higher. A tone row containing a symmetric (palindromic) list of musical intervals (including the directions) will be described as possessing the *symmetric interval property*.

Any row that has each complimentary pair summing to n will always have this property. This follows by construction. If a and b are consecutive notes in the first half of the row, then $n - b$ and $n - a$ are consecutive notes in the second half of the row. The interval between a and b, $b - a$, is equal to the distance between $n - b$ and $n - a$, since $(n - a) - (n - b) = b - a$.

A tone row that has the symmetric interval property will always have its retrogrades and inversions agreeing. An inversion of the original tone row will feature the same list of intervals with the u's replaced by d's and vice versa. On the other hand, because of the symmetry in the list of intervals, the retrograde of the row will also feature the same list of intervals with the u's and d's flipped. (Going backward through a tone row means reading the list of intervals backward, but also flipping the u's and d's of each interval.) Therefore, the inversions and retrogrades of the original row will have the exact same list of intervals. Once the correct number of half steps is determined to match two particular rows, the equations identifying specific rows then follow quickly.

Note that having the symmetric interval property is not enough to ensure that the tone row is invariant. As explained in Theorem 7.3.2, the distance between the first and last notes must be an odd number of half steps, because the sum of the complimentary pairs must be odd. In Example 7.3.1 the gap between the opening and closing notes of the row is five half steps (a perfect fourth), while for a chromatic scale, that gap is 11 half steps (a major seventh). Each of these numbers is odd.

For comparison, suppose that we want to mimic the invariance of the tone row in Example 7.3.1, but we adjust the last note to a B♭, which is 10 half steps above the starting note C (or two below it). Keeping the same symmetric interval structure (except for

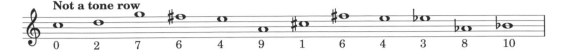

Not a tone row

0   2   7   6   4   9   1   6   4   3   8   10

FIGURE 7.11. Although the set of notes above has the same symmetric interval structure as Example 7.3.1, it is not a legitimate tone row because E and F♯ are repeated.

the middle interval, which has no effect on the symmetry) yields the set of pitches in Figure 7.11. However, in this case, the 12 pitches do not form a true tone row because both E and F♯ are repeated. This will always happen if the interval between the opening and closing notes of the row consists of an even number of half steps. Also note that each complimentary pair in Figure 7.11 sums to 10, but since this is an even number, Theorem 7.3.2 does not apply.

To count the total number of invariant tone rows with the symmetric interval property, we return to some combinatorial ideas from our discussion of change ringing. Recall that multiplication is the key operation when counting combinations with multiple choices (e.g., $n!$ represents the total number of permutations of n items).

Suppose that the first and last pitches of a symmetric row differ by n half steps, where $n \in \{1,3,5,7,9,11\}$. There are 12 choices for the starting note, but since n is fixed, the last note is now determined. The second note has 10 possibilities remaining since it cannot duplicate the first and last notes. Making the intervals symmetric, or ensuring that each complimentary pair sums to n, now fixes the penultimate note of the row as well. Continuing in this fashion, we have eight choices for the third note, six for the fourth note, four for the fifth note, and only two for the sixth note. The last six notes of the tone row are all determined by the first six, since the palindromic interval structure must be maintained. Finally, the same count will be achieved regardless of which value of n is selected. Since there are six possible values for n, we conclude that the number of invariant tone rows exhibiting the symmetric interval property is

$$6 \cdot (12 \cdot 10 \cdot 8 \cdot 6 \cdot 4 \cdot 2) = 276{,}480.$$

7.3.3   Tritone symmetry

The arguments above explain why the only tone rows having their retrogrades and inversions equivalent are those possessing the symmetric interval property. However, there is another type of invariant tone row, with a different kind of symmetry, that will also have 24 equivalencies between rows. In this case, the retrogrades and transpositions agree, and the inversions and retrograde-inversions align. The key to this type of symmetry is the tritone interval (six half steps), which divides the octave equally in half.

In Figure 7.12 we demonstrate an invariant tone row that satisfies $P_6 = R_0$. As with our previous invariant examples, this row only generates 24 new tone rows under the four usual transformations. The key feature to this invariant row is that complimentary pairs of notes are always a tritone apart. This is easily seen by examining the numbers under the row, as each complimentary pair has a difference equal to 6. For example, the 3rd and 10th notes in P_0 are C and F♯, respectively, which are a tritone apart since $7 - 1 = 6$. Or, the 4th and 9th notes are B♭ and E, respectively, which are also a tritone

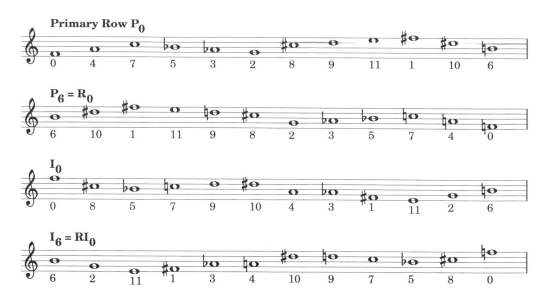

FigURE 7.12. The tone row P_0 has tritone symmetry and consequently satisfies the relations $P_6 = R_0$ and $I_6 = RI_0$. It is another example of a primary tone row that only generates 24 new rows under the usual symmetry transformations.

apart $(11 - 5 = 6)$. We will describe a tone row that has each complementary pair a tritone apart as having *tritone symmetry*. In terms of numerical values, a tone row has tritone symmetry if the difference between all complimentary pairs is 6.

Returning to the row P_0 in Figure 7.12, note that if we transpose P_0 up six half steps, the first six notes of P_6 will match the last six notes of P_0, but in reverse order. At the same time, because the tritone equally divides the octave, the last six notes of P_6 will agree with the first six of P_0, also in the reverse order. Combining these two observations leads to the key identity $P_6 = R_0$. As before, shifting each side of this equation by a different number of half steps leads to the following 12 relations between the primary forms of the row and its retrogrades:

$$
\begin{array}{lll}
P_6 = R_0 & P_{10} = R_4 & P_2 = R_8 \\
P_7 = R_1 & P_{11} = R_5 & P_3 = R_9 \\
P_8 = R_2 & P_0 = R_6 & P_4 = R_{10} \\
P_9 = R_3 & P_1 = R_7 & P_5 = R_{11}.
\end{array}
\tag{7.2}
$$

Since rows P_6 and R_0 are equivalent, we might expect that rows I_6 and RI_0 are as well. This is indeed the case (see Figure 7.12) and can be verified by applying the inversion I_0 to both sides of $P_6 = R_0$. Specifically, we have

$$
P_6 = R_0 \implies I_0 * P_6 = I_0 * R_0 \implies I_6 = RI_0,
$$

where the last equation follows from Equation (7.1). The remaining 12 equivalencies are found by replacing P with I and R with RI in each of the 12 equations in System (7.2).

The tritone is the only interval that will yield this special type of invariance. In other words, the equation $P_n = R_0$ can only be satisfied if $n = 6$. While it is possible to match the first half of the row P_n with the last half of P_0 (in reverse order), to ensure that this same trait also occurs with the remaining halves of the rows requires the choice $n = 6$.

Theorem 7.3.3. *Suppose that P_0 is a primary tone row with tritone symmetry, that is, every complimentary pair is six half steps apart. Then P_0 is invariant and satisfies the relations $P_6 = R_0$ and $I_6 = RI_0$. A primary row that has every complimentary pair n half steps apart will not be invariant unless $n = 6$.*

Proof: Using the numerical method of notating tone rows, suppose that P_0 is a tone row with tritone symmetry, and let a and $a + 6$ (mod 12) represent an arbitrary complimentary pair in the row. We then compute that

$$
\begin{array}{ccccccc}
P_0 & = & 0 & a & \cdots & a+6 & 6 \\
P_6 & = & 6 & a+6 & \cdots & a+12 & 12 \\
& = & 6 & a+6 & \cdots & a & 0 & = R_0.
\end{array}
$$

Notice that the complimentary pair $(a, a + 6)$ becomes $(a + 6, a)$ under the transformation P_6. Since every complimentary pair is of this form, it follows that all complimentary pairs in the original row will be sent to themselves under the transformation P_6, but in the reverse order. This proves that $P_6 = R_0$, and thus P_0 is an invariant tone row.

Next, suppose that each complimentary pair is n half steps apart. Is it possible that the relation $P_m = R_0$ could hold for some integer m? Repeating the same argument as above, we find that

$$
\begin{array}{ccccccc}
P_0 & = & 0 & a & \cdots & a+n & n \\
P_m & = & m & a+m & \cdots & a+n+m & n+m \\
R_0 & = & n & a+n & \cdots & a & 0.
\end{array}
$$

In order for $P_m = R_0$, each entry in the last two rows must agree. This is the case if and only if $m = n$ and $m + n = 0$ (mod 12). But these equations are both satisfied only for $m = n = 0$ or $m = n = 6$. The first solution is impossible because each complimentary pair would consist of the same note, violating the definition of a tone row. The last case is precisely tritone symmetry ($n = 6$) and confirms that the invariance is found by shifting $m = 6$ half steps. $\qquad\square$

How common are invariant tone rows?

We are now ready to determine the likelihood of running across an invariant tone row, one that only generates 24 new rows. By Theorems 7.3.2 and 7.3.3, the only invariant rows are those possessing the symmetric interval property or tritone symmetry. All other tone rows will generate a full set of 48 new rows using the transformations P, I, R, and RI. We have already counted the number of possible tone rows with the symmetric interval property (276,480). The number of different tone rows with tritone symmetry is

$$ 12 \cdot 10 \cdot 8 \cdot 6 \cdot 4 \cdot 2 = 46{,}080. $$

This value is obtained by counting the number of choices for each of the first six notes in the tone row, since the last six notes must be a tritone away. There are 10 choices for the second note (as opposed to 11) because once the first note is chosen, the last note is determined as well, leaving a possible 10 notes to select from. Thus, there are a total of $276{,}480 + 46{,}080 = 322{,}560$ invariant tone rows. Since the total number of tone rows is 12!, the chances of randomly composing a tone row that generates only 24 new rows under the four symmetry operations is

$$\frac{322{,}560}{12!} = \frac{1}{1485} \approx 0.0006734,$$

or about 0.07%. Although they are fun to play around with, the special invariant tone rows require a good degree of mathematical intuition to construct and are quite rare in the world of 12-tone music.

7.3.4   The number of distinct tone rows

When counting items, mathematicians distinguish between those that are related to each other by some transformation and those that are not. For example, $\mathbb{Z}_{12}$, the integers modulo 12, only needs 12 representative elements, known as *equivalence classes*. Each integer differs by a multiple of 12 from one of the members of the set $\{0, 1, 2, 3, \ldots, 11\}$. Hence, this set is sufficient to describe the full space. For example, the integers 22, 70, and -14 all belong to the equivalence class for 10, denoted as $[[10]]$, since each is a multiple of 12 away from 10. Every integer in $[[10]]$ can be written as $10 + k \cdot 12$ for some choice of $k \in \mathbb{Z}$.

It is also important to realize that two different equivalence classes never share an element in common. For instance, the intersection of $[[10]]$ and $[[7]]$ is empty because if the equation $10 + 12k_1 = 7 + 12k_2$ were true for some integers k_1 and k_2, then $3 = 12(k_2 - k_1)$ implies that 3 would be a multiple of 12, a clear fallacy.

In the spirit of equivalence classes, we now determine the number of "distinct" tone rows by eliminating those that can be obtained from another row under one of our four symmetry transformations. Specifically, given a primary row P_0, the equivalence class $[[P_0]]$ will consist of all the rows generated by the transformations P_n, R_n, I_n, or RI_n. This is typically 48 different tone rows, but as we have seen, if P_0 is invariant, then the number of generated rows reduces to 24. In either case, the tone rows associated with P_0 all count as "one."

Given this setup, the number of distinct tone rows is 9,985,920. This fact was apparently first proven by David Reiner [1985] using group theory and Burnside's counting lemma. We give a simpler but less rigorous proof below.

Theorem 7.3.4. *Suppose that two tone rows are called equivalent if there is a symmetry transformation (P_n, R_n, I_n, or RI_n) mapping one to the other. Then, the number of distinct 12-tone rows is 9,985,920.*

Proof: First note that any given tone row P_0 will generate either 24 or 48 total rows (including itself) under the four symmetry transformations P_n, R_n, I_n, or RI_n. This follows because there are 24 rows determined by P_n and I_n which are always different, regardless of the value of n. Moreover, if there is a special relation between two rows, for example, $I_9 = R_0$, then there will be 24 such equations, obtained by shifting the subscript and/or applying R or I to each side of the equation. This automatically reduces the number of generated rows from 48 to 24, but never any number in between.

When we count the number of distinct tone rows, we are really counting the number of equivalence classes $[[P_0]]$. Two rows belong to the same class if there is some symmetry operation transforming one row into the other. Since we don't want to include these extra rows in our count, we should divide the total number of rows by 48. However, the invariant tone rows only generate 24 new tone rows under symmetry. It follows that we should perform the count by first splitting the set of 12-tone rows into two groups: the invariant tone rows, which generate 24 new additional rows, and the non-invariant tone rows, which generate 48 new rows. Fortunately, we have argued above that the only invariant tone rows possess either the symmetric interval property or tritone symmetry. This allowed us to count the total number of invariant rows, yielding 322,560 such rows.

Therefore, the total number of distinct tone rows is the number of non-invariant rows divided by 48, plus the number of invariant rows divided by 24. This yields

$$\frac{12! - 322{,}560}{48} + \frac{322{,}560}{24} = 9{,}985{,}920.$$

$\square$

7.3.5 Twelve-tone music and group theory

We close this section with a brief investigation of 12-tone music from the perspective of group theory. Suppose we collect the 48 symmetry operations P_n, I_n, R_n, and RI_n into a set G. Does G form a group under composition? Surprisingly, if we adhere to our precise definitions for each transformation, then the answer is no.

There are two primary reasons for this disappointing fact. First, certain elements will compose in different ways depending on the tone row they are being applied to. From a mathematical perspective this is problematic, because it means that we need to keep track of the underlying tone row. When we studied the dihedral group D_4 in Section 5.3.3, the actual numbering of the original square was irrelevant; it was how those numbers were permuted under the symmetry transformations that mattered. But with our notation and definitions for G, the underlying tone row does affect the result of certain compositions. In mathematical jargon, we say that composition in G is not *well defined*.

Here is an example illustrating the fact that composition is not well defined. Suppose that P_0 is the row 0 2 $\cdots$ 9 5. We claim that $R_3 * I_0 = RI_1$. In other words, taking the R_3 tone row and reflecting it about its starting pitch will produce the row RI_1. This can be checked as follows:

$$
\begin{array}{rcccccc}
P_0 & = & 0 & 2 & \cdots & 9 & 5 \\
R_0 & = & 5 & 9 & \cdots & 2 & 0 \\
R_3 & = & 8 & 0 & \cdots & 5 & 3 \\
R_3 * I_0 & = & 8 & 4 & \cdots & 11 & 1,
\end{array}
$$

and

$$
\begin{array}{rcccccc}
P_0 & = & 0 & 2 & \cdots & 9 & 5 \\
I_0 & = & 0 & -2 & \cdots & -9 & -5 \\
I_1 & = & 1 & -1 & \cdots & -8 & -4 \\
RI_1 & = & -4 & -8 & \cdots & -1 & 1 \\
& = & 8 & 4 & \cdots & 11 & 1.
\end{array}
$$

On the other hand, suppose that we adjust the last note of P_0 from 5 to 7. Then we have $R_3 * I_0 - RI_5$, and the value of the composition has changed. We check:

$$
\begin{array}{rcccccc}
P_0 & = & 0 & 2 & \cdots & 9 & 7 \\
R_0 & = & 7 & 9 & \cdots & 2 & 0 \\
R_3 & = & 10 & 0 & \cdots & 5 & 3 \\
R_3 * I_0 & = & 10 & 8 & \cdots & 3 & 5,
\end{array}
$$

and

$$
\begin{array}{rcccccc}
P_0 & = & 0 & 2 & \cdots & 9 & 7 \\
I_0 & = & 0 & -2 & \cdots & -9 & -7 \\
I_5 & = & 5 & 3 & \cdots & -4 & -2 \\
RI_5 & = & -2 & -4 & \cdots & 3 & 5 \\
& = & 10 & 8 & \cdots & 3 & 5.
\end{array}
$$

The fact that $R_3 * I_0$ can vary depending on the underlying tone row demonstrates that composition is not well defined on our set G.

Composition is not associative on G

The other issue preventing G from being a group is that the group operation, composition of transformations, is not associative. This can be seen from the following identities. Using the numerical method to notate tone rows, suppose that a primary row begins on 0 and ends on b. For any m or n, we have the following relations between inversions, retrogrades, and retrograde-inversions:

$$
\begin{array}{rclcrcl}
R_n * I_m & = & RI_{m+n+2b}, & \quad & RI_n * I_m & = & R_{m+n-2b}, \\
I_m * R_n & = & RI_{m+n}, & \quad & I_m * RI_n & = & R_{m+n}.
\end{array} \tag{7.3}
$$

Note that two of these equations depend on the value of b in different ways.

Let us compare $(I_0 * R_0) * I_0$ with $I_0 * (R_0 * I_0)$, using a tone row with $b = 5$. By the identities in (7.3), we have

$$
(I_0 * R_0) * I_0 = RI_0 * I_0 = R_{-10} = R_2,
$$

while

$$
I_0 * (R_0 * I_0) = I_0 * RI_{10} = R_{10}.
$$

This means that $*$ is not associative, and therefore G cannot be a group under composition.

The basic problem with our setup is that we have defined inversions to occur about the starting note of a tone row. This creates problems when inversions and retrogrades are combined, as the note of reflection changes from the opening note to the last note of the row. This is apparent in a composition such as $R_0 * I_0$, where the expected result is simply RI_0 rather than the actual RI_{2b}. Similarly, the composition of an inversion with itself ought to return the identity. However, since the reflections are about *different* notes, we have $I_n * I_n = P_{2n}$, rather than P_0.

For a mathematical fun fact, note that if $b = 6$, then G does form a group under composition. In other words, if we restrict to only those tone rows whose first and last notes are a tritone apart, then G is a group of order 48. In this special case, G is a commutative group (since $2b = 0 \pmod{12}$) and is isomorphic to $\mathbb{Z}_{12} \times \mathbb{Z}_2 \times \mathbb{Z}_2$, the Cartesian product of the integers mod 12 and the four musical symmetries.

To rectify the issues preventing G from forming a group, it is necessary to specify that an inversion always be taken with respect to the same note, say, the first note of the primary tone row (0). Thus, the operation I_0 will always imply a reflection about 0, rather than the starting note of the particular row. With this setup, the operations P and I no longer commute, as $I_0 * P_n = P_{-n} * I_0$, although R and I will now commute with each other. Using this definition of I makes G a group under composition with 48 elements. For group theory enthusiasts, G is isomorphic to $D_{12} \times \mathbb{Z}_2$, the Cartesian product of the dihedral group of order 12 and the integers mod 2.[2]

Exercises for Section 7.3

1. Consider the primary tone row P_0 shown below.

Primary Row P_0

  a. Using the numbers $0, 1, \ldots, 11$, indicate the value of each pitch with respect to the starting note. In other words, $D = 0$, $D\sharp = 1$, $E = 2$, and so forth.

  b. Using your answer to part **a.**, construct the entire tone row matrix for P_0 using numbers instead of pitches.

  c. The excerpt below is based on tone row P_0. Each of the four parts follows a different row. Identify the names of each row used (P_5, R_4, I_2, etc.). What do you notice when comparing the rhythm in the treble clef with that of the bass clef?

2. Suppose that P_0 is an invariant tone row satisfying the relation $I_9 = R_0$.

  a. Which type of symmetry does the row possess, the symmetrical interval property or tritone symmetry? Explain.

  b. Which row is equivalent to I_0? Which row is equivalent to I_4?

  c. Which row is equivalent to the primary row P_0? Which row is equivalent to P_6?

---

[2]See Sections 9.8–9.10 of Benson, *Music* for an explanation of this fact and for the definition of the Cartesian product of two groups.

3. Write down an invariant tone row, other than one from the text, that satisfies the symmetric interval property. List all 24 equations that relate one type of tone row to another.

4. Write down an invariant tone row, other than one from the text, that satisfies the tritone property. Give a few of the equations that relate one type of tone row to another.

References for Chapter 7

Benson, D. J.: 2007, *Music: A Mathematical Offering*, Cambridge University Press.

Cross, J.: 2003, Composing with Numbers: Sets, Rows and Magic Squares, *in* J. Fauvel, R. Flood and R. Wilson (eds), *Music and Mathematics: From Pythagoras to Fractals*, Oxford University Press, pp. 130–146.

Lansky, P. and Perle, G.: 2013, Twelve-Note Composition, *Grove Music Online*, Oxford University Press.

Reiner, D.: 1985, Enumeration in Music Theory, *American Mathematical Monthly* **92**(1), 51–54.

Schoenberg, A.: 1952, *Suite Für Klavier (Suite for Piano)*, Universal Edition, no. 7627.

Stein, L. (ed.): 1975, *Style and Idea: Selected Writings of Arnold Schoenberg*, St. Martin's Press, New York.

Wright, D.: 2009, *Mathematics and Music*, American Mathematical Society.

Chapter 8

Mathematical Modern Music

Our final chapter considers the mathematically inspired works of three modern composers: Peter Maxwell Davies, Steve Reich, and Iannis Xenakis. Davies has written pieces that use *magic squares* to derive the tonal and rhythmic material in a fashion similar to the 12-tone method developed by Schoenberg (and later Babbitt). A magic square is a special array of numbers where each column, row, and diagonal sum to the same number. Davies uses an 8×8 magic square (depicting the planet Mercury) to determine the notes and durations in his piece *A Mirror of Whitening Light*.

After Davies we consider the highly rhythmic, pulsating work of the great American composer Steve Reich. Reich's creative use of rhythmic phase shifting and minimalism in works such as *It's Gonna Rain* and *Clapping Music* is a splendid example of how simple mathematical ideas (in this case, phase shifting and cyclic groups) can generate provocative and memorable music. We also examine the highly mathematical Greek composer Iannis Xenakis. Trained in engineering and mathematics, and working briefly as an architect with Le Corbusier, Xenakis brings a highly unique and distinctly mathematical approach to his compositions. We discuss his use of glissandi in the score for *Metastasis* to create a ruled surface, a technique that later inspired his design of the Philips Pavilion. We also describe Xenakis's stochastic music, using his piece *Pithoprakta* as a model, where the pitches, timbres, durations, and dynamics are all determined using probability density functions.

The chapter closes with a sample final project inviting the reader to compose their own mathematical piece of music. This culminating assignment, one that asks students to combine their artistic and analytic skills, has worked particularly well in the author's courses on math and music. Some sample compositions of former students are described in detail in order to provide inspiration for the project.

8.1 Sir Peter Maxwell Davies: Magic Squares

A magic square in a musical composition is not a block of numbers—it is a generating principle, to be learned and known intimately, perceived inwardly as a multidimensional projection into that vast (chaotic!) area of the internal ear—the space/time crucible—where music is conceived. ... Projected onto the page, a magic square is a

FIGURE 8.1. Composer Peter Maxwell Davies. © www.johnbattenphotography.co.uk.

dead, black conglomeration of digits; tune in, and one hears a powerful, orbiting dy-
namo of musical images, glowing with numen and lumen.

— Peter Maxwell Davies[1]

One of the most interesting uses of a mathematical tool for constructing a piece of music, a process that extends the techniques of Schoenberg's 12-tone method, is the innovative application of *magic squares* by the British composer Sir Peter Maxwell Davies (see Figure 8.1). As indicated in the opening quote, Davies views a magic square, a square grid of whole numbers where all the rows, columns, and diagonals sum to the same number, as an architectural blueprint for the compositional process. By moving through the square in various patterns, the resulting sequence of numbers determines the pitches in the music, and even their durations, in much the same way that a tone row determines the pitches in the 12-tone technique. Davies uses magic squares as a compositional device in many of his pieces, beginning with *Ave Maris Stella* (1975), a work that utilizes the 9×9 magic square associated with the moon to permute notes of a plainchant melody and to determine the durations of the notes. Other works employing magic squares include the operas *The Lighthouse* (1979) and *Resurrection* (1987), *Strathclyde Concerto No. 3 for Horn and Trumpet* (1989), many of his symphonies, and *A Mirror of Whitening Light* (1977), a work we explore in great detail below.

Born in Salford, Lancashire, England in 1934, Davies studied at the Manchester University and Royal Manchester College of Music. He helped found the group *New Music Manchester*, committed to contemporary music, in 1953. Davies studied in Rome, Princeton, and Australia before settling on the Orkney Islands off the coast of Scotland in 1971. From 1992 to 2002 Davies served as the associate conductor/composer with the BBC Philharmonic Orchestra. He has conducted many of the world's great orchestras, including the Cleveland and Boston Symphony Orchestras and the Russian National Orchestra. Davies was knighted in 1987 (thus the Sir honorific) and appointed the Master of the Queen's Music in 2004. He was awarded an Honorary Doctorate of Music by Oxford University in 2005. Interestingly, in 1996 Davies became one of the first classical composers to open a music download website, www.maxopus.com, to promote his works.

---

[1]Davies, Composition Questions Answered.

8.1.1 Magic squares

Before we learn how Davies applied magic squares in his compositions, we explore some of the mathematical theory and history regarding magic squares. An $n \times n$ *matrix* is a square array of numbers with n rows and n columns. For instance,

$$\begin{bmatrix} 1 & 3 & 5 \\ 2 & 4 & 6 \\ 7 & 8 & 9 \end{bmatrix} \tag{8.1}$$

is an example of a 3×3 matrix. The tone row matrix with numerical values is an example of a 12×12 matrix. In general, a matrix can be rectangular, that is, it can have a different number of rows and columns. However, in this section we are only interested in $n \times n$ square matrices. A square matrix is called *magic* if the sums of its rows, columns, and two main diagonals are all equivalent. Matrix (8.1) is not magic because the first two rows sum to 9 and 12, respectively.

Definition 8.1.1. *An $n \times n$ square array containing the numbers*

$$1, 2, 3, 4, \cdots, n^2 - 1, n^2,$$

each occurring precisely once, and having each row, column, and main diagonal sum to the same amount, is called a <u>magic square</u>. The <u>order</u> of a magic square is the number of rows or columns in the square.

For clarity, there are two main diagonals, each running from one corner of the square to the opposite corner. The diagonal between the upper left and lower right corners will be called the *leading* or *left* diagonal, while the diagonal stretching from the upper right to lower left corner will be called the *secondary* or *right* diagonal.

8	1	6
3	5	7
4	9	2

TABLE 8.1. The *Lo Shu*, a magic square of order 3.

Table 8.1 shows an example of a magic square with three rows and columns. Note that each row, column, and main diagonal sums to 15, so the square is indeed a magic one. This magic square was known to the ancient Chinese, who called it the *Lo Shu*, and was first recorded in a manuscript around 2200 BCE. According to legend, while walking along the bank of the Lo River (or Yellow River), Emperor Yu watched a mystical turtle crawl out of the water. There was nothing particularly interesting about the turtle except that it contained a series of dots in each panel of its shell. Remarkably, Yu noticed that the total number of dots in any row, column, or diagonal was always the same (see Figure 8.2).

FIGURE 8.2. The mystical turtle that inspired the legend of the Lo Shu.

The emperor brought his new mathematical pet back to his palace, where it spent the rest of its years as the most famous turtle in the world, earning visits from mathematicians and kings alike. The pattern on the turtle shell became known as the *Lo Shu* (the word "shu" means *book* or *scroll*) and was frequently used on charms and magic stones. Feng Shui formulas of astrology and I-Ching are each based on the Lo Shu magic square.

Magic squares have a long and storied history, beginning with the discovery of the 3×3 Lo Shu magic square by the ancient Chinese. Various cultures have pondered their magic and assigned mystical properties to their existence.[2] The Lo Shu magic square was also well known and used by many other cultures, both past and present, including the Mayan Indians, the modern Hausa people of northwestern Nigeria, the ancient Babylonians, and prehistoric cave dwellers in northern France. It had important religious meaning in the Islamic tradition and to the ancient Chinese, who associated the even numbers with the *yin*, the female principle, and the odd numbers with the *yang*, the male principle. The five in the central square represented the earth. Notice that the eight outside numbers in the Lo Shu magic square alternate between the yin and the yang, symbolizing an equal balance among the key elements. Also observe that if the square is rotated by a multiple of $90°$, or reflected about any of the four symmetry axes, it still remains magic.

8.1.2 Some examples

Is it possible to have a magic square of order less than 3? If there is just one row and column, then $\boxed{1}$ is a magic square since its rows, columns, and diagonals all sum to 1. This is a trivial example of a magic square, although the astrologer and theologian Heinrich Cornelius Agrippa viewed the magic square of order 1 as a symbol of God's everlasting perfection.

What about a magic square of order 2? Is it possible to construct a 2×2 magic square using the numbers $1, 2, 3, 4$? Suppose that we place a 1 in the upper left corner. Figure 8.3 shows the six possible 2×2 squares that have a 1 in the upper left corner. Note that each

---

[2]For a detailed history, classification, and construction of magic squares, see Pickover, *Zen of Magic Squares, Circles, and Stars.*

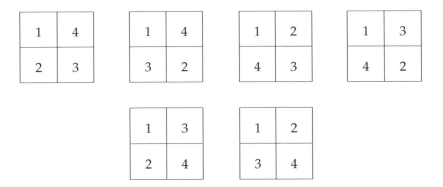

FIGURE 8.3. There is no magic square of order 2.

row in the first two squares (top left of figure), each column in the second set of squares (top right of figure), and each diagonal in the bottom set of squares sums to 5. However, none of these squares are magic because each one contains rows, columns, or diagonals that do not sum to the same amount. If we try to find a magic square of order 2 with a different number in the upper left corner, we can rotate that square to obtain one of the examples in Figure 8.3. Consequently, we conclude that there is no magic square of order 2.

Ignoring its symmetric counterparts, the Lo Shu is the only magic square of order 3. In other words, any 3×3 magic square is either the Lo Shu magic square shown in Table 8.1 or one of its seven symmetric cousins (three from the rotations R_{90}, R_{180}, and R_{270}, and four from the reflections H, V, D_L, and D_R). Table 8.2 shows a version of the Lo Shu obtained from reflecting about the left diagonal (applying the transformation D_L).

Magic squares have inspired artists for centuries. A famous example of a 4×4 magic square appears in the 1514 engraving *Melancholia I* by German artist and mathematician Albrecht Dürer (see Figure 8.4). The sum of any row, column, or main diagonal is 34, so the square is indeed a magic one. Notice that Dürer managed to include the date of his work in the middle of the bottom row of the magic square. Also observe that the sum of the four corners, and the sum of the four middle squares, both equal 34. This fact is valid for any magic square of order 4 (see Exercise 5). Unlike the case $n = 3$, magic squares of order 4 are far more plentiful; there are 880 geometrically distinct 4×4 magic squares.[3]

8	3	4
1	5	9
6	7	2

TABLE 8.2. A reflected version of the Lo Shu magic square. The Lo Shu is the only magic square of order 3.

[3]For a complete list of the order 4 magic squares, as well as some interesting background and fun facts, see Harvey Heinz's webpage http://www.magic-squares.net/order4list.htm.

16	3	2	13
5	10	11	8
9	6	7	12
4	15	14	1

FIGURE 8.4. Dürer's *Melancholia I* featuring a magic square of order 4 in the upper right corner (enlargement to the right). The date of the carving, 1514, can be found in the middle of the last row of the magic square. © The Trustees of the British Museum.

The famous Spanish architect Antoni Gaudí featured a symbolic "magic-like" square on the front facade of his remarkable church the Sagrada Família in Barcelona (left image, Figure 8.5). Here, the numbers 10 and 14 are used twice, while 12 and 16 are absent, but every row, column, and main diagonal sums to 33, a number with great religious symbolism. The right image in Figure 8.5 shows a print by mathematically trained artist Margaret Kepner that is based on the 8×8 magic square known as the *Gwalior Square*.

8.1.3   The magic constant

It turns out that every magic square of the same size (the same n) will have each row, column, or main diagonal summing to the same amount, called the *magic constant*. We explain and justify this below. For the case $n = 3$, the magic constant is 15, while for $n = 4$, it is 34.

Definition 8.1.2. *The magic constant, denoted as M_n, is given by the formula*

$$M_n = \frac{n(n^2 + 1)}{2}.$$

FIGURE 8.5. The pseduo-magic square on the front of Gaudí's Sagrada Família (*left*; author photo), and *Magic Square 8 Study: A Breeze over Gwalior* (*right*; 2013), a work based on an 8×8 magic square by mathematically inspired artist Margaret Kepner.

Note that the value of the magic constant depends on n; in fact, it is an increasing cubic function of n. The values of M_n for some small values of n are shown in Table 8.3.

n	2	3	4	5	6	7	8	9	10
M_n	5	15	34	65	111	175	260	369	505

TABLE 8.3. The magic constant M_n for different values of n.

Theorem 8.1.3 (Magic Constant). *The rows, columns, and main diagonals of any $n \times n$ magic square always sum up to the same value, the magic constant.*

Before proving this theorem, we derive an important identity about the sum of the first n natural numbers. There is a wonderful story about the great German mathematician Carl Friedrich Gauss (1777–1855), who as a young boy was told, perhaps as a punishment for goofing off during class, to sum up all of the integers between 1 and 100. After a few moments of thinking, Gauss quickly declared that the answer was 5050. The method the young genius supposedly used is quite illustrative.

Instead of adding the first few numbers $1 + 2 + 3 + 4$, Gauss combined the first and last numbers in the list, $1 + 100 = 101$. He then combined the second number with the penultimate number in the list, $2 + 99 = 101$. The pattern continues with $3 + 98 = 101$, $4 + 97 = 101$, etc., as shown below:

$$1 + 2 + 3 + \cdots + 98 + 99 + 100.$$

$$101$$

It is clear that each grouping always yields a sum of 101. Since there are 50 such groupings (the last pairing will be $50 + 51 = 101$), it follows that the sum is $50 \cdot 101 = 5050$, which is the answer provided by the young prodigy.

The above argument generalizes to any sum as long as the difference between successive numbers remains constant (such a sum is called an *arithmetic series*). Suppose we wanted to sum the first n natural numbers beginning with 1 and ending with n. Using the method described above, we pair the first and last numbers together, $1 + n = n + 1$; the second and penultimate numbers, $2 + (n - 1) = n + 1$; the third and antepenultimate numbers, $3 + (n - 2) = n + 1$; and so on,

$$1 + 2 + 3 + \cdots + n - 2 + n - 1 + n.$$

$$n+1$$

As before, there are $n/2$ such groupings, so the total sum will be $(n/2) \cdot (n + 1) = n(n + 1)/2$. This is an important formula, one that can be proven more rigorously with a technique called the *Principle of Induction*:

$$1 + 2 + 3 + \cdots + n - 2 + n - 1 + n = \frac{n(n + 1)}{2}. \tag{8.2}$$

Proof of the Magic Constant Theorem: Suppose that we have an $n \times n$ magic square (so it has n rows and columns). Let S represent the common sum of any row, column, or main diagonal. First, let's compute the sum of *all* the squares in the magic square. There are two ways to write down this larger sum. One way is to note that every natural number from 1 to n^2 appears exactly once. Thus, the total sum of all the numbers in the magic square is

$$1 + 2 + 3 + \cdots + n^2 - 2 + n^2 - 1 + n^2.$$

Using Formula (8.2), we can replace n by n^2 to obtain an expression for this sum. It is

$$\frac{n^2(n^2 + 1)}{2}. \tag{8.3}$$

On the other hand, by definition, each row of the magic square has to sum to S. Since there are n rows, the sum of all the entries in the magic square must be $n \cdot S$. (A similar argument works for the n columns.) Equating this with the value given in Equation (8.3), we find that

$$nS = \frac{n^2(n^2 + 1)}{2} \qquad \text{or} \qquad S = \frac{n(n^2 + 1)}{2}.$$

Since the formula for S depends only on n, and not on the actual location of the numbers in the magic square, it follows that the sum S will be the same for any two magic squares of the same size. Moreover, since $S = M_n$, we have also shown that the magic constant is indeed the sum of any row, column, or main diagonal in a magic square. This completes the proof. $\qquad\qquad\qquad\qquad\qquad\qquad\qquad\qquad\qquad\qquad\qquad\qquad\qquad\qquad\qquad\qquad\qquad$ $\square$

8.1.4 *A Mirror of Whitening Light*

A Mirror of Whitening Light, op. 75 was composed by Davies in 1977. It was commissioned by the instrumental ensemble London Sinfonietta and dedicated to one of his teachers, the American composer Roger Sessions, with whom Davies studied at Princeton. The title of the piece, originally in Latin as *Speculum Luminis Dealbensis*, refers to alchemy, in particular the process of "whitening" a base metal into gold. Davies was inspired by the remarkable properties of sunlight, specifically those he observed from his window while composing on the Orkney Islands, where most of his work, spanning five-plus decades, has been written. The "mirror" in the title refers to the reflection of the sunlight off the bay outside Davies's window.

In the composer's note in the score for the piece, Davies [1978b] writes,

> *Fancifully perhaps, I often see the great cliff-bound bay before my window where the Atlantic and the North Sea meet as a huge alchemical crucible, rich in speculative connotations, and at all times a miracle of ever-changing reflected light, and it is this which is the physical Mirror of the title.*

The work is written for a small chamber orchestra consisting of a flute, piccolo, oboe, English horn, clarinet, bassoon, French horn, trumpet, trombone, celesta, two violins, viola, cello, double bass, and three percussion instruments: crotales (antique cymbals), glockenspiel, and marimba. It was first performed by the London Sinfonietta at Queen Elizabeth Hall in London in 1977, with Davies conducting. The piece is approximately 22 minutes in length.

The remarkable aspect of this work is how the notes and their lengths are created entirely from an 8×8 magic square.[4] In the introductory lecture for the opening performance, Davies states that he uses the magic square of the sun, and he proceeds to explain the properties of a magic square with entries ranging from 1 to 64. However, the magic square of the sun is actually a 6×6 square, not 8×8. It is clear from an analysis of the music that Davies was in fact using the magic square of the planet Mercury, shown in Table 8.4. Although this magic square is associated with Mercury (ostensibly because the number 8 and the magic constant $M_8 = 260$ have numerological connections to the planet) rather than the sun, the interesting alchemical properties of the *element* mercury serve to connect this particular magic square with the title of the piece.

8	58	59	5	4	62	63	1
49	15	14	52	53	11	10	56
41	23	22	44	45	19	18	48
32	34	35	29	28	38	39	25
40	26	27	37	36	30	31	33
17	47	46	20	21	43	42	24
9	55	54	12	13	51	50	16
64	2	3	61	60	6	7	57

TABLE 8.4. The 8×8 magic square of Mercury.

---

[4]Our analysis of the piece is based on Cross, Composing with Numbers.

G $_1$	E $_2$	F $_3$	D $_4$	F♯ $_5$	A $_6$	G♯ $_7$	C $_8$
E $_9$	C♯ $_{10}$	D $_{11}$	B $_{12}$	D♯ $_{13}$	F♯ $_{14}$	F $_{15}$	A $_{16}$
F $_{17}$	D $_{18}$	E♭ $_{19}$	C $_{20}$	E $_{21}$	G $_{22}$	G♭ $_{23}$	B♭ $_{24}$
D $_{25}$	B $_{26}$	C $_{27}$	A $_{28}$	C♯ $_{29}$	E $_{30}$	E♭ $_{31}$	G $_{32}$
F♯ $_{33}$	D♯ $_{34}$	E $_{35}$	C♯ $_{36}$	F $_{37}$	G♯ $_{38}$	G $_{39}$	B $_{40}$
A $_{41}$	F♯ $_{42}$	G $_{43}$	E $_{44}$	A♭ $_{45}$	B $_{46}$	B♭ $_{47}$	D $_{48}$
G♯ $_{49}$	F $_{50}$	F♯ $_{51}$	D♯ $_{52}$	G $_{53}$	A♯ $_{54}$	A $_{55}$	C♯ $_{56}$
C $_{57}$	A $_{58}$	A♯ $_{59}$	G $_{60}$	B $_{61}$	D $_{62}$	C♯ $_{63}$	F $_{64}$

FIGURE 8.6. The eight-note phrase (left) used by Davies in *A Mirror of Whitening Light*, derived from the plainchant *Veni Sancte Spiritus*. On the right are the notes of the plainchant transposed repeatedly to fill an 8×8 grid.

To utilize the magic square as a compositional device, Davies begins with the notes from the plainchant *Veni Sancte Spiritus* (Come Holy Spirit), often called the "Golden Sequence," to create a melodic phrase with eight distinct notes (see Figure 8.6). These eight notes are used to generate a *transposition square*, a term coined by music analyst David Roberts [2000]. To form the transposition square, Davies places the eight notes across the top row and leftmost column of an 8×8 matrix. He then transposes the opening phrase seven times so that it begins on the first note of each row. For example, since the main melodic phrase begins with the notes G, E, and F, the second row of the matrix is the original melody transposed down a minor third to begin on E instead of G, while the third row is transposed down a whole step to begin on F instead of G (see the table in Figure 8.6). This is precisely the procedure used to create a 12-tone matrix, except that Davies does not invert the original melody; he simply copies it down the left-hand column. By construction, the resulting square will be symmetric about the main diagonal (running from the upper left to lower right corners). After the transposition square is complete, the numbers 1 through 64 are placed consecutively throughout the square, as shown in the table in Figure 8.6.

The final step is to use the plainchant matrix as a map onto the magic square of Mercury. For instance, the top row of the magic square begins with $8, 58, 59$. Referring to the table in Figure 8.6, we see that square 8 corresponds to the note C, square 58 corresponds to A, and square 59 corresponds to A♯. Continuing the process, each entry in the magic square of Mercury is identified with a particular note, as shown in Table 8.5.

This special array of 64 notes is the architectural blueprint used by the composer to create the music. Davies traverses his magic square in a variety of ways, going across rows, up and down columns, using zigzag patterns, or even spiraling around the square. Each path generates a sequence of notes for a particular instrument, and just as with the 12-tone technique, the register (which octave) used for each note can vary widely. Davies is careful to disguise the plainchant by utilizing different octaves within a particular instrument's part.

C	A	A♯	F♯	D	D	C♯	G
G♯	F	F♯	D♯	G	D	C♯	C♯
A	G♭	G	E	A♭ (D)	E♭	D	D (A♭)
G	D♯	E	C♯	A (D)	G♯	G	D (A)
B	B	C	F	C♯	E	E♭	F♯
F	B♭	B	C	E	G	F♯	B♭
E	A	A♯	B	D♯	F♯	F	A
F	E	F	B	G	A	G♯	C

TABLE 8.5. Davies's mapping of the plainchant notes onto the 8×8 magic square of Mercury. The notes in parentheses are adjustments to the square that were used by the composer.

Finding the various paths describing each instrument is a bit like finding words in a word search puzzle. Sometimes the mathematical patterns dissolve at the composer's discretion, for musical reasons. To make matters even more confusing when analyzing the piece, it appears that Davies altered four of the numbers in the magic square of Mercury [Cross, 2003]. It is not clear why this was done, but the numbers 45 and 48 are interchanged (third row) in Davies's magic square, as well as the numbers 25 and 28 (fourth row). Interestingly, since $45 + 28 = 48 + 25$, Davies's transformed square nearly remains a magical one, as the sum of the rows and columns remains unchanged under this transformation, equal to the magic constant $M_8 = 260$. However, the sum on the diagonal from the upper right to lower left corner becomes 257, three shy of the magic constant. The four different notes corresponding to this interchange of numbers are indicated by parentheses in Table 8.5.

The composer's adjustment to the magic square of Mercury is apparent right from the beginning of the piece. Just as the start of a 12-tone composition helps identify the underlying tone row, the opening lines in *A Mirror of Whitening Light* confirm the interchanging of notes described above. The starting notes for the first five instruments to play are given below, with Davies's alterations from the magic square indicated in parentheses:

> Trumpet: C A B♭ G♭ D
> Flute: D C♯ G A♭ F G♭ E♭ G
> Clarinet: A F♯ G E (D) E♭ D (A♭) G E♭ E C♯
> Crotales: (D) A♭
> Celesta: G (A) B B C F C♯ E D♯ F♯ F B♭ B C.

Beginning with the trumpet and flute, the first 13 notes of the magic square are sounded (using enharmonic equivalents where necessary), reading across the first and second rows from left to right. The clarinet part starts with the third row of Davies's magic square, with

FIGURE 8.7. The flute part starting at rehearsal letter Z in *A Mirror of Whitening Light*. Dynamics and articulations have been removed for easier reading.

the crotales and celesta continuing the line until the middle of the sixth row. Assuming that the notes in parentheses in Table 8.5 are selected, the pattern is an exact match with the magic square.

Another interesting example is the flute entrance at rehearsal letter Z (p. 51 of the score [Davies, 1978b]), about halfway through the piece (see Figure 8.7). Just before the flute enters, the viola plays a G, corresponding to the upper right corner of the magic square. The flute then plays the following sequence of notes: C♯, D, D, F♯, A♯, A, C, which correspond to the first row of the magic square moving from right to left. (Sometimes we must consider two consecutive notes as the same note, as with the opening two C♯s at rehearsal letter Z, while on other occasions we consider them as distinct notes, as must be done with the following pair of Ds.) Over the next 32 measures, concluding at the double bar line (p. 59) just before rehearsal letter D1, the flute plays the remaining 56 notes of the magic square by spiraling around the square in the counterclockwise direction, moving inward each revolution to avoid forming a cycle. The final four flute notes of this passage are D, C♯, F, C♯, in the center of the magic square. As with the opening, all the notes correspond precisely to those in the square, assuming that the parenthetical choices are utilized.

An even more startling observation is that the durations of the notes correspond to the same counterclockwise path through Davies's magic square, except that the numbers of the square have been reduced modulo 8. Suppose that we measure duration in terms of eighth notes, so that a length of 1 means an eighth note, a length of 2 corresponds to

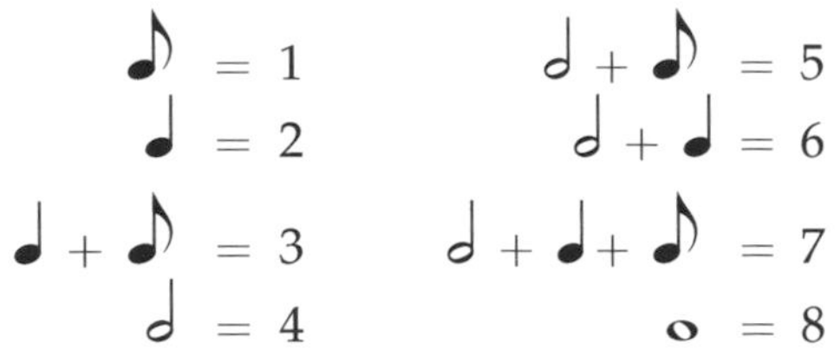

$$\eighthnote = 1 \qquad \halfnote + \eighthnote = 5$$

$$\quarternote = 2 \qquad \halfnote + \quarternote = 6$$

$$\quarternote + \eighthnote = 3 \qquad \halfnote + \quarternote + \eighthnote = 7$$

$$\halfnote = 4 \qquad \wholenote = 8$$

FIGURE 8.8. Identifying the integers 1 through 8 with the corresponding number of eighth notes.

two eighth notes or a quarter note, a length of 3 corresponds to a dotted quarter note (equivalent to three eighth notes), and so on (see Figure 8.8).

Next, consider the magic square of Mercury reduced modulo 8 (see Table 8.6). In other words, take each number in the square, divide it by 8, and record the remainder. Normally we would identify a multiple of 8 with the number 0; however, in order to obtain a positive length, we identify a multiple of 8 with the number 8 (a whole note). If we follow the same counterclockwise spiral pattern in the reduced magic square, including the adjustments made by the composer (shown in parentheses in Table 8.6), then the path traced out yields the precise durations of each note, assuming that a rest following a note counts toward the length of the note preceding it. For the consecutive notes that Davies regards as one "note," we must sum the durations to obtain the correct value in Table 8.6.

For instance, the flute part starting at rehearsal letter Z begins with a C♯ played over the span of seven eighth notes (two eighth notes each tied with a half note, plus the additional eighth note in the second measure). Then follows a D half note tied to a quarter note (equivalent to six eighth notes), another D half note (four eighth notes), and an F♯ held for a duration equivalent to five eighth notes. Next is an A♯ eighth note followed by a quarter rest. Counting the quarter rest as two eighth notes, the A♯ thus counts as

8	2	3	5	4	6	7	1
1	7	6	4	5	3	2	8
1	7	6	4	5 (8)	3	2	8 (5)
8	2	3	5	4 (1)	6	7	1 (4)
8	2	3	5	4	6	7	1
1	7	6	4	5	3	2	8
1	7	6	4	5	3	2	8
8	2	3	5	4	6	7	1

TABLE 8.6. The 8×8 magic square of Mercury reduced modulo 8, except that any number equivalent to 0, that is, any multiple of 8, has been written as an 8.

three eighth notes. Finally, we have an A quarter note (two eighth notes) followed by a C quarter note tied to a dotted half note (equivalent to eight eighth notes). The resulting sequence of durations, $7, 6, 4, 5, 3, 2, 8$, is precisely the first row of Table 8.6 (excluding the upper right corner) moving from right to left. In sum, the 63 flute notes and their durations between rehearsal letters Z and D1 correspond to a counterclockwise spiral pattern through Davies's magic square.

It is worth pointing out that the inclusion of rests in the counting method described above gives the composer more freedom to choose the durations of each note. For example, there are six different ways to achieve a count of 6: $6 + 0, 5 + 1, 4 + 2, 3 + 3, 2 + 4, 1 + 5$, where the second number corresponds to the number of eighth notes equivalent to the length of the rest. Thus, the apparent restriction imposed by the magic square reduced modulo 8 is not as confining as it first seems. Nonetheless, Davies seems to exploit this rest "loophole" minimally, and there are clearly less options available when the integers in the reduced square are small.

Davies's use of magic squares in constructing the musical material for *A Mirror of Whitening Light* is intentional, creative, clever, and mathematical. In many ways, it extends the ideas of Schoenberg's 12-tone technique by allowing for more varied and imaginative ways to arrange a particular class of pitches. It also leaves the student or music theorist with hours of enjoyment locating the patterns in Davies's version of a musical word search. We close this section with some insightful words of the composer delivered at the premiere, with the gleeful observation that Davies also recognizes the abstract algebra inherent in his work, as evidenced by his reference to change ringing (emphasis added).

> *And if you go across that square of the numbers arranged in a particular way, they make very interesting patterns. And I see these patterns, in the first place, possibly as dance patterns; and one gets to know them by heart. One doesn't in fact deal with numbers at all. One deals rather as somebody who is dealing with* **bell-changes, with actual patterns with changes***. . . . I firmly believe that the more one controls the flow of one's wildest inspiration, the wilder it sounds. And so when I really wanted to be wild towards the climax of this work, I imposed very rigid rhythmic and tonal controls derived from the plainsong, and from that magic square; and the result is really quite extraordinary I find, even now.*[5]

Exercises for Section 8.1

1. Generalize the supposed technique of Gauss used to derive Equation (8.2) to find the sums of the following arithmetic series:

   a. $1 + 3 + 5 + 7 + \cdots + 93 + 95 + 97 + 99$

   b. $3 + 7 + 11 + 15 + \cdots + 187 + 191 + 195 + 199$

2. By writing the specific letter names of the notes and their durations (in terms of eighth notes), confirm that the flute part shown in Figure 8.7 corresponds to a counterclockwise spiral through both Davies's magic square (notes) and the magic square reduced modulo 8 (durations). Remember that some pairs of consecutive notes should be considered the same note, with their durations combined.

---

[5]Davies, Composer's Note on *A Mirror of Whitening Light*.

FIGURE 8.9. The first violin part in Davies's *A Mirror of Whitening Light*, starting at rehearsal letter F1.

3. Figure 8.9 shows the first violin part for *A Mirror of Whitening Light*, beginning at rehearsal letter F1.

 a. List the letter names of the notes consecutively and find the pattern in Table 8.5 used by the composer to create the given music.

 b. Consider the durations of each note, including rests. Find the pattern in the reduced magic square (Table 8.6) corresponding to the lengths of each note. Be sure to explain your calculations for those measures with time signatures having an 8 or 16 in the denominator. *Hint:* For the entire excerpt, count using the quarter note as one beat rather than an eighth note.

4. What famous eighteenth-century American created some beautiful magic squares? This individual was so proud of a particular 16×16 example that he described it as the "most magically magical of any magic square ever made."[6]

5. Prove that the sum of the four corners in *any* 4×4 magic square is equal to 34, the magic constant. Conclude that the sum of the central four squares is also 34.

6. Complete the following 4×4 magic square. There is one, unique solution. Be sure to show your work or describe how you arrived at your solution.

	13	11	
7			1
14			12
	8	2	

---

[6]Pickover, *Zen of Magic Squares, Circles, and Stars*, p. 151.

7. Complete the following 4×4 magic square. In this case, there are two different solutions. Find them both. Be sure to show your work or describe how you arrived at your solutions.

1		13	
	9		5
	7		11
15		3	

1		13	
	9		5
	7		11
15		3	

8. Complete the following 6×6 magic square. Be sure to show your work or describe how you arrived at your solution.

	1	6		19	24
3	32	7	21		25
31		2		27	20
8	28	33		10	
30	5	34	12		16
4	36		13	18	

9. If the sum of the numbers in each 2×2 corner of a 4×4 magic square equals the magic constant, then the square is an example of a *gnomon magic square*. Of the four 4×4 magic squares shown in this section (Dürer's square in Figure 8.4 and the magic squares in Exercises 6 and 7), which ones are gnomon magic squares?[7]

8.2 Steve Reich: Phase Shifting

He didn't reinvent the wheel so much as he showed us a new way to ride.

— John Adams[8]

---

[7]This problem was inspired by the observations of former student John Kane. Of the 880 geometrically distinct 4×4 magic squares, 432 are gnomon.

[8]John Adams, from the liner notes to the CD collection *Steve Reich: Works 1965–95*, Nonesuch Records, 1997, p. 17.

FIGURE 8.10. Composer Steve Reich. Photo by Wonge Bergmann.

One of the most basic yet frequently applied mathematical transformations is a translation or shift. By taking the graph of a function and shifting it in either the vertical or horizontal direction, we create a new function that is easily related to the original. If the function is periodic, such as the trigonometric function $y = \sin x$ with period 2π, then shifting it horizontally will not alter its period or its general shape; however, it does change *when* the wave starts. Such a shift is called a *phase shift*, a term we encountered earlier in our discussion of sine waves in Section 3.3.2. In music, the use of phase shifts has been thoroughly explored in a variety of contexts and with different instruments by the innovative American composer Steve Reich (see Figure 8.10).

Born in New York City in 1936 to the Broadway lyricist June Stillman, Reich's parents divorced early in his life, resulting in shared time between New York and California. The long train rides across the country inspired his later work *Different Trains*. Although he studied piano at a young age, his interest in music did not blossom until the age of 14 when he began studying drums with Roland Kohloff. Before his 16th birthday, Reich enrolled at Cornell University, where he majored in philosophy and dabbled with some music courses. After graduating from Cornell in 1957, Reich returned to New York City to study musical composition, beginning privately with Hall Overton (1957–1958) and continuing on at the famous Juilliard School with Bergsma and Persichetti (1958–1961). He later earned a master's degree in composition from Mills College, where he studied with Luciano Berio and Darius Milhaud (1961–1963).

Early works

Reich eventually settled in San Francisco and wrote his first acknowledged work, *It's Gonna Rain*, in 1965. The piece featured what would become defining traits of Reich's music: short repeating patterns and stationary harmony. Reich's notion of *phasing*, where rhythmic patterns move in and out of step with each other, was introduced in this piece using recordings of an end-of-the-world sermon by a black Pentecostal street preacher named Brother Walter. Reich generated phasing by using two recording tapes that moved in and out of phase. He discussed his motivation for the use of phase in an interview with Jonathan Cott [1997]. Note the reference to Bartók's *Mikrokosmos*, discussed in Section 5.1.2.

Phase really has to do with the canon ... I picked it up mostly from some of the simpler piano pieces in Bartók's "Mikrokosmos" ... In my early tape pieces "It's Gonna Rain" and "Come Out," you have one tape loop going and another identical loop slipping slightly behind the first one, and what you really have is a unison canon or round where the rhythmic interval between the first and second voices is variable and constantly changing. "Phase" was just a technical word I used at the time to refer to the function of the tape recorders.[9]

Another early work that demonstrates Reich's use of phasing is the piece *Violin Phase*, written in 1967. The work can be performed with one violinist playing against prere-corded tracks that are phased in and out, or with four violinists who re-create the phasing through their individual parts. A primary melodic theme is stated at the outset, which generates all of the music. This key theme is continuously repeated by one player while the other players join in with the same musical idea, but shifted ever so slightly for con-trast. Each player repeats the same phrase, gradually shifting in and out of agreement with the original and with the other players. The result is an oral demonstration of a mathematical phase shift. Since the music is repetitive with gradual changes, it has often been characterized as *minimalist*. (Reich was influenced by minimalist composer Terry Riley.) The same effect can be obtained using an audio device with a prerecorded version of the main melody that the performer plays against.

Musical mayhem

In 1966, Reich created his own ensemble called "Steve Reich and Musicians," which started small (three members), but soon grew to a size of 20. For many years, this was the only ensemble allowed to perform Reich's works because he didn't publish any music until the mid-1970s. Initially, Reich's ensemble performed in museum halls and art galleries instead of concert halls (e.g., the Guggenheim Museum in New York). Based on the encouragement of the well-known conductor Michael Tilson Thomas, Reich agreed to have his piece *Four Organs* (1970) performed as part of a Boston Symphony Orchestra series of "new music." The performance of *Four Organs* in Carnegie Hall, New York, was legendary, featuring booing, heckling, and widespread unrest among the audience. Some attendees rushed the stage begging for an end to the music. If this had occurred in a mosh pit at a Nirvana show, no one would have been surprised; but this was classical music being performed in one of the most respected concert halls in the world!

As word spread of the musical mayhem at Carnegie Hall, Reich became an overnight sensation and a sought-after composer. He continued to explore his phasing ideas, em-phasizing rhythmic pulsation, repetitive patterns, and slight harmonic changes. His music soon became classified as minimalist (à la Philip Glass), where small changes or gradual shifts perpetuated over time create a piece. Reich proceeded to become a popular and award-winning composer, winning Grammy awards in 1990 for *Different Trains* (1988) and in 1999 for *Music for 18 Musicians* (1974–1976). In 2007, Reich won the Polar Music Prize, and two years later, he received the Pulitzer Prize for Music for his composition *Double Sextet*. Thomas provided the following explanation for the power of Reich's music:

My estimation of his music was greatly enhanced soon after when I heard "It's Gonna Rain" and "Violin Phase." The pieces were long, witty, spiritual, swinging; and best

---

[9]Cott, Interview with Steve Reich, p. 27.

of all, the notes were great. Hearing those first instrumental pieces was joy like that of hearing Monteverdi, Pérotin, or James Brown for the first time. It was amazing that someone could be discovering so much music with such economy of means. There was something streetwise and at the same time enormously innocent about it.[10]

8.2.1 *Clapping Music*

The notion of generating large amounts of music from a simple idea harkens back to our discussion of musical symmetry in Chapter 5. There, it was the symmetry that produced new versions of old material; with Reich's music, symmetry is replaced by the phase shift. A wonderful example demonstrating this idea is the whimsical *Clapping Music*, which incorporates several interesting mathematical ideas.

In 1972, Reich wrote *Clapping Music*, a work for two players using only their hands as instruments. The composer's intention was to create a piece featuring his phasing technique which could be performed easily, requiring no instruments or electronic devices. However, instead of the gradual phasing employed in Reich's earlier works such as *Violin Phase*, a feat that is difficult to accomplish with hand claps, *Clapping Music* features phasing that always shifts a whole number of beats from the original.

The first two measures for *Clapping Music* are shown in Figure 8.11. Notice the lack of a time signature, although a fast tempo is indicated at the start. (Recall that ♩ = 160 means that there should be 160 quarter-note beats per minute.) The first clapper repeats the same rhythmic pattern throughout the piece, a pattern that we will numerically describe as 3, 2, 1, 2, referring to the number of eighth notes clapped between rests. This rhythmic pattern is a variation of a fundamental African bell pattern. (Reich studied African drumming at the University of Ghana in 1970.) Since the eighth rests are the same duration as the "notes," each measure contains 12 eighth-note beats.

The second clapper initially is in unison with the first, playing together for 12 repetitions. Then, the fundamental rhythmic pattern undergoes a left cyclic shift (σ) by one eighth note. In other words, the first of the opening three eighth notes now becomes the last beat of the measure, and the measure opens with two consecutive eighth notes as opposed to three. Numerically, the second measure for clapper 2 is 2, 2, 1, 2, 1. The

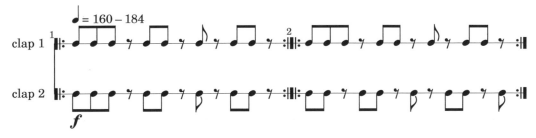

FIGURE 8.11. The first two bars of Steve Reich's *Clapping Music*. Clapper 1 repeats the same rhythmic pattern throughout the whole piece, while clapper 2 performs a left cyclic shift of the pattern in the previous bar. Each pattern is to be repeated 12 times before proceeding on to the next measure. © Copyright 1980 by Universal Edition (London) Ltd., London/UE 16182.

---

[10]Michael Tilson Thomas, from the liner notes to the CD collection *Steve Reich: Works 1965–95*, Nonesuch Records, 1997, p. 18.

second clapper, now phase shifted one beat to the left, repeats this pattern 12 times, juxtaposed against the first clapper playing the original pattern. Then, the shifting process continues as clapper 2 plays the pattern 1, 2, 1, 2, 2 in the third measure. After repeating each new pattern 12 times, the second clapper applies another cyclic shift. In mathematical notation, beginning with the second measure, clapper 2 plays the rhythmic patterns $\sigma, \sigma^2, \sigma^3, \ldots, \sigma^{11}$, until returning back to the original with σ^{12}. The piece concludes when the two clappers are back in unison again, clapping out the original pattern in measure 13.

There are many clever aspects of *Clapping Music*, and mathematics can help us gain a deeper understanding of their significance. First, by using a *cyclic* shift, one that moves the entire pattern to the left and places the first note at the end, Reich preserves the number of beats per measure. Instead of reducing the number of beats from 12 to 11 to 10 and so on, the original pattern and its cyclic shifts all have 12 beats. This is similar to the idea of closure from group theory (see Property **(i)** in Definition 5.3.1).

Second, the 3, 2, 1, 2 fundamental rhythmic pattern is asymmetric in the sense that each new measure created by the one-beat cyclic shift is distinct from every other measure, including the original. Thus, clapper 2 always has a new pattern to play after each stage of the shift. A contrasting example is the rhythmic pattern 3, 1, 3, 1, which repeats itself after six shifts. The 12 distinct patterns arising from the shifts also serve to create interesting and appealing musical structure depending on how the original and shifted patterns match up. When the original pattern and its shifted version are completely out of phase (such as the second measure), the music consists of repeated eighth notes. On the other hand, when simultaneous rests occur (e.g., measures numbered 4, 6, and 10), the music takes on a dance-like character.

Reich is very clear about the importance of the contrasting patterns. In the "Directions for Performance" given at the start of the score for *Clapping Music*, he explains that the lack of a time signature is intended to discourage any accents at the start of a measure. He warns clapper 2 to keep the downbeat on the first beat of the measure, not on the first beat of the three claps in the main pattern. This is critical to the success of the piece, as it blends the rhythms together, making it difficult for the listener to discern a clear downbeat at the start of each measure, one of the hallmark traits of minimalist music. As is typical with much of Reich's music, he also gives very precise instructions regarding the duration of the piece (approximately 5 minutes), the method of clapping in order to achieve a uniform sound, and directions for electronic amplification if needed.

A third mathematical feature of the work is its clear connection with the group $\mathbb{Z}_{12}$, the cyclic group of order 12, or the integers $\{0, 1, \ldots, 11\}$ under addition modulo 12. The repeated cyclic shift to the left by one eighth-note beat produces a group structure identical to $\mathbb{Z}_{12}$. The operation of adding 1 (the group generator) in $\mathbb{Z}_{12}$ corresponds to the one-beat left shift of the original rhythmic pattern. The fact that all 12 patterns are distinct means that we are dealing with the full group $\mathbb{Z}_{12}$ and not a subgroup.

On the uniqueness of the 3, 2, 1, 2 **rhythmic pattern**

Following the outline and ideas of *Clapping Music*, what are the possible rhythmic patterns that a composer could choose from to create such a piece? Among these choices, what is special about Reich's 3, 2, 1, 2 pattern? These questions are answered in Joel Haack's instructive article "Clapping Music—A Combinatorial Problem" [Haack, 1991].

FIGURE 8.12. A *Clapping Music* comic from *Cat and Girl*.

Before we count the possible number of patterns, we first need to establish a few ground rules. We will restrict our attention to measures with 12 beats, under the following two assumptions:

1. Rests occur for precisely four beats (so the remaining eight beats contain claps).

2. Rests cannot occur on consecutive beats.

To count the number of possible 12-beat patterns that satisfy these two rules, we will distinguish those patterns that begin with a clap from those that do not. As Haack explains, if we begin with a clap, then we are allowed four clap-rest combinations and four claps that can be arranged in any order. For example, we could start with four clap-rest combinations in a row and conclude with four successive claps. This yields the sequence $1, 1, 1, 1, 4$ (recall that a comma represents a rest). Or, if we played the pattern clap, clap-rest, clap, clap-rest, etc., we would create the sequence $2, 2, 2, 2$. To count the total number of 12-beat patterns beginning with a clap, we need to choose four arbitrary spots out of a possible eight for the clap-rest combinations. Then the remaining four spots automatically receive claps. The number we seek is well known in probability theory and combinatorics. It is called 8 *choose* 4, denoted $\binom{8}{4}$.

Definition 8.2.1. *The number of ways to choose k items from a set of size n, without regard to the order in which they are chosen, is* <u>n choose k</u>*, denoted* $\binom{n}{k}$*, and is found using the formula*[11]

$$\binom{n}{k} = \frac{n!}{k!(n-k)!}. \tag{8.4}$$

---

[11]The number $\binom{n}{k}$ is also called a *binomial coefficient*. For a proof of Formula 8.4, see Gerstein, *Discrete Mathematics and Algebraic Structures*, pp. 265–267.

For example, there are $\binom{4}{2} = \dfrac{4!}{2! \cdot 2!} = 6$ ways to choose a subset of two items from a set of four:

$$\{a,b,c,d\} \implies \begin{cases} \{a,b\} & \{b,c\} \\ \{a,c\} & \{b,d\} \\ \{a,d\} & \{c,d\}. \end{cases}$$

Note that the order within a subset is irrelevant, so the subset $\{a,b\}$ is considered equivalent to the subset $\{b,a\}$. For card players, the number of five-card poker hands from a 52-card deck is $\binom{52}{5} = 52!/(5! \cdot 47!) = 2{,}598{,}960$.

Thus, the total number of 12-beat rhythmic patterns that satisfy our two rules and begin with a clap is $\binom{8}{4} = 8!/(4! \cdot 4!) = 70$. Half of these possibilities, or 35, will end with a rest, since there is an equal chance of placing a clap-rest combination in the last of the eight spots as there is a clap. If we right shift each of these 35 patterns by one beat, then we create all of the 35 possible patterns that begin with a rest. Alternatively, we could assume that the first beat is a rest and then count the number of ways to choose three random spots from among the remaining seven for the three clap-rest combinations, (three not four since the opening rest leaves only three rests remaining). This number is $\binom{7}{3} = 7!/(3! \cdot 4!) = 35$, as expected.

In sum, there are 105 possible "Reich patterns" satisfying the two rules above, 70 that begin with a clap and 35 that begin with a rest. A further reduction is possible by identifying those rhythmic patterns that are equivalent under some number of cyclic shifts. For example, after one left shift, the fundamental pattern $3, 2, 1, 2$ becomes $2, 2, 1, 2, 1$, and after another shift, it changes to $1, 2, 1, 2, 2$. We now consider these patterns to be the "same" since they are related by some number of cyclic shifts. This is similar to creating a left coset within a group (see Section 6.2.3). The 12 possible cyclic shifts of Reich's fundamental pattern are collected together as one set and count as a single rhythmic pattern represented by the sequence $3, 2, 1, 2$. Applying a mathematical formula called *Burnside's counting lemma*, a technique for counting the number of identifications in a group with symmetry, Haack shows that under this particular identification, there are only 10 possible rhythmic patterns. If we list the largest number of consecutive claps first, these patterns are:

$$\begin{array}{ll}
5,1,1,1 & 3,1,3,1 \\
4,2,1,1 & 3,2,2,1 \\
4,1,2,1 & 3,2,1,2 \\
4,1,1,2 & 3,1,2,2 \\
3,3,1,1 & 2,2,2,2.
\end{array} \qquad (8.5)$$

This list can be created by finding all of the ways to decompose the number 8 into the sum of four whole numbers. Note that there is no overlap in the list above; each sequence generates its own special set of rhythmic patterns under shifting. Also note that the sequence $3, 1, 1, 3$ is equivalent to $3, 3, 1, 1$, since each will generate the same set of 12 patterns via shifting. The corresponding rhythmic patterns for some of these new sequences are shown in Figure 8.13.

Although Burnside's lemma is beyond the scope of this text, it is clear that the 10 patterns listed in (8.5) will generate the 105 we counted previously. This follows because all of the sequences, except for $3, 1, 3, 1$ and $2, 2, 2, 2$, will produce a total of 12 distinct patterns under shifting. This gives a total of $8 \cdot 12 = 96$ patterns. Then, since $3, 1, 3, 1$ and

FIGURE 8.13. The corresponding rhythmic patterns for the sequences $4, 1, 2, 1$ (*left*) and $3, 1, 3, 1$ (*right*).

$2, 2, 2, 2$ produce six and three new patterns, respectively, we have found $96 + 6 + 3 = 105$ total patterns, as expected.

Out of the eight possible sequences in (8.5) that produce 12 distinct measures of music, only two of them do not have the same number repeated consecutively in the list: Reich's pattern of $3, 2, 1, 2$ and the pattern $4, 1, 2, 1$. This is noteworthy, for it distinguishes Reich's choice from the group. A pattern with the same number of claps heard consecutively, such as $3, 3, 1, 1$, is too predictable. It and its ensuing shifts are not as interesting rhythmically or musically. Try clapping out the different rhythms corresponding to the sequences in (8.5). The syncopation and lack of regularity in the $3, 2, 1, 2$ pattern are what makes Reich's clever construction a success. Our mathematical excursion into the field of combinatorics has helped explain why this is the case.

Retrogrades

Some interesting symmetry occurs within the set of 12 different rhythmic patterns of *Clapping Music*. Focusing on the part of the second clapper, measure 4 will be the pattern $2, 1, 2, 3$, which is a retrograde of the original $3, 2, 1, 2$ pattern. If we ignore the lack of a tonal center or pitch, then measure 4 is actually a retrograde-inversion ($180°$ rotation) of measure 1. Even more surprising is the fact that *every* measure of the second clapper is a retrograde of some other measure.[12] The retrograde pairs occur in the following sets of measures: $\{1, 4\}, \{2, 3\}, \{5, 12\}, \{6, 11\}, \{7, 10\}, \{8, 9\}$ (see Exercise 6). Notice that each pair sums to 5 (modulo 12). It turns out that this same phenomenon occurs for other possible arrangements of eight claps and four rests.

Theorem 8.2.2 (Retrograde Clapping Theorem). *Suppose that a piece constructed in the same fashion as "Clapping Music" has a retrograde of the opening measure in measure n. Then each measure of the piece will have a retrograde somewhere else in the piece. Measures a and b will be retrogrades of each other if and only if $a + b = n + 1$ (mod 12).*

Proof: The key fact in the argument is that moving forward in the piece corresponds to a left shift, while moving backward is a right shift. For instance, measure 3 of the piece is both a left shift by two beats and a right shift by 10 beats.

Suppose that the retrograde of the opening measure is found in measure n. The second measure of the piece is obtained from the first by a left shift of one beat, so its retrograde would be found by taking the retrograde of the first measure and shifting it right by one beat. In other words, if measures 1 and n are retrogrades of each other, then so are measures 2 and $n - 1$. Likewise, the argument continues and measures 3 and $n - 2$ are retrogrades. The argument also works in the other direction, so measures 12 and $n + 1$ are retrogrades, 11 and $n + 2$ are retrogrades, and so on. In this fashion, the sum of each retrograde pair is always the same, namely, $n + 1$ (mod 12), and each measure a is guaranteed a retrograde partner $b = n + 1 - a$ (mod 12). $\square$

---

[12]Thanks to Julia Lam, a former student, for bringing this to my attention.

8.2.2 Phase shifts

In Section 3.3.2, we discussed phase shifts in the context of the graphs of sine waves. For example, the function $y = \sin(t - \pi)$ is a sine wave that begins at $t = \pi$, instead of $t = 0$. The graph of $\sin t$ shifts right by π units. By setting $t = \pi$, the angle inside the parentheses of $\sin(t - \pi)$ becomes $\pi - \pi = 0$. Thus, the new "0" has become π. In general, if t is replaced by $t - c$, then the graph of the new function $f(t - c)$ will be a right shift of the old one by c units. Replacing t by $t + c$ will cause the graph to shift left by c units.

In contrast, vertical shifts are obtained by adding or subtracting a number to the output of a function. In other words, the function $f(t) + c$ is the graph of f shifted up c units. Notice that the c is located *outside* the parentheses. This is an example of a *range change*, where the range of the function (the output) is adjusted. A function such as $f(t + c)$ or $\sin(t + 3\pi)$ represents a *domain change* because we are shifting what goes into the function, not what comes out. On the left in Figure 8.14, we show the graphs of $f(x)$ and two vertical shifts, $f(x) + 2$ and $f(x) - 2$, which are the graphs of f shifted up and down two units, respectively. On the right in Figure 8.14 are the graphs of $f(x)$ and $f(x - 1)$, which is a right shift of f by one unit. In essence, a domain change such as $x \mapsto x - 1$ is moving the x-axis (the domain) one unit to the *left*. This has the effect of making the original graph move to the *right* by one unit. A domain change typically has the opposite effect of what we would expect.

Returning to *Clapping Music*, the structure of the piece is to perform consecutive left shifts by one eighth-note pulse. Mathematically speaking, this corresponds to the sequence of functions $f(x), f(x + 1), f(x + 2), \ldots, f(x + 11)$, where $f(x + 12) = f(x)$. Thus, shifting the graph 12 units to the left returns the original graph, which is another way of saying that the period of the graph is 12.

For example, the equation $\sin(t + 2\pi) = \sin(t)$ can be understood in two different ways. From one viewpoint, 2π has been added to the angle t. Since this returns us to the same spot on the unit circle, the y-coordinate is unchanged, and hence the value of sine remains the same. On the other hand, the graph of $\sin(t + 2\pi)$ is a left shift of the graph of $\sin(t)$. Since the period of the sine function is 2π, this left shift has no effect on the graph, that is, $\sin(t + 2\pi) = \sin(t)$.

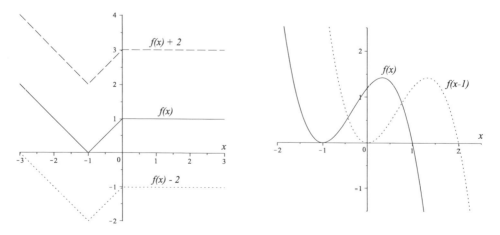

FIGURE 8.14. Some vertical shifts of $f(x)$ (*left*) and a horizontal shift (*right*).

Exercises for Section 8.2

1. In the game of Powerball, players choose five different numbers from a collection of 59. How many possible different choices are there? Assume that the order of the numbers chosen is irrelevant.

2. An ice cream store offers 30 different flavors. You are ordering the extra large cone, which comes with four scoops.

    a. Assuming that you choose four different flavors, how many possible ice cream cones can you create?

    b. Suppose that the first two flavors you choose are chocolate and vanilla. How many different cones can you create?

3. What does $\binom{n}{1}$ simplify to? Explain why this makes sense in terms of counting.

4. Show that $\binom{n}{2}$ simplifies to $\dfrac{n(n-1)}{2}$.

5. Suppose that a and b are two whole numbers less than n such that $a + b = n$. Show that $\binom{n}{a} = \binom{n}{b}$. Explain why this makes sense in terms of counting.

6. Write out the remaining measures of *Clapping Music*, continuing the cyclic shift begun in Figure 8.11. The very last measure (after 12 total shifts) should be the same as the first. Check that the following pairs of bars are retrogrades of each other: $\{1,4\}$, $\{2,3\}$, $\{5,12\}$, $\{6,11\}$, $\{7,10\}$, $\{8,9\}$. Find a rhythmically adept friend and perform the piece together.

7. Write out an entire version of *Clapping Music* using the $4,1,2,1$ pattern as the primary rhythmic figure. The first measure is shown to the left in Figure 8.13. Which pairs of measures are retrogrades of each other? Confirm that Theorem 8.2.2 is satisfied here.

8. Figure 8.15 shows the graphs of the cosine and sine functions. Explain the identity $\sin(t + \pi/2) = \cos t$ in two different ways, one by using a phase shift and the other by applying the trig addition formula $\sin(\alpha + \beta) = \sin\alpha\cos\beta + \cos\alpha\sin\beta$.

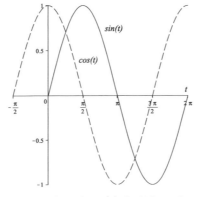

FIGURE 8.15. The graphs of $\sin(t)$ (*solid*) and $\cos(t)$ (*dashed*).

8.3   Xenakis: Stochastic Music

Linear polyphony destroys itself by its very complexity; what one hears is in reality nothing but a mass of notes in various registers. ... In fact, when linear combinations and their polyphonic superpositions no longer operate, what will count will be the statistical mean of isolated states and of transformations of sonic components at a given moment. The macroscopic effect can then be controlled by the mean of the movements of elements which we select. The result is the introduction of the notion of probability, which implies, in this particular case, combinatory calculus. Here, in a few words, is the possible escape route from the "linear category" in musical thought.

— Iannis Xenakis[13]

8.3.1   A Greek architect

Iannis Xenakis was a transformative modern-era composer who applied mathematics and physics to music in ways that had never been seen before. One of his defining beliefs was that probability theory could be applied to music in order to determine the length and frequency of the pitches played. By focusing on the overall structure (or mass) of sound produced by a group of instruments, mathematical calculations were developed to build the music.

Xenakis had a colorful and at times tramatic early life, one that shaped his ensuing musical works. He was born in 1922 in Braila, Romania, the eldest son of a Greek businessman. In 1932 he began studying at a boarding school in Greece on the Aegean island of Spetsai. Six years later he graduated from boarding school and began his studies in architecture and engineering at the National Technical University of Athens. He also studied music, particularly harmony and counterpoint. His work at the university was interrupted in 1940 when Italy invaded Greece. Xenakis served as a member of the Greek resistance during World War II, joining the communist National Liberation Front (EAM) to fight the occupation by Germany and Italy. He later joined the Greek People's Liberation Army (ELAS).

In December 1944, during the Greek Civil War and Britain's involvement, he was severely injured in street fighting with British tanks, losing eyesight in his left eye. Despite the war and his experiences fighting, he eventually graduated from the Technical University in 1947. However, Xenakis was forced to flee from Greece to Paris, when a new, non-communist government came to power. He was subsequently sentenced to death (in absentia) by the right-wing administration, a sentence that was not lifted until 1974.

While in Paris, Xenakis worked as an engineering assistant in the architectural studio of Le Corbusier. Although he started small, he was soon given a lead role as a design engineer. During this period, Xenakis tried to study music in Paris with Boulanger, Honegger, and Milhaud, but was rejected, viewed as a rebel with little training. An important moment in Xenakis's development occurred when the famous composer Olivier Messiaen encouraged him to incorporate his training in engineering and mathematics into his music. "[Y]ou have the good fortune of being Greek, of being an architect and having studied special mathematics. Take advantage of these things. Do them in your music."[14] In 1959, Xenakis left Le Corbusier's studio and began to make a living as a composer and teacher.

---

[13]Xenakis, *Crisis of Serial Music*.
[14]Matossian, *Xenakis*, p. 48.

Figure 8.16. The Philips Pavilion from the 1958 Brussels World Fair, a building inspired by the score of Xenakis's *Metastasis*. Photo by Louis Warzée.

8.3.2 *Metastasis* and the Philips Pavilion

One interesting feature of Xenakis's earlier compositions was the interplay between music and architecture. A great example of this was his collaboration with Le Corbusier on the Philips Pavilion at the 1958 Brussels World Fair (see Figure 8.16). The design of the building used mathematical surfaces called conoids and hyperbolic paraboloids (saddles). Remarkably, the architectural ideas for the pavilion originated from a piece of music that Xenakis had composed in 1953–1954 called *Metastasis* (Transformations).

Metastasis is an orchestral work for 61 players (46 strings), with each string player on a *different* part. This is a particularly unique aspect of the piece, as most composers (past and present) write the same part for each member of a section (e.g., all first violin parts are identical). The piece features multiple *glissandi* (straight lines) in the music for the string and horn parts. These indicate for the musician to begin at a certain pitch and slide through all the frequencies on the way to a different pitch (the new pitch can be higher or lower). This is simple enough to accomplish on an instrument like a slide trombone, but it is also possible on string instruments, where sliding a finger along the string can change the frequency continuously. One of Xenakis's interesting insights was the realization that drawing glissandi in the score can create a special surface of straight lines, called a *ruled surface* (see Figure 8.17). This later inspired his design of the Philips Pavilion. Notice the lines on the outside of the building in Figure 8.16. Interestingly, Xenakis thought of the glissandi in the score as graphs of straight lines (time on the horizontal axis, pitch on the vertical), where different slopes correspond to different "sound spaces."

The focus of *Metastasis* is on the totality of sound and timbre, rather than individual pitches. Xenakis strove to portray "sound events made out of a large number of individual sounds [that] are not separately perceptible, ... [to] reunite them again ... [so that] a new sound is formed which may be perceived in its entirety."[15] The notes he chose were based on 12-note tone rows, with durations assigned using the Fibonacci sequence and

---

[15]Matossian, *Xenakis*, p. 58.

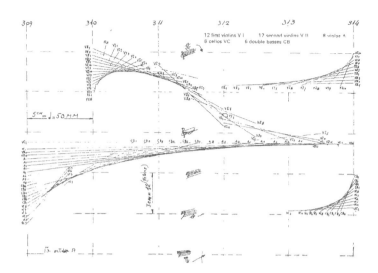

FIGURE 8.17. Glissandi in the score of *Metastasis* form a special shape called a ruled surface.

the golden ratio, a concept we encountered earlier with Bartók's music (see Section 5.2) and one that Xenakis also employed in some of his other architectural designs.

8.3.3 *Pithoprakta*: Continuity versus discontinuity

After *Metastasis*, Xenakis began to develop his "stochastic music" as a way of further exploring the collective sound over the individual parts. A *stochastic* process is random and nondeterministic; the next state of the environment is not fully determined by the previous one. Xenakis decided to apply probability theory to govern all aspects of his music. The duration, pitch, timbre, and dynamics were all determined by a *probability density function*, a special function used in probability theory to predict the likelihood of a particular event occurring. For Xenakis, this was like hearing "the sound of science." He strove to create a musical embodiment of the laws of nature and physics.

One example of a work demonstrating these ideas is the orchestral piece *Pithoprakta* (Actions through Probabilities; 1955–1956), which drew inspiration from Maxwell-Boltzmann's kinetic theory of gases. The work was dedicated to Hermann Scherchen, who conducted its premiere in March 1957 in Munich. It is scored for 46 strings (again, all playing *different* parts), two trombones, one xylophone, and one wood block.

As with *Metastasis*, Xenakis focuses on the overall block of sound in *Pithoprakta*, using density functions to build different "clouds of sounds."[16] For Xenakis, the individual notes and timbres are "unpredictable" and less relevant; it is the combined mass of sound created by the ensemble that matters, a mass that can be controlled mathematically. The work creatively explores the conflict between continuity and discontinuity, terms with precise mathematical meanings. The continuous sounds are produced by glissandi in the strings and trombones, as well as short bow strokes. Discontinuous sounds are obtained by pizzicati plucking in the strings, tapping the bridge of the stringed instruments with the opposite side of the bow, and a rather jarring use of the wood block.

---

[16]Xenakis, *Formalized Music*, p. 12.

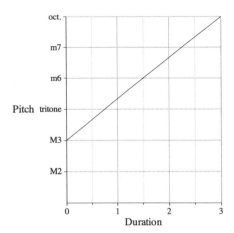

FIGURE 8.18. A glissando traveling eight half steps, from a major third (M3) to an octave, in three time units, will have a slope or "speed" of 8/3.

To calculate the notes for each player, different "speeds" are determined based on a set of rules. According to Xenakis, speed is measured by the slope of a line. If a continuous glissando is graphed as a line, then the change in pitch can be associated with a particular speed, measured as the slope of the line. For example, a glissando that rises from a major third to an octave higher (eight half steps), over three time units, will have a speed (slope) of 8/3 (see Figure 8.18). The goals that Xenakis tried to achieve in determining how to distribute speeds among all the musicians are given below.

1. The density of speeds should be constant. This means that two regions of the same length along the pitch axis average the same number of glissandi.

2. The distribution of speeds is uniform. This is accomplished by using a bell curve, known as a *Gaussian normal distribution*. The likelihood that a particular speed is contained in a given interval equals the area under the curve over that interval.

3. Isotropy, a notion of balance among physical attributes, is obtained by having an equal number of sounds ascending and descending.

4. Pitches are freely distributed, that is, no one pitch is favored over any other. Note how this generalizes Schoenberg's 12-tone method.

Figure 8.19 demonstrates the result of Xenakis's calculations over a few measures of *Pithoprakta*. The 24 different violin parts reflect many of the goals outlined above. At any particular moment, the notes are freely distributed across the staff, there are roughly equal numbers of rising and falling glissandi, and the slopes of the glissandi are constantly changing from beat to beat. This particular excerpt captures a continuous moment of the piece. Other instances, particularly when the wood block enters with striking accents and the string players are tapping their bridges with the opposite side of their bows, embody a discontinuous phase. Listening to this fascinating work, it becomes fairly clear that Xenakis's experiences with warfare had a strong influence on the piece. The slow rise and fall of the trombone glissandi are reminiscent of a WWII fighter plane, while the

FIGURE 8.19. The 24 different violin parts from measures 52–57 of Xenakis's *Pithoprakta*.

unexpected smack of the wood block conjures up sounds of gunfire. The battle between continuous and discontinuous sounds is an apt metaphor for the tragedy of war.

Xenakis continued to compose mathematical music and wrote several articles and essays on stochastic processes, game theory, and computer programming in music. In 1966 he helped establish the school *Equipe de Mathématique et Automatique Musicales* (EMAMu) for the study of computer-assisted composition. Eventually his work was translated and expanded into the well-known text *Formalized Music: Thought and Mathematics in Composition* [Xenakis, 1992]. In 1999 he was awarded the Swedish Polar Music Prize along with the great Stevie Wonder. After several years fighting a serious illness, Xenakis passed away in February of 2001. His legacy lives on through his expository writings and his uniquely mathematical style of composing.

Exercise for Section 8.3

1. Find the speed (or slope) of each glissando shown in Figure 8.20. Notice the difference between the labels on the vertical axes.

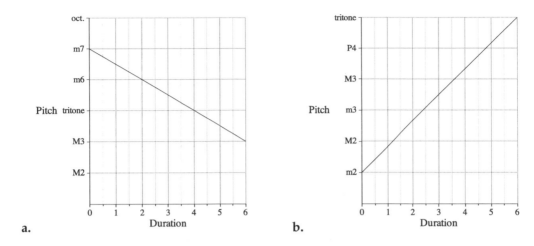

FIGURE 8.20. Two sample speeds envisioned by Xenakis.

8.4   Final Project: A Mathematical Composition

We have explored several musical examples where mathematical techniques are an important aspect of the compositional process. From the use of symmetry to build additional music out of a given theme, to the 12-tone technique of Schoenberg, to the phase shifts of Steve Reich, to the stochastic calculation-based works of Xenakis, we have witnessed a wide variety of compositional styles, each infused with its own mathematical character. To close this chapter (and the book), we invite the reader to create their own mathematical piece of music.

The final exercise for our course in math and music is to compose and perform a short piece of music based on some of the concepts discussed in the text. Whether you consider yourself an advanced musician or a neophyte, composing a piece of music can be a fun and informative endeavor. The aim is to combine your creative and analytic abilities and gain a deeper understanding of some of the ideas we have been investigating.

To appreciate the full compositional experience, your piece should be performed in front of an audience (e.g., your classmates). You can be the performer, or you can ask a peer or peers to play your piece. Before the performance, you should give a brief presentation explaining the mathematical connections utilized in your work. You may also discuss your inspirations for the piece, creative impulses, special notation, etc.

Your final project should consist of the following four items:

1. A brief presentation about your composition.

2. A performance of your musical work.

3. The musical composition itself, turned in on staff paper.

4. A brief written report explaining the mathematical connections in your piece and the ideas you are trying to convey.

Some advice

- There are several good computer programs available that can be a valuable aid for composing and generating the notes on the staff. One free option is the program *MuseScore*,[17] which is easy to use and has the capacity to play back your composition, as well as produce high-quality printed music.

- *I've never composed before. Where do I begin?* There is no easy answer to this question. First, decide what instrument(s) you are writing for (e.g., piano, voice, violin, tuba, percussion, etc.). Are you going to have more than one part? Your choice may depend on your musical abilities or those of your peers. Next, try writing a simple melody (say, four to eight measures long) by sitting at the piano and playing the notes. Write your melody down on staff paper. Is there any obvious symmetry apparent? Do you want there to be? From here, you could compose a countermelody or play around with your original to create more music. This is a good place to utilize symmetry (see Chapter 5). Deciding on the overall structure of the work first may help with the process of composing. Be sure to be conscious of the rhythms you use, as this is a central ingredient. What time signature do you envision your work in?

- Varying the tempo, rhythm, dynamics, tonality, feeling, etc., is a great way to make your piece come alive. Be sure to think about the overall structure of your work. How do you musically distinguish different sections of your piece?

- *What are some of the mathematical ideas I could use in my composition?* Here are some of the relevant concepts from the second half of the book: musical group theory, change ringing, 12-tone technique, magic squares, minimalism, and the golden ratio. Some of the specific mathematical concepts available include symmetry (reflections, translations, rotations), group theory, permutations, patterns, phasing, creative use of rhythm, and the Fibonacci numbers. Use your imagination, be creative, and apply some of the mathematics you've learned from this text. Do not try to throw in a little of everything. Focus on one or two primary concepts to generate your music.

- Be sure to rehearse your piece beforehand, several times. If you are not playing your own piece, make sure that your performer(s) have rehearsed sufficiently ahead of time. Part of the challenge of this project is to do the necessary logistical work beforehand to ensure a smooth performance. If you have more than one player, then you will have to write out the music for each part and find time for your performers to get together for rehearsals. This takes extra time and planning. Moreover, you may find that your piece needs some minor tweaking after hearing it played for the first time. Dress appropriately and don't forget to take a bow after your performance.

Sample student compositions

We now describe some sample mathematical pieces written by the author's past students which have been particularly memorable.

---

[17]Available at http://musescore.org/ .

1. *Life of Pi* by Jacob Miller (piano). Miller first assigned pitches and durations to the numbers $0, 1, 2, \ldots, 9$. He then used the first 30 digits of π to generate a melody and incorporated symmetry to construct more music out of his main theme.

2. *Joey 12 Tone* by Joseph Kramkowski (vibraphone). Kramkowski composed a genuine 12-tone piece based on a single principal tone row. He used the standard symmetry operations (transpositions, retrogrades, inversions, and retrograde-inversions) to construct the work. Having the piece performed on the vibraphone added an ethereal quality that worked particularly well given its atonal nature.

3. *Magic Numbers* by Emely Ventura (piano). Ventura began by assigning the numbers $1, 2, \ldots, 24, 25$ to the white keys on the piano symmetrically located around middle C. (She chose the white keys because they are easier to play.) Inspired by the use of magic squares in Davies's *A Mirror of Whitening Light*, Ventura used a 5×5 magic square (see Table 8.7) to generate her music, traveling through the magic square horizontally, vertically, and diagonally. She admitted to "cheating" in a few places for "harmonic reasons," taking her own artistic liberty. The piece was exactly 65 measures long because this is the magic constant for $n = 5$.

11	24	7	20	3
4	12	25	8	16
10	18	1	14	22
17	5	13	21	9
23	6	19	2	15

TABLE 8.7. The 5×5 pseudo-magic square used by Ventura in her piece *Magic Numbers*. While the rows and columns sum to the magic constant 65, the left and right diagonals sum to 60 and 40, respectively.

4. *Bartalk* by Christina Catalano (piano). The title of the piece is a homonym of the composer that inspired the work, as well as a play on the popular NPR show *Car Talk*. Inspired by Bartók's *Subject and Reflection* and *Music for Strings, Percussion and Celesta*, Catalano used exact inversions, retrogrades, and the Fibonacci and Lucas numbers to construct her piece. She wrote five- and eight-note motifs and placed symmetric versions of these motifs at measure numbers corresponding to the Lucas numbers. Certain key structural points occur at Fibonacci-numbered measures, such as a tempo change and dynamical climax at bar 55. Catalano managed to incorporate the mathematics while also writing a genuine piece of music. The tonality was neither major nor minor, but it was not atonal either, mimicking a key aspect of Bartók's music.

5. *Star in D₅* by Alexandra Gitto (piano). In Gitto's own words, the piece was "inspired by group theory and the dihedral group." She first computed the group multiplication table for D_5, the dihedral group of degree five (the 10 symmetries of a regular pentagon). Each element of the group was written down numerically according to the order of the five vertices in the pentagon. Then, after assigning each number a particular note (1 ↔ A, 2 ↔ B, 3 ↔ C, 4 ↔ D, 5 ↔ F), Gitto used the group elements and the multiplication table to construct her piece.

Star in D₅ was written in $\frac{5}{8}$ so that each group element fits precisely into one measure. The piece opens with each element played once, announcing the group. Then, a fixed element g is chosen (e.g., $g = R_{72}$) to be played in the right hand while the left hand cycles through all elements except for e and g^{-1}. The product of the right- and left-hand group elements (essentially a left coset) are played in unison. In essence, the group multiplication table is being portrayed musically. The piece concludes with each rotation composed with its inverse in order to obtain the identity in both hands (a recognizable ascending scale). This mimics the practice of sounding rounds in change ringing to announce the start and close of an extent. Gitto's love of abstract algebra was quite evident in her piece. She later went on to become a math major.

6. *Newcomer's Waltz* by Megan Whitacre (piano and dance). The final piece we discuss was a true gem, a joy to hear and watch. Whitacre, a member of the College's Ballroom Dance Team, was inspired by her experiences as a dancer. Her work was intended to reflect the challenges and inner struggles of being a dancer and a first-year college student. As a result, the music is "a little eerie" (composer's words), employing a minor scale with a sharp fourth scale degree (e.g., F♯ in the key of C minor), and emphasizing tritones and minor seconds. Inspired by Bach, a haunting opening melody in the right hand is transposed and inverted multiple times (see Figure 8.21). A countersubject enters in the right hand in measure 9, while the melody shifts to the left hand. A simple phase shift is used in the right-hand melody, à la Steve Reich, inserting one-measure rests at various locations in order to create different juxtapositions of the melody and countersubject. Whitacre remarked that "[t]his phase shift changes the way the entire piece comes together."

What truly distinguished her piece was the addition of a remarkable accompanying dance choreographed by Whitacre. While a friend played the work on the piano (off to the side), Whitacre and her dance partner performed a waltz (the piece is in $\frac{3}{4}$) choreographed to match both the music and the mathematics. For instance, when both hands play in unison, the dancers' movements are identical. When the melody is inverted, the dancers face each other to "reflect" one another's movements. When a phase shift happens in the music, a time delay takes place with one dancer three beats behind the other. The choreography was a beautiful and memorable illustration of the mathematical concepts present in the work.

FIGURE 8.21. The eerie opening melody from Whitacre's *Newcomer's Waltz*, featuring the tritone and minor second intervals.

References for Chapter 8

Cott, J.: 1997, Interview with Steve Reich, *Steve Reich: Works 1965–1995*, Nonesuch Records. CD Liner Notes.

Cross, J.: 2003, Composing with Numbers: Sets, Rows and Magic Squares, *in* J. Fauvel, R. Flood and R. Wilson (eds), *Music and Mathematics: From Pythagoras to Fractals*, Oxford University Press, pp. 130–146.

Cross, J.: 2007–2015, Manchester School, *Grove Music Online*, Oxford University Press.

Davies, P. M.: 1978a, Composer's Note on *A Mirror of Whitening Light*. http://www.maxopus.com.

Davies, P. M.: 1978b, *A Mirror of Whitening Light: Speculum Luminis Dealbensis*, Boosey & Hawkes Music Publishers Ltd.

Davies, P. M.: January 2006, Composition Questions Answered—1 of 4, Interviews & Speeches. http://www.maxopus.com.

Davies, P. M.: n.d., The Official Website for Sir Peter Maxwell Davies. http://www.maxopus.com.

Gerstein, L. J.: 1987, *Discrete Mathematics and Algebraic Structures*, W. H. Freeman and Company.

Haack, J. K.: 1991, Clapping Music—a Combinatorial Problem, *College Math. J.* **22**(3), 224–227.

Kepner, M.: 2013, Magic Square 8 Study: A Breeze over Gwalior. Archival inkjet print.

Matossian, N.: 1986, *Xenakis*, Kahn and Averill, London.

McGregor, R.: 2000, Compositional Processes in Some Works of the 1980s, *in* R. McGregor (ed.), *Perspectives on Peter Maxwell Davies*, Ashgate Publishing Ltd., pp. 93–114.

Pickover, C. A.: 2002, *The Zen of Magic Squares, Circles, and Stars: An Exhibition of Suprising Structures across Dimensions*, Princeton University Press.

Reich, S.: 1980, *Clapping Music*, Universal Edition (London) Ltd., no. 16182.

Reich, S.: n.d., The Steve Reich Website. http://www.stevereich.com/.

Roberts, D.: 2000, Alma Redemptoris Mater, *in* R. McGregor (ed.), *Perspectives on Peter Maxwell Davies*, Ashgate Publishing Ltd., pp. 1–22.

Warnaby, J.: 2003, Davies, Peter Maxwell, *Grove Music Online*, Oxford University Press.

Weisstein, E. W.: n.d.a, Gnomon Magic Squares, MathWorld: A Wolfram Web Resource. http://mathworld.wolfram.com/GnomonMagicSquare.html.

Weisstein, E. W.: n.d.b, Magic Square, MathWorld: A Wolfram Web Resource. http://mathworld.wolfram.com/MagicSquare.html.

Xenakis, I.: 1955, The Crisis of Serial Music, *Gravesaner Blätter* **1**, 2–4.

Xenakis, I.: 1967a, *Metastasis*, Boosey & Hawkes Music Publishers Ltd.

Xenakis, I.: 1967b, *Pithoprakta*, Boosey & Hawkes Music Publishers Ltd.

Xenakis, I.: 1992, *Formalized Music: Thought and Mathematics in Music* (revised edition), Pendragon Press.

Credits

Chapter 1

Figures 1.3 and 1.4, *Messa da Requiem* by Giuseppe Verdi. © Copyright 1937 (renewed 1965) by C. F. Peters Corporation. Used by permission. All Rights Reserved.

Figure 1.6, *America* (from *West Side Story*) by Leonard Bernstein and Stephen Sondheim. © Copyright 1956, 1957, 1958, 1959 by Amberson Holdings LLC and Stephen Sondheim. Copyright renewed. Leonard Bernstein Music Publishing Company LLC, publisher. Boosey & Hawkes, agent for rental. International copyright secured. Reprinted by permission of Boosey & Hawkes, Inc.

Figure 1.8, *Take Five* by Paul Desmond. © Copyright 1960 (renewed 1988) by Desmond Music Company (U.S.A.) and Derry Music Company (World except U.S.A.). Reprinted by permission. All Rights Reserved.

Figure 1.10, *The Rite of Spring* by Igor Stravinsky. Reprinted by generous permission of Dover Publications, Inc.

Figure 1.18, *Fake Empire*, words and music by Matt Berninger and Bryce Dessner. © 2007 Val Jester Music and Hawk Ridge Songs. This arrangement © 2014 Val Jester Music and Hawk Ridge Songs. All rights administered by Bug Music, Inc., a BMG Chrysalis company. All Rights Reserved. Used by permission. Reprinted by permission of Hal Leonard Corporation.

Chapter 2

Figures 2.8, 2.12, 2.26, and 2.41, as well as the figures for Exercises 5 and 6 in Section 2.1, were created by graphic artist Carrie Peck.

Figure 2.14, *Blue Monk* by Thelonious Monk. © Copyright 1962 (renewed 1990) by Thelonious Music Corp. International copyright secured. All Rights Reserved. Reprinted by permission of Hal Leonard Corporation and Thelonious Music Corp.

Figure 2.17, *Harlem Nocturne*, words by Earle Hagen, music by Dick Rogers. © Copyright 1940, 1946, 1951 Shapiro, Bernstein & Co., Inc., New York. Copyright renewed. This arrangement © Copyright 2014 Shapiro, Bernstein & Co., Inc., New York. International copyright secured. All Rights Reserved. Used by permission. Reprinted by permission of Hal Leonard Corporation.

Figure 2.44, *"Tolkien" Circle of Fifths* by Joshua Wells. Reprinted by permission from Joshua Wells, www.oddquartet.com. © 2012.

Chapter 3

Figures 3.3 and 3.8 were created by graphic artist Carrie Peck.

Chapter 4

Figures 4.1 and 4.4 were created by graphic artist Carrie Peck. Figure 4.4 is based on Figure 5 in "Faggot's Fretful Fiasco" by Ian Stewart, in J. Fauvel, R. Flood, and R. Wilson (eds), *Music and Mathematics: From Pythagoras to Fractals*, Oxford University Press, 2003, pp. 60–75.

Figure 4.7, *Bosanquet's Enharmonic Harmonium*. Image used by permission of Science & Society Picture Library, Science Museum Group, London.

Chapter 5

Figure 5.1 was created by graphic artist Carrie Peck.

Figures 5.7 and 5.8, *Mikrokosmos*, SZ107 by Béla Bartók. © Copyright 1987 by Hawkes & Son (London) Ltd. Reprinted by permission of Boosey & Hawkes, Inc.

Figure 5.9, *Contrapunctus XI* from *The Art of Fugue* by Johann Sebastian Bach. Reprinted by generous permission of Dover Publications, Inc.

Figures 5.13 and 5.14, the Royal theme and the *Crab Canon*, respectively, from the *Musical Offering* by Johann Sebastian Bach. Reprinted by generous permission of Dover Publications, Inc.

Figure 5.16, *Ludus Tonalis* by Paul Hindemith. © Copyright 1942 by Schott Music GmbH & Co. KG. © Copyright renewed. All Rights Reserved. Used by permission of European American Music Distributors Company, sole U.S. and Canadian agent for Schott Music GmbH & Co. KG, Mainz, Germany.

Figure 5.21 is a reproduction of Figure 4: Fugue from *Music for Strings, Percussion and Celesta*, from the article "Bartók, Lendvai and the Principle of Proportional Analysis," by Roy Howat, in *Music Analysis*, vol. 2, no. 1, March 1983, pp. 69–95, published by Wiley. Reproduction created by graphic artist Carrie Peck. Reprinted by permission of John Wiley and Sons, Inc. © 1983, Roy Howat. *Music Analysis* © 1983, Blackwell Publishing Ltd.

Figures 5.22 and 5.23 from *Music for Strings, Percussion and Celesta*, SZ106 by Béla Bartók. © Copyright 1937 by Boosey & Hawkes, Inc., for the U.S.A. Copyright renewed. Reprinted by permission of Boosey & Hawkes, Inc. World rights (excluding the U.S.A.) obtained from Universal Edition A.G. © Copyright 1937 by Universal Edition A.G., Wien/UE 34129. Reprinted by permission of Universal Edition A.G.

Chapter 6

Figure 6.1 was created by graphic artist Carrie Peck. It is based on a figure located at `http://www.phillyringers.com/stmarks/newwhat.htm`.

In Figure 6.2, the photograph of the Swan Bell Tower is used courtesy of The Bell Tower in Perth, Western Australia. © Copyright 2014 by Swan Bells, the Bell Tower, Perth, Western Australia. Image located at `http://www.thebelltower.com.au/photo-gallery/`.

In Figure 6.2, the photograph of the floating belfry is used by permission of PA Photos Limited. © PA.

Chapter 7

Figures 7.2, 7.4, 7.5, and 7.6 from *Suite für Klavier*, op. 25 by Arnold Schoenberg. © Copyright 1925, 1952 by Universal Edition A.G., Wien/UE 7627. Reprinted by permission of Universal Edition A.G.

Figure 7.10 was created by graphic artist Carrie Peck.

Chapter 8

Figure 8.2 was created by graphic artist Carrie Peck.

The image to the right in Figure 8.5, *Magic Square 8 Study: A Breeze over Gwalior* by Margaret Kepner, is used by generous permission of the artist. More information is available at `http://MEKvisysuals.yolasite.com`.

Figures 8.7 and 8.9 from *A Mirror of Whitening Light* by Peter Maxwell Davies. © Copyright 1978 by Boosey & Hawkes Music Publishers Ltd. Reprinted by permission.

Figure 8.11, *Clapping Music* by Steve Reich. © Copyright 1980 by Universal Edition (London) Ltd., London/UE 16182. Reprinted by permission.

Figure 8.12, *The Rise and Fall of Steve Reich* by Dorothy Gambrell (*Cat and Girl*, April 2009), is reprinted by permission of Dorothy Gambrell, `http://catandgirl.com/?p=1992`.

In Figure 8.16 the photograph of the Philips Pavilion was taken by Louis Warzée with a camera using Kodachrome 24×36 rolls. The image is reprinted here with kind permission of his son, Guy Warzée.

Figure 8.17, showing the score of *Metastasis* by Iannis Xenakis, is reprinted from page 3 of *Formalized Music: Thought and Mathematics in Music* by Iannis Xenakis, Pendragon Press, 1992. The image is reproduced here with kind permission of Pendragon Press.

Figure 8.19, *Pithoprakta* by Iannis Xenakis. © Copyright 1967 by Boosey & Hawkes Music Publishers Ltd. Reprinted by permission of Boosey & Hawkes, Inc. The image is reproduced from page 17 of *Formalized Music: Thought and Mathematics in Music* by Iannis Xenakis, Pendragon Press, 1992, with kind permission of Pendragon Press.

Index